3/84 CR

QC
718
.N53
1983

Introduction to Plasma Theory

WILEY SERIES IN PLASMA PHYSICS

SANBORN C. BROWN ADVISORY EDITOR
RESEARCH LABORATORY OF ELECTRONICS
MASSACHUSETTS INSTITUTE OF TECHNOLOGY

BEKEFI • RADIATION PROCESSES IN PLASMAS
BROWN • ELECTRON-MOLECULE SCATTERING
BROWN • INTRODUCTION TO ELECTRICAL DISCHARGES IN GASES
GILARDINI • LOW ENERGY ELECTRON COLLISIONS IN GASES
GOLANT, ZHILINSKY, AND SAKHAROV • FUNDAMENTALS OF PLASMA PHYSICS
HEALD AND WHARTON • PLASMA DIAGNOSTICS WITH MICROWAVES
HUXLEY AND CROMPTON • THE DIFFUSION AND DRIFT OF ELECTRONS IN GASES
LICHTENBERG • PHASE-SPACE DYNAMICS OF PARTICLES
MACDONALD • MICROWAVE BREAKDOWN IN GASES
McDANIEL • COLLISION PHENOMENA IN IONIZED GASES
McDANIEL AND MASON • THE MOBILITY AND DIFFUSION OF IONS IN GASES
MITCHNER AND KRUGER • PARTIALLY IONIZED GASES
NASSER • FUNDAMENTALS OF GASEOUS IONIZATION IN PLASMA ELECTRONICS
NICHOLSON • INTRODUCTION TO PLASMA THEORY
TIDMAN AND KRALL • SHOCK WAVES IN COLLISIONLESS PLASMAS

Introduction to Plasma Theory

Dwight R. Nicholson
University of Iowa

John Wiley & Sons
New York • Chichester • Brisbane • Toronto • Singapore

Copyright © 1983, by John Wiley & Sons, Inc.

All rights reserved. Published simultaneously in Canada.

Reproduction or translation of any part of
this work beyond that permitted by Sections
107 and 108 of the 1976 United States Copyright
Act without the permission of the copyright
owner is unlawful. Requests for permission
or further information should be addressed to
the Permissions Department, John Wiley & Sons.

Library of Congress Cataloging in Publication Data:

Nicholson, Dwight R. (Dwight Roy)
Introduction to plasma theory.

(A Wiley series in plasma physics, ISSN 0271-602X)
Includes bibliographies and index.
1. Plasma (Ionized gases) I. Title. II. Series.

QC718.N53 1983 530.4′4 82-13658
ISBN 0-471-09045-X

Printed in the United States of America

10 9 8 7 6 5 4 3 2 1

To my wife Jane and my parents Forrest and Johanna

RAMSEY LIBRARY
UNC-Asheville
Asheville, NC

Preface

The purpose of this book is to teach the basic theoretical principles of plasma physics. It is not intended to be an encyclopedia of results and techniques. Nor is it intended to be used primarily as a reference book. It is intended to develop the basic techniques of plasma physics from the beginning, namely, from Maxwell's equations and Newton's law of motion. Absolutely no previous knowledge of plasma physics is assumed. Although the book is primarily intended for a one year course at the first or second year graduate level, it can also be used for a one or two semester course at the junior or senior undergraduate level. Such an undergraduate course would make use of that half of the book which assumes a knowledge only of undergraduate electricity and magnetism. The other half of the book, suitable for the graduate level, requires familiarity with complex variables, Fourier transformation, and the Dirac delta function.

The book is organized in a logical fashion. Although this is not the standard organization of an introductory course in plasma physics, I have found that students at the graduate level respond well to this organization. After the introductory material of Chapters 1 and 2 (single particle motion), the exact theories of Chapters 3 to 5 (Klimontovich and Liouville equations), which are equivalent to Maxwell's equations plus Newton's law of motion, are replaced via approximations by the Vlasov equation of Chapter 6. Further approximations lead to the fluid theory (Chapter 7) and magnetohydrodynamic theory (Chapter 8). The book concludes with two chapters on discrete particle effects (Chapter 9) and weak turbulence theory (Chapter 10). Chapter 6, and Chapters 7 and 8, are meant to be self-contained, so that the book can easily be used by instructors who wish the standard organization. Thus, the introductory material of Chapters 1 and 2 can be immediately followed by Chapters 7 and 8. This would be enough material for a one semester undergraduate course, while the first half of a two semester graduate course could continue with Chapter 6 on Vlasov theory, followed in the second semester by Chapters 3 to 5 on kinetic theory and then by Chapters 9 and 10.

It is a pleasure to acknowledge the help of many individuals in writing this book. My views on plasma physics have been shaped over the years by dozens of plasma physicists, especially Allan N. Kaufman and Martin V. Goldman. The students in graduate plasma physics courses at the University of Colorado and the University of Iowa have contributed many useful suggestions (Sun Guo-Zheng deserves special mention). The manuscript was professionally typed and edited by Alice Conwell Shank, Gail Maxwell, Susan D. Imhoff, and Janet R. Kephart. The figures were skillfully drafted by John R. Birkbeck, Jr. and Jeana K. Wonderlich. The preparation of this book was supported by the University of Colorado, the University of Iowa, the United States Department of Energy, the United States National Aeronautics and Space Administration, and the United States National Science Foundation.

Dwight R. Nicholson

Contents

CHAPTER

CHAPTER 1

Introduction

1.1 INTRODUCTION

A *plasma* is a gas of charged particles, in which the potential energy of a typical particle due to its nearest neighbor is much smaller than its kinetic energy. The *plasma state* is the fourth state of matter: heating a solid makes a liquid, heating a liquid makes a gas, heating a gas makes a plasma. (Compare the ancient Greeks' earth, water, air, and fire.) The word plasma comes from the Greek plásma, meaning "something formed or molded." It was introduced to describe ionized gases by Tonks and Langmuir [1]. More than 99% of the known universe is in the plasma state. (Note that our definition excludes certain configurations such as the electron gas in a metal and so-called "strongly coupled" plasmas which are found, for example, near the surface of the sun. These need to be treated by techniques other than those found in this book.)

In this book, we shall always consider plasma having roughly equal numbers of singly charged ions ($+e$) and electrons ($-e$), each with average density n_0 (particles per cubic centimeter). In nature many plasmas have more than two species of charged particles, and many ions have more than one electron missing. It is easy to generalize the results of this book to such plasmas.

EXERCISE Name a well-known proposed source of energy that involves plasma with more than one species of ion.

1.2 DEBYE SHIELDING

In a plasma we have many charged particles flying around at high speeds. Consider a special test particle of charge $q_T > 0$ and infinite mass, located at the origin of a three-dimensional coordinate system containing an infinite, uniform plasma. The test charge repels all other ions, and attracts all electrons. Thus, around our test charge the electron density n_e increases and the ion density decreases. The test ion gathers a *shielding cloud* that tends to cancel its own charge (Fig. 1.1).

Consider Poisson's equation relating the electric potential φ to the charge density ρ due to electrons, ions, and test charge,

$$\nabla^2\varphi = -4\pi\rho = 4\pi e(n_e - n_i) - 4\pi q_T\,\delta(\mathbf{r}) \tag{1.1}$$

where $\delta(\mathbf{r}) \equiv \delta(x)\delta(y)\delta(z)$ is the product of three Dirac delta functions. After the introduction of the test charge, we wait for a long enough time that the electrons with temperature T_e have come to thermal equilibrium with themselves, and the ions with temperature T_i have come to thermal equilibrium with themselves, but not so long that the electrons and ions have come to thermal equilibrium with each other at the same temperature (see Section 1.6). Then equilibrium statistical mechanics predicts that

$$n_e = n_0 \exp\left(\frac{e\varphi}{T_e}\right), \qquad n_i = n_0 \exp\left(\frac{-e\varphi}{T_i}\right) \tag{1.2}$$

where each density becomes n_0 at large distances from the test charge where the potential vanishes. Boltzmann's constant is absorbed into the temperatures T_e and T_i, which have units of energy and are measured in units of electron-volts (eV).

Assuming that $e\varphi/T_e \ll 1$ and $e\varphi/T_i \ll 1$, we expand the exponents in (1.2) and write (1.1) away from $\mathbf{r} = 0$ as

$$\nabla^2\varphi = \frac{1}{r^2}\frac{d}{dr}\left(r^2\frac{d\varphi}{dr}\right) = 4\pi n_0 e^2\left(\frac{1}{T_e} + \frac{1}{T_i}\right)\varphi \tag{1.3}$$

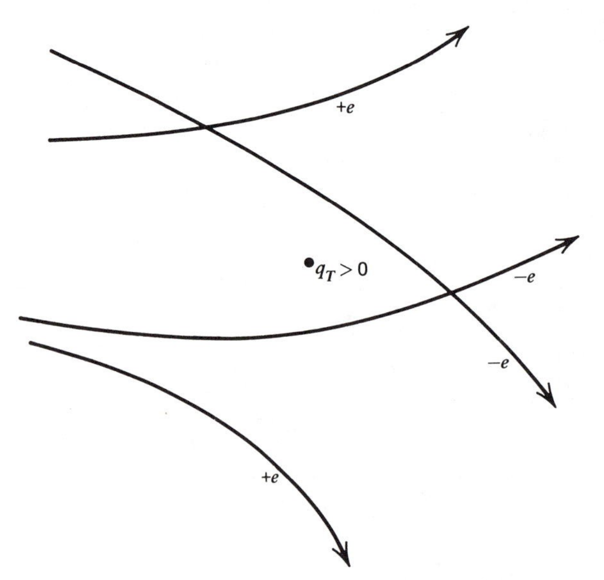

Fig. 1.1 A test charge in a plasma attracts particles of opposite sign and repels particles of like sign, thus forming a shielding cloud that tends to cancel its charge.

If we define the electron and ion *Debye lengths*

$$\boxed{\lambda_{e,i} \equiv \left(\frac{T_{e,i}}{4\pi n_0 e^2}\right)^{1/2}} \tag{1.4}$$

and the *total Debye length*

$$\lambda_D^{-2} = \lambda_e^{-2} + \lambda_i^{-2} \tag{1.5}$$

Eq. (1.3) then becomes

$$\frac{1}{r^2}\frac{d}{dr}\left(r^2 \frac{d\varphi}{dr}\right) = \lambda_D^{-2}\varphi \tag{1.6}$$

Trying a solution of the form $\varphi = \tilde{\varphi}/r$, we find

$$\frac{d^2\tilde{\varphi}}{dr^2} = \lambda_D^{-2}\tilde{\varphi} \tag{1.7}$$

The solution that falls off properly at large distances is $\tilde{\varphi} \propto \exp(-r/\lambda_D)$. From elementary electricity and magnetism we know that the solution to (1.1) at locations very close to $\mathbf{r} = 0$ is $\varphi = q_T/r$; thus, the desired solution to (1.1) at all distances is

$$\boxed{\varphi = \frac{q_T}{r}\exp\left(\frac{-r}{\lambda_D}\right)} \tag{1.8}$$

The potential due to a test charge in a plasma falls off much faster than in vacuum. This phenomenon is known as *Debye shielding*, and is our first example of plasma *collective behavior*. For distances $r >>$ the Debye length λ_D, the shielding cloud effectively cancels the test charge q_T. Numerically, the Debye length of species s with temperature T_s is roughly $\lambda_s \approx 740[T_s(\text{eV})/n(\text{cm}^{-3})]^{1/2}$ in units of cm.

EXERCISE Prove that the net charge in the shielding cloud exactly cancels the test charge q_T.

It is not necessary that q_T be a special particle. In fact, each particle in a plasma tries to gather its own shielding cloud. However, since the particles are moving, they are not completely successful. In an equal temperature plasma ($T_e = T_i$), a typical slowly moving ion has the full electron component of its shielding cloud and a part of the ion component, while a typical rapidly moving electron has a part of the electron component of its shielding cloud and almost none of the ion component.

1.3 PLASMA PARAMETER

In a plasma where each species has density n_0, the distance between a particle and its nearest neighbor is roughly $n_0^{-1/3}$. The average potential energy Φ of a particle due to its nearest neighbor is, in absolute value,

$$|\Phi| \sim \frac{e^2}{r} \sim n_0^{1/3} e^2 \tag{1.9}$$

Our definition of a plasma requires that this potential energy be much less than the typical particle's kinetic energy

$$\frac{1}{2} m_s \langle v^2 \rangle = \frac{3}{2} T_s = \frac{3}{2} m_s v_s^2 \tag{1.10}$$

where m_s is the mass of species s, $\langle \ \rangle$ means an average over all particle velocities at a given point in space, and we have defined the *thermal speed* v_s of species s by

$$v_s \equiv \left(\frac{T_s}{m_s} \right)^{1/2} \tag{1.11}$$

For electrons, $v_e \approx 4 \times 10^7 \, T_e^{1/2}$ (eV) in units of cm/s. Our definition of a plasma requires

$$n_0^{1/3} e^2 << T_s \tag{1.12}$$

or

$$n_0^{2/3} \left(\frac{T_s}{n_0 e^2} \right) >> 1 \tag{1.13}$$

Raising each side of (1.13) to the 3/2 power, and recalling the definition (1.4) of the Debye length, we have (dropping factors of 4π, etc.)

$$\boxed{\Lambda_s \equiv n_0 \lambda_s^3 >> 1} \tag{1.14}$$

where Λ_s is called the *plasma parameter of species s*. (*Note:* Some authors call Λ_s^{-1} the plasma parameter.) The plasma parameter is just the number of particles of species s in a box each side of which has length the Debye length (a Debye cube). Equation (1.14) tells us that, by definition, a plasma is an ionized gas that has many particles in a Debye cube. Numerically, $\Lambda_s \approx 4 \times 10^8 \, T_s^{3/2}(\text{eV})/n_0^{1/2}(\text{cm}^{-3})$. We will often substitute the total Debye length λ_D in (1.14), and define the result $\Lambda \equiv n_0 \lambda_D^3$ to be the *plasma parameter*.

EXERCISE Evaluate the electron thermal speed, electron Debye length, and electron plasma parameter for the following plasmas.

(a) A tokamak or mirror machine with $T_e \approx 1$ keV, $n_0 \approx 10^{13}$ cm^{-3}.

(b) The solar wind near the earth with $T_e \approx 10$ eV, $n_0 \approx 10$ cm^{-3}.

(c) The ionosphere at 300 km above the earth's surface with $T_e \approx 0.1$ eV, $n_0 \approx 10^6$ cm^{-3}.

(d) A laser fusion, electron beam fusion, or ion beam fusion plasma with $T_e \approx 1$ keV, $n_0 \approx 10^{20}$ cm^{-3}.

(e) The sun's center with $T_e \approx 1$ keV, $n_0 \approx 10^{23}$ cm^{-3}.

It is fairly easy to see why many ionized gases found in nature are indeed plasmas. If the potential energy of a particle due to its nearest neighbor were greater than its kinetic energy, then there would be a strong tendency for electrons and ions to bind together into atoms, thus destroying the plasma. The need to keep ions and electrons from forming bound states means that most plasmas have temperatures in excess of one electron-volt.

EXERCISE The temperature of intergalactic plasma is currently unknown, but it could well be much lower than 1 eV. How could the plasma maintain itself at such a low temperature? (*Hint:* $n_0 \approx 10^{-5}\ \text{cm}^{-3}$).

Of course, it is possible to find situations where a plasma exists jointly with another state. For example, in the lower ionosphere there are regions where 99% of the atoms are neutral and only 1% are ionized. In this *partially ionized plasma*, the ionized component can be a legitimate plasma according to (1.14), where Λ_s should be calculated using only the parameters of the ionized component. Typically, there will be a continuous exchange of particles between the unionized gas and the ionized plasma, through the processes of atomic recombination and ionization.

We can now evaluate the validity of the assumption made before (1.3), that $e\varphi/T_s << 1$. This assumption is most severe for the nearest neighbor to the test charge (which we now take to have charge $q_T = +e$). Using the unshielded form of the potential, we require

$$\frac{e}{T_s}\left(\frac{e}{r}\right) \approx \frac{e}{T_s}\left(\frac{e}{n_0^{-1/3}}\right) << 1 \tag{1.15}$$

or

$$n_0^{1/3}\, e^2 << T_s \tag{1.16}$$

which is just the condition (1.12) required by the definition of a plasma. Thus, our derivation of Debye shielding is correct for any ionized gas that is indeed a plasma.

1.4 PLASMA FREQUENCY

Consider a hypothetical slab of plasma of thickness L, where for the present we consider the ions to have infinite mass, but equal density n_0 and opposite charge to the electrons while the electrons are held rigidly in place with respect to each other, but can move freely through the ions. Suppose the electron slab is displaced a distance δ to the right of the ion slab and then allowed to move freely (Fig. 1.2). What happens?

An electric field will be set up, causing the electron slab to be pulled back toward the ions. When the electrons exactly overlap the ions, the net force is zero, but the electron slab has substantial speed to the left. Thus, the electron slab overshoots, and the net result is harmonic oscillation. The frequency of the oscillation is called the *electron plasma frequency*. It depends only on the electron density, the electron charge, and the electron mass. Let's calculate it.

Poisson's equation in one dimension is ($\partial_x \equiv \partial/\partial x$)

$$\partial_x E = 4\pi\rho \tag{1.17}$$

where E is the electric field. Referring to Fig. 1.3, we take the boundary condition $E(x = 0) = 0$, and assume throughout that $\delta << L$. From (1.17) the electric field over most of the slab is $4\pi n_0 e\delta$, and the force per unit area on the electron slab is (electric field) $\times$ (charge per unit area) or $-4\pi n_0^2 e^2 \delta L$. Newton's second law is

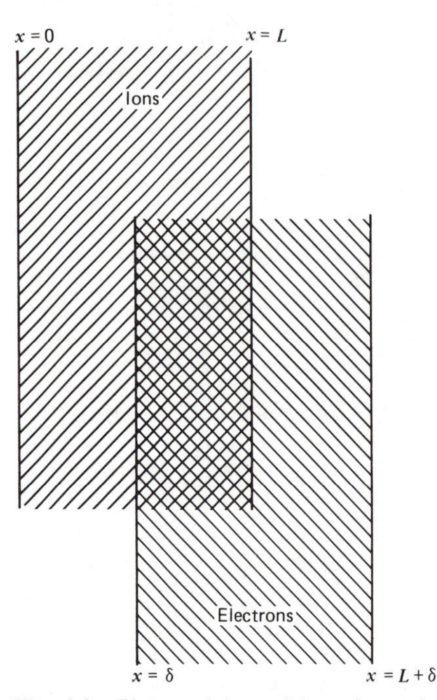

Fig. 1.2 Plasma slab model used to calculate the plasma frequency.

(force per unit area) = (mass per unit area) × (acceleration), or

$$(-4\pi n_0^2 e^2 \delta L) = (n_0 m_e L)(\ddot{\delta}) \tag{1.18}$$

where an overdot is a time derivative. Equation (1.18) is in the standard form of a harmonic oscillator equation,

$$\ddot{\delta} + \left(\frac{4\pi n_0 e^2}{m_e}\right) \delta = 0 \tag{1.19}$$

with characteristic frequency

$$\boxed{\omega_e \equiv \left(\frac{4\pi n_0 e^2}{m_e}\right)^{1/2}} \tag{1.20}$$

which is called the *electron plasma frequency*. Numerically, $\omega_e = 2\pi \times 9000\, n_e^{1/2}$ (cm^{-3}) in units of s^{-1}.

EXERCISE Calculate the electron plasma frequency ω_e and $\omega_e/2\pi$ (e.g., in MHz and kHz) for the five plasmas in the exercise below (1.14).

By analogy with the electron plasma frequency (1.20) we define the ion plasma frequency ω_i for a general ion species with density n_i and ion charge Ze as

$$\omega_i \equiv \left(\frac{4\pi n_i Z^2 e^2}{m_i}\right)^{1/2} \tag{1.21}$$

The total plasma frequency ω_p for a two-component plasma is defined as

$$\omega_p^2 \equiv \omega_e^2 + \omega_i^2 \tag{1.22}$$

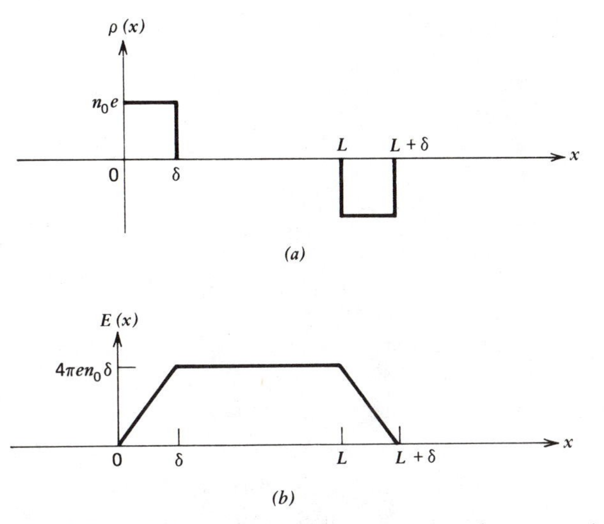

Fig. 1.3 Calculation of the electron plasma frequency. (*a*) Charge density. (*b*) Electric field.

(See Problem 1.3.) For most plasmas in nature $\omega_e >> \omega_i$, so $\omega_p^2 \approx \omega_e^2$. We will see in a later chapter that the general response of an unmagnetized plasma to a perturbation in the electron density is a set of oscillations with frequencies very close to the electron plasma frequency ω_e.

The relation among the Debye length λ_s, the plasma frequency ω_s, and the thermal speed v_s, for the species s, is

$$\boxed{\lambda_s = v_s/\omega_s} \tag{1.23}$$

EXERCISE Demonstrate (1.23).

1.5 OTHER PARAMETERS

Many of the plasmas in nature and in the laboratory occur in the presence of magnetic fields. Thus, it is important to consider the motion of an individual charged particle in a magnetic field. The Lorentz force equation for a particle of charge q_s and mass m_s moving in a constant magnetic field $\mathbf{B} = B_0\hat{z}$ is

$$m_s\ddot{\mathbf{r}} = \frac{q_s}{c}(\dot{\mathbf{r}} \times B_0\hat{z}) \tag{1.24}$$

For initial conditions $\mathbf{r}(t = 0) = (x_0,y_0,z_0)$ and $\dot{\mathbf{r}}(t = 0) = (0, v_\perp,v_z)$ the solution of (1.24) is

$$x(t) = x_0 + \frac{v_\perp}{\Omega_s}(1 - \cos \Omega_s t)$$

$$y(t) = y_0 + \frac{v_\perp}{\Omega_s}\sin \Omega_s t$$

$$z(t) = z_0 + v_z t \tag{1.25}$$

where we have defined the *gyrofrequency*

$$\boxed{\Omega_s \equiv \frac{q_s B_0}{m_s c}} \tag{1.26}$$

EXERCISE Verify that (1.25) is the solution of (1.24) with the desired initial conditions.

Numerically, $\Omega_e = -2 \times 10^7 B_0$ (gauss, abbreviated G) in units of s^{-1}, and $\Omega_i = 10^4 B_0$ (gauss) in units of s^{-1} if the ions are protons.

The nature of the motion (1.25) is a constant velocity in the $\hat{z}$-direction, and a circular gyration in the x-y plane with angular frequency $|\Omega_s|$ and center at the *guiding center* position $\mathbf{r}_{gc}$ given by

$$\mathbf{r}_{gc} = (x_0 + v_\perp/\Omega_s, y_0, z_0 + v_z t) \tag{1.27}$$

The radius of the circle in the x-y plane is the *gyroradius* $v_\perp/|\Omega_s|$. The *mean gyroradius* r_s of species s is defined by setting $v_\perp$ equal to the thermal speed, so

$$\boxed{r_s \equiv v_s/|\Omega_s|} \tag{1.28}$$

EXERCISE In the exercise below (1.14), calculate and order the frequencies ω_e, ω_i, $|\Omega_e|$, Ω_i; also calculate the gyroradii r_e and r_i; take $T_i = T_e$ and use the following parameters.

(a) Protons, $B_0 = 10$ kG.
(b) Protons, $B_0 = 10^{-5}$ G.
(c) O^+ ions, $B_0 = 0.5$ G.
(d) Deuterons, $B_0 = 0$ and $B_0 = 10^6$ G.
(e) Protons, $B_0 = 100$ G.

At this point, let us briefly mention relativistic and quantum effects. For simplicity, we shall always treat nonrelativistic plasmas. In principle, there is no difficulty in generalizing any of the results of this course to include special relativistic effects; these are discussed at length in the book by Clemmow and Dougherty [2].

EXERCISE To what regime of electron temperature are we limited by the nonrelativistic assumption? How about ion temperature if the ions are protons?

There are, of course, many plasmas in which special relativistic effects do become important. For example, cosmic rays may be thought of as a component of the interstellar and intergalactic plasma with relativistic temperature.

We shall also neglect quantum mechanical effects. For most of the laboratory and astrophysical plasmas in which we might be interested, this is a good assumption. There are, of course, plasmas in which quantum effects are very important. An example would be solid state plasmas. As a rough criterion for the neglect of quantum effects, one might require that the typical de Broglie length $h/m_s v_s$ be much less than the average distance between particles $n_0^{-1/3}$.

EXERCISE What is the maximum density allowed by this criterion for electrons with temperature
(a) 10 eV?
(b) 1 keV?
(c) 100 keV?

In other applications, such as collisions (see next section), one might require the de Broglie length to be much smaller than the distance of closest approach of the colliding particles.

In addition to these assumptions, we shall also neglect the magnetic field in many of the sections of this book. This neglect is made for simplicity, in order that the basic physical phenomena can be elucidated without the complications of a magnetic field. In practice, the magnetic field can usually be ignored when the typical frequency (inverse time scale) of a phenomenon is much larger than the gyrofrequencies of both plasma species.

1.6 COLLISIONS

A typical charged particle in a plasma is at any instant interacting electrostatically (see Problem 1.5) with many other charged particles. If we did not know about Debye shielding, we might think that a typical particle is simultaneously having Coulomb collisions with all of the other particles in the plasma. However, the field of our typical particle is greatly reduced from its vacuum field at distances greater than a Debye length, so that the particle is really not colliding with particles at large distances. Thus, we may roughly think of each particle as undergoing Λ simultaneous Coulomb collisions.

From our definition of a plasma, we know that the potential energy of interaction of each particle with its nearest neighbor is small. Since the potential energy is a measure of the effect of a collision, this means that the strongest one of its Λ simultaneous collisions (the one with its nearest neighbor) is relatively weak. Thus, a typical charged particle in a plasma is simultaneously undergoing Λ weak collisions. We shall soon see that even though Λ is a large number for a plasma, the total effect of all the simultaneous collisions is still weak. Of course, a weak effect can still be a very important effect. In the magnetic bottles like tokamaks and mirror machines currently being used to study controlled thermonuclear fusion plasmas, ion-ion collisions are one of the most important loss mechanisms.

Mathematically, the importance of collisions is contained in an expression called the *collision frequency*, which is the inverse of the time it takes for a particle to suffer a collision. Exactly what is meant by a collision of a charged particle depends upon the definition, and we will consider two different definitions with different physical content. Our mathematical derivation of the collision frequency is an approximate one, intended to be simple but yet to yield the correct results within factors of two or so. A more rigorous development can be found in the book by Spitzer [3]. (See Problem 1.6.)

Consider the situation shown in Fig. 1.4. A particle of charge q, mass m is incident on another particle of charge q_0 and infinite mass with incident speed v_0.

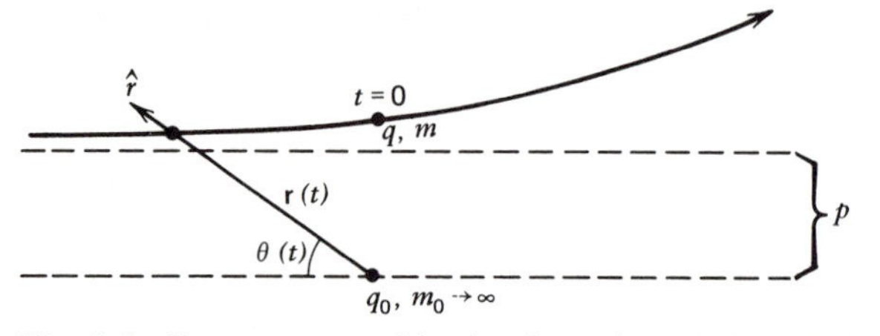

Fig. 1.4 Parameters used in the discussion of the collision frequency in Section 1.6.

If the incident particle were undeflected, it would have position $x = v_0 t$ along the upper dashed line in Fig. 1.4, being at $x = 0$ directly above the scattering charge q_0 at $t = 0$. The separation p of the two dashed lines is the *impact parameter*. If the scattering angle is small, the final parallel speed (parallel to the dashed lines) will be quite close to v_0. The perpendicular speed $v_\perp$ can be obtained by calculating the total perpendicular impulse

$$mv_\perp = \int_{-\infty}^{\infty} dt\, F_\perp(t) \tag{1.29}$$

where $F_\perp$ is the perpendicular force that the particle experiences in its orbit. Since the scattering angle $v_\perp / v_0$ is small, we can to a good approximation use the *unperturbed orbit* $x = v_0 t$ to evaluate the right side of (1.29). This approximation is a very useful one in plasma physics. In Fig. 1.4, Newton's second law with the Coulomb force law is

$$m\ddot{\mathbf{r}} = \frac{qq_0}{r^2}\,\hat{r} \tag{1.30}$$

where $\hat{r}$ is a unit vector in the **r**-direction. Then

$$F_\perp = \frac{qq_0}{r^2}\sin\theta = \frac{qq_0 \sin\theta}{(p/\sin\theta)^2} = \frac{qq_0}{p^2}\sin^3\theta \tag{1.31}$$

where we have used $p = r\sin\theta$ since the particle is assumed to be traveling along the upper dashed line. Equation (1.29) then reads

$$v_\perp = \frac{qq_0}{mp^2}\int_{-\infty}^{\infty} dt\, \sin^3\theta(t) \tag{1.32}$$

The relation between θ and t is obtained from

$$x = -r\cos\theta = \frac{-p\cos\theta}{\sin\theta} = v_0 t \tag{1.33}$$

so that

$$dt = \frac{p}{v_0}\,\frac{d\theta}{\sin^2\theta} \tag{1.34}$$

EXERCISE Verify (1.34).

Using (1.34) in (1.32), we find

$$v_\perp = \frac{qq_0}{mv_0 p}\int_0^{\pi} d\theta\, \sin\theta = \frac{2qq_0}{mv_0 p} \tag{1.35}$$

Defining the quantity

$$p_0 \equiv \frac{2qq_0}{mv_0^2} \tag{1.36}$$

we have

$$\frac{v_\perp}{v_0} = \frac{p_0}{p} \tag{1.37}$$

which is strictly valid only when $v_\perp << v_0, p >> p_0$. In some books, the parameter p_0 is called the *Landau length.*

EXERCISE Show that if $qq_0 > 0$, then p_0 is the distance of closest possible approach for a particle of initial speed v_0.

Although (1.37) is not valid for large angle collisions, let us use it to get a rough idea of the impact parameter p which yields a large angle collision; we do this by setting $v_\perp$ equal to v_0 in (1.37) to obtain $p = p_0$. Thus, any impact parameter $p \leq p_0$ will yield a large angle collision. Suppose the incident particle is an electron, and the (almost) stationary scatterer is an ion. (Although Fig. 1.4 shows a repulsive collision, our development is equally valid for attractive collisions.) The *cross section* for scattering through a large angle by one ion is πp_0^2. Consider an electron that enters a gas of ions. It will have a large angle collision after a time given roughly by setting (the total cross section of the ions in a tube of unit cross-sectional area, and length equal to the distance traveled) equal to (the unit area), or (time) $\times$ (velocity) $\times$ (number per unit volume) $\times$ (cross section) $= 1$. The inverse of this time gives us the collision frequency ν_L for *l*arge angle collisions; thus

$$\nu_L = \pi n_0 v_0 p_0^2 = \frac{4\pi n_0 q^2 q_0^2}{m^2 v_0^3} = \frac{4\pi n_0 e^4}{m_e^2 v_0^3} \tag{1.38}$$

Note that ν_L is proportional to the inverse third power of the particle speed.

Recall that a typical charged particle in a plasma is simultaneously undergoing Λ collisions. Only a very few of these are of the large angle type that lead to (1.38), since a large angle collision involves a potential energy of interaction comparable to the kinetic energy of the incident particle and, by the definition of a plasma, the potential energy of a particle due to its nearest neighbor is small compared to its kinetic energy. Thus, a particle undergoes many more small angle collisions than large angle collisions. It turns out that the cumulative effect of these small angle collisions is substantially larger than the effect of the large angle collisions, as we shall now show.

Unlike the large angle collisions, the many small angle collisions can produce a large effect only after many of them occur. But these small angle collisions produce velocity changes in random directions, some up, some down, some left, some right. We need to know how to measure the cumulative effect of many small random events.

Consider a variable Δx that is the sum of many small random variables Δx_i, $i = 1, 2, \ldots, N$,

$$\Delta x = \Delta x_1 + \Delta x_2 + \ldots + \Delta x_N \tag{1.39}$$

Suppose $\langle \Delta x_i \rangle = 0$ for each i and $\langle (\Delta x_i)^n \rangle$ is the same for each i, where $\langle \ \rangle$ indicates ensemble average [4]. Furthermore, suppose $\langle \Delta x_i \, \Delta x_j \rangle = 0$ if $i \neq j$, so that Δx_i is uncorrelated with Δx_j, $i \neq j$. Then by (1.39) we have $\langle \Delta x \rangle = 0$, and

$$\begin{aligned} \langle (\Delta x)^2 \rangle &= \left\langle \left(\sum_{i=1}^{N} \Delta x_i \right)^2 \right\rangle \\ &= \sum_{i=1}^{N} \langle (\Delta x_i)^2 \rangle \\ &= N \langle (\Delta x_1)^2 \rangle \end{aligned} \tag{1.40}$$

Consider a typical particle moving in the z-direction through a gas of scattering centers. As it moves, it suffers many small angle collisions given by $v_\perp$ which can be decomposed into random variables Δv_x and Δv_y. These latter have just the properties of our random variable Δx_i above. For one collision, with a given impact parameter p (Fig. 1.5), we have from (1.37)

$$\langle v_\perp{}^2 \rangle = \langle (\Delta v_x)^2 \rangle + \langle (\Delta v_y)^2 \rangle = \frac{v_0{}^2 p_0{}^2}{p^2} \tag{1.41}$$

Since Δv_x must have the same statistical properties as Δv_y, we must have

$$\langle (\Delta v_x)^2 \rangle = \langle (\Delta v_y)^2 \rangle = \frac{1}{2} \frac{v_0{}^2 p_0{}^2}{p^2} \tag{1.42}$$

Then by (1.40) we have, for the total x velocity $\Delta v_x{}^{\text{tot}}$,

$$\langle (\Delta v_x{}^{\text{tot}})^2 \rangle = N \langle (\Delta v_x)^2 \rangle = \frac{N}{2} \frac{v_0{}^2 p_0{}^2}{p^2} \tag{1.43}$$

Since we are considering a particle moving through a gas of scattering centers, it is more useful for our purposes to have the time derivative of (1.43), where on the

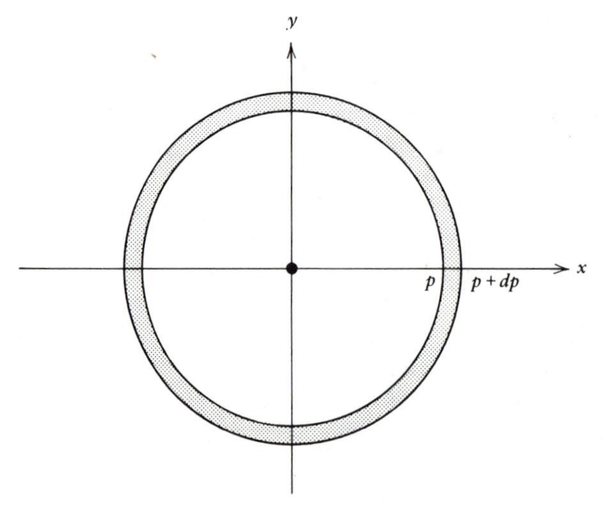

Fig. 1.5 The incident particle is located at the origin and is traveling into the paper. It makes simultaneous small angle collisions with all of the scattering centers randomly distributed with impact parameters between p and $p + dp$.

right we shall have $dN/dt = 2\pi p\, dp\, n_0 v_0$ as the number of scattering centers, with impact parameter between p and $p + dp$, which our incident particle encounters per unit time. The time derivative of (1.43) is then

$$\frac{d}{dt} \langle (\Delta v_x^{\text{tot}})^2 \rangle = \pi n_0 v_0^3 p_0^2 \frac{dp}{p} \tag{1.44}$$

We have calculated (1.44) for only one set of impact parameters between p and $p + dp$. The same logic that led to (1.40) also allows us to sum (integrate) the right side of (1.44) over all impact parameters to obtain a total change in mean square velocity in the $\hat{x}$-direction. Likewise, we can add the total $\hat{x}$-direction and the total $\hat{y}$-direction mean square velocities to obtain a total mean square perpendicular velocity $\langle (\Delta v_\perp^{\text{tot}})^2 \rangle$. With this final factor of two we have

$$\frac{d}{dt} \langle (\Delta v_\perp^{\text{tot}})^2 \rangle = 2\pi n_0 v_0^3 p_0^2 \int_{p_{\min}}^{p_{\max}} \frac{dp}{p} \tag{1.45}$$

What should we use for $p_{\max}$ and $p_{\min}$? Recall that our derivation of the scattering angle $v_\perp / v_0$ in (1.37) uses the Coulomb force law. However, we know from Section 1.2 that the true force law is modified by Debye shielding and is essentially negligible at distances (impact parameters) much greater than a Debye length. Thus, it is consistent with the approximate nature of the present calculation to replace $p_{\max}$ with λ_D. In the case of $p_{\min}$, we use the fact that our scattering formula (1.37) is not valid for impact parameters $p < |p_0|$ to replace $p_{\min}$ by $|p_0|$. Equation (1.45) is then

$$\frac{d}{dt} \langle (\Delta v_\perp^{\text{tot}})^2 \rangle = 2\pi n_0 v_0^3 p_0^2 \ln \left(\frac{\lambda_D}{|p_0|} \right) \tag{1.46}$$

Since the logarithm is such a slowly varying function of its argument, it will suffice to make a very rough evaluation of λ_D / p_0. In the definition of p_0 in (1.36) we take $q = -e$, $q_0 = +e$, $m = m_e$, and for this rough calculation replace v_0 by the electron thermal speed v_e to obtain

$$\frac{\lambda_D}{|p_0|} \approx \frac{\lambda_D m_e v_e^2}{2e^2} \approx \frac{m_e \lambda_D^3 \omega_e^2}{2e^2} \approx 2\pi n_0 \lambda_D^3 = 2\pi\Lambda \tag{1.47}$$

where we have ignored the difference between λ_D and λ_e. Dropping the small factor 2π compared to the large plasma parameter Λ, and using the definition (1.36) of p_0, we find that (1.46) becomes

$$\frac{d}{dt} \langle (\Delta v_\perp^{\text{tot}})^2 \rangle = \frac{8\pi n_0 e^4}{m_e^2 v_0} \ln \Lambda \tag{1.48}$$

A reasonable definition for the scattering time due to small angle collisions is the time it takes $\langle (\Delta v_\perp^{\text{tot}})^2 \rangle$ to equal v_0^2 according to (1.48); the inverse of this time is the *collision frequency* ν_c due to small-angle collisions:

$$\boxed{\nu_c = \frac{8\pi n_0 e^4 \ln \Lambda}{m_e^2 v_0^3}} \tag{1.49}$$

Note again the inverse cube dependence on the velocity v_0. One important aspect of ν_c is that it is a factor $2 \ln \Lambda$ larger than the collision frequency ν_L for large angle collisions given by (1.38). This is a substantial factor in a plasma ($\ln \Lambda = 14$ if $\Lambda = 10^6$). Thus, the deflection of a charged particle in a plasma is predominantly due to the many random small angle collisions that it suffers, rather than the rare large angle collisions.

Throughout one's study of plasma physics, it is useful to identify each phenomenon as a collective effect or as a single particle effect. The oscillation of the plasma slab in Section 1.4, characterized by the plasma frequency ω_e, is a collective effect involving many particles acting simultaneously to produce a large electric field. The collisional deflection of a particle, represented by the collision frequency ν_c in (1.49), is a single particle effect caused by many collisions with individual particles that do not act cooperatively.

EXERCISE Is the Debye shielding described in Section 1.2 a collective effect or a single particle effect?

It is instructive to calculate the ratio of ν_c to ω_e, which is, taking a typical speed $v_0 = v_e$ in (1.49),

$$\frac{\nu_c}{\omega_e} \approx \frac{8\pi n_0 e^4 \ln \Lambda}{m_e^2 v_e^3 \omega_e} = \frac{\ln \Lambda}{2\pi n_0 \lambda_e^3} = \frac{\ln \Lambda}{2\pi \Lambda_e} \tag{1.50}$$

By crudely dropping the factor $\ln \Lambda/2\pi$ and replacing Λ_e by Λ, we have the easily remembered but very approximate expression

$$\boxed{\frac{\nu_c}{\omega_e} \approx \frac{1}{\Lambda}} \tag{1.51}$$

Thus, the collision frequency in a plasma is very much smaller than the plasma frequency. In this respect, single particle effects are less important than collective effects. A wave with frequency near ω_e will oscillate many times before being substantially damped because of collisions.

EXERCISE What is the ratio of the collisional mean free path, for a typical electron, to the electron Debye length?

The collision frequency ν_c that we calculated in (1.49) is the one appropriate to the collisions of electrons with ions, ν_{ei}. The collision frequency ν_{ee} of electrons with electrons could be calculated in the same way, by moving to the center-of-mass frame rather than taking the scattering center to have infinite mass. This procedure would only introduce factors of two or so, so that within such factors we have $\nu_{ee} \approx \nu_{ei}$. Next, consider ion-ion collisions between ions having the same temperature as the electrons that have collision frequency ν_{ee}. Equation (1.49) yields, with m_e replaced by m_i and $v_i = (m_e/m_i)^{1/2} v_e$ instead of v_0, $\nu_{ii} \approx (m_e/m_i)^{1/2} \nu_{ee}$. Finally, consider ions scattered by electrons (or Mack trucks scattered by pedestrians). This calculation in the center-of-mass frame would introduce another factor of $(m_e/m_i)^{1/2}$, so that $\nu_{ie} \approx (m_e/m_i)\nu_{ee}$.

Suppose an electron–proton plasma is prepared in such a way that the electrons and protons have arbitrary velocity distributions, and comparable but not equal temperatures. On the time scale $\nu_{ee}^{-1} \approx \nu_{ei}^{-1} \approx \Lambda\, \omega_e^{-1}$, the electrons will thermalize via electron-electron and electron-ion collisions and obtain a Maxwellian distribution. On a time scale 43 times longer, the ions will thermalize and obtain a Maxwellian at the ion temperature via ion–ion collisions. Finally, on a time scale 43 times longer still, the electrons and ions will come to the same temperature via ion–electron collisions.

This completes our brief introduction to the important basic concepts of plasma physics. In the next chapter, we shall consider the motion of single charged particles in electric and magnetic fields.

REFERENCES

[1] L. Tonks and I. Langmuir, *Phys. Rev.*, *33*, 195 (1929).

[2] P. C. Clemmow and J. P. Dougherty, *Electrodynamics of Particles and Plasmas*, Addison-Wesley, Reading, Mass., 1969.

[3] L. Spitzer, Jr., *Physics of Fully Ionized Gases*, 2nd ed., Wiley-Interscience, New York, 1962.

[4] F. Reif, *Fundamentals of Statistical and Thermal Physics*, McGraw-Hill, New York, 1965.

PROBLEMS

1.1 Debye Shielding

In the discussion of Debye shielding in Section 1.2, suppose that the ions are infinitely massive and thus cannot respond to the introduction of the test charge. How does the answer change?

1.2 Potential Energy (Birdsall's Problem)

A sphere of plasma has equal uniform densities n_0 of electrons and infinitely massive ions. The electrons are moved to the surface of the sphere, which they cover uniformly. What is the potential energy in the system? Sketch the electric field and electric potential as a function of radius. If the electrons initially had temperature T_e, and it is found that the potential energy is equal to the total initial electron kinetic energy, what is the radius of the sphere in terms of the electron Debye length?

1.3 Total Plasma Frequency

In the discussion of the plasma frequency in Section 1.4, suppose the ions are not infinitely massive but have mass m_i. Modify the discussion to show that the slabs oscillate with the total plasma frequency defined in (1.22).

1.4 Plasma in a Gravitational Field

Consider an electron–proton plasma with equal temperatures $T = T_e = T_i$, no magnetic field, and a gravitational acceleration g in the $-\hat{z}$-direction. We desire the densities $n_e(z)$ and $n_i(z)$, where $z = 0$ can be thought of as the surface of a planet. If the electrons and ions were neutral, their densities would be given by the Boltzmann law $n_{e,i} \propto \exp(-m_{e,i}\, gz/T)$. Then the scale height $T/m_{e,i}\, g$ would be quite different for electrons and ions. However, this would give rise to huge electric fields that would tend to move ions up and electrons down. Taking into account the electric field, use the Boltzmann law and the initial guess that $n_e(z) \approx n_i(z)$, to be checked at the end of the calculation, to find self-consistent electron and ion density distributions.

1.5 Electrostatic Interaction

Show that in nonrelativistic plasma, the Coulomb force between two typical particles is much more important than the magnetic field part of the Lorentz force.

1.6 Collisions

Read Sections 5.1, 5.2, and 5.3 of Spitzer [3] and compare his treatment of collisions to our Section 1.6. Watch out for differences in notation, and explain all apparent differences of factors of two.

CHAPTER 2

Single Particle Motion

2.1 INTRODUCTION

A plasma consists of many charged particles moving in self-consistent electric and magnetic fields. The fields affect the particle orbits, and the particle orbits affect the fields. The general solution of any problem in plasma physics can be quite complicated. In this chapter, we consider the motion of a single charged particle moving in prescribed fields. After studying this part of the problem in isolation, we can proceed in following chapters to include these particle orbits in the self-consistent determination of the fields.

2.2 E × B DRIFTS

Consider a particle with $v_z = 0$ gyrating in a magnetic field $\mathbf{B}_0$ in the $\hat{z}$-direction, with an electric field $\mathbf{E}_0$ in the $-\hat{y}$-direction perpendicular to the magnetic field as in Fig. 2.1. (The symbol $\hat{a}$ always means a unit vector in the **a**-direction.) The electric field $\mathbf{E}_0$ cannot accelerate the particle indefinitely, because the magnetic field will turn the particle. (The component of electric field E_z, which we ignore here, can accelerate particles indefinitely. In a plasma, the resulting current usually acts to cancel the charge that caused the electric field in the first place. There are, however, important cases where this cancellation is hindered; for example, the earth's aurora, and tokamak runaway electrons.) What does happen? When the charge q_s is positive, the ion is accelerated on the way down. This gives it a larger local gyroradius at the bottom of its orbit than at the top; recall that the gyroradius is $r_s = v_\perp/\Omega_s$. Thus, the motion will be a spiral in the x-y plane as shown in Fig. 2.2, where we have used the symmetry of the situation to draw the upward part of each orbit. We see that the orbit does not connect to itself, but has jumped a

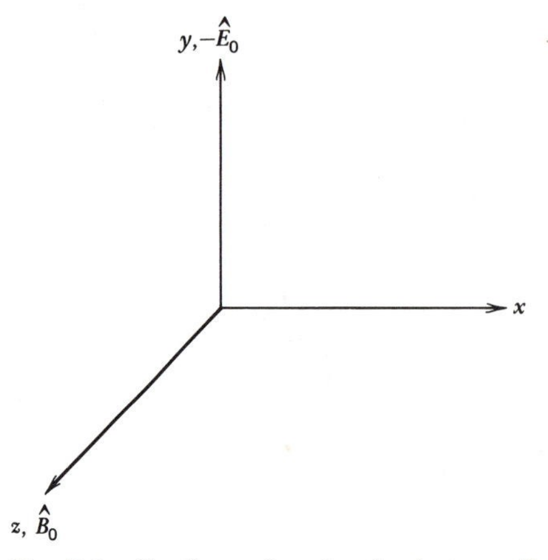

Fig. 2.1 Configuration that leads to an **E** × **B** drift.

certain distance to the left during one orbit. The net result is that the particle has a *drift* velocity $\mathbf{v}_d$ to the left. Let us guess how big the drift speed is. If we average over many gyroperiods, we see that the average accleration is zero. Thus, the net force must be zero. The force downward is $q_s\mathbf{E}_0$, while the force upward is $(q_s/c)\mathbf{v}_d \times \mathbf{B}_0$. We must have

$$\overline{m_s\dot{\mathbf{v}}} = 0 = q_s\mathbf{E}_0 + \frac{q_s}{c}\mathbf{v}_d \times \mathbf{B}_0 \tag{2.1}$$

where $\overline{(\quad)}$ indicates an average over one gyroperiod. Taking the cross product of (2.1) with $\mathbf{B}_0$, and assuming $\mathbf{v}_d \cdot \mathbf{B}_0 = 0$, we find

$$\boxed{\mathbf{v}_d = c\,\frac{\mathbf{E}_0 \times \mathbf{B}_0}{B_0^{\,2}}} \tag{2.2}$$

Note that the drift velocity does not depend on the particle's charge or mass.

EXERCISE What is the drift speed of an electron in the earth's magnetosphere if $|\mathbf{B}_0| = 0.1$ G and $|\mathbf{E}_0| = 10^{-3}$ V cm^{-1}? (Remember 1 sV = 300 V.) What if the particle were a uranium atom with charge $q_s = +57\,e$?

Let us now make sure that our guess is correct, and that the kind of motion we have in mind, namely a gyration about the magnetic field lines, accompanied by a drift, is an exact solution to the equation of motion. This equation is

$$m_s\dot{\mathbf{v}} = q_s\mathbf{E}_0 + \frac{q_s}{c}\mathbf{v} \times \mathbf{B}_0 \tag{2.3}$$

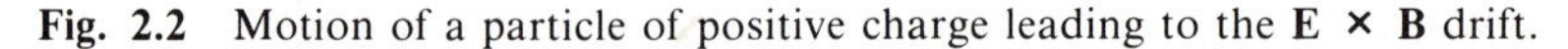

Fig. 2.2 Motion of a particle of positive charge leading to the **E** × **B** drift.

We define a new variable $\tilde{\mathbf{v}}$ by splitting $\mathbf{v}$ into a piece $\tilde{\mathbf{v}}$ (which will turn out oscillatory) and a piece $\mathbf{v}_d$ as given by (2.2). Taking

$$\mathbf{v} = \tilde{\mathbf{v}} + \mathbf{v}_d \tag{2.4}$$

(2.3) becomes

$$\begin{aligned} m_s\dot{\tilde{\mathbf{v}}} &= q_s\mathbf{E}_0 + \frac{q_s}{c}\,\tilde{\mathbf{v}} \times \mathbf{B}_0 + \frac{q_s}{c}\,\mathbf{v}_d \times \mathbf{B}_0 \\ &= q_s\mathbf{E}_0 + \frac{q_s}{c}\,\tilde{\mathbf{v}} \times \mathbf{B}_0 + \frac{q_s}{B_0^2}\,(\mathbf{E}_0 \times \mathbf{B}_0) \times \mathbf{B}_0 \\ &= \frac{q_s}{c}\,\tilde{\mathbf{v}} \times \mathbf{B}_0 \end{aligned} \tag{2.5}$$

But the final form of (2.5) is an equation that we have already solved in (1.25) with a solution that represents gyromotion about the magnetic field. Adapting that solution we find

$$\tilde{\mathbf{v}} = v_\perp(\sin \Omega_s t, \cos \Omega_s t, 0) \tag{2.6}$$

where $v_\perp$ is any constant. Thus, the total solution is

$$\boxed{\mathbf{v} = v_\perp(\sin \Omega_s t, \cos \Omega_s t, 0) + c\,\frac{\mathbf{E}_0 \times \mathbf{B}_0}{B_0^2}} \tag{2.7}$$

Note that any constant $v_\perp$ is acceptable, including $v_\perp = 0$.

EXERCISE What is the initial velocity implied by the solution (2.7) with $v_\perp = 0$? Sketch orbits with $v_\perp = 0$, $v_\perp < v_d$, $v_\perp = v_d$, and $v_\perp > v_d$.

Note that this entire discussion would apply if an arbitrary (temporally and spatially constant) force $\mathbf{F}_\perp$ such that $\mathbf{F}_\perp \cdot \mathbf{B}_0 = 0$ were to replace $q_s\mathbf{E}_0$ in the force equation (2.1). Thus, instead of the drift velocity (2.2), we obtain

$$\boxed{\mathbf{v}_d = \frac{c}{q_s}\,\frac{\mathbf{F}_\perp \times \mathbf{B}_0}{B_0^2} = \frac{1}{\Omega_s}\left(\frac{\mathbf{F}_\perp}{m_s} \times \hat{B}_0\right)} \tag{2.8}$$

EXERCISE What is the gravitational drift speed of an electron in a tokamak, with $|\mathbf{B}_0| = 10$ kG? How about a proton? Does either of these drifts make it hard to confine a plasma in a volume of order $(1\text{ m})^3$ for a time of order 1 s?

We proceed to discuss other kinds of drifts. We have already seen that any real force gives a drift according to (2.8). We shall now see that any so-called "fictitious" force also gives a drift. For example, it is sometimes said that magnetic fields exert a "pressure." This is , of course, not a real pressure, yet we shall find that corresponding to magnetic "pressure" is a drift related to ∇B_0. Likewise, when an existing drift speed changes, the resulting acceleration is experienced as an "inertial" force, which then gives rise to its own drift.

2.3 GRAD-B DRIFT

First, let us calculate the so-called *grad-B* drift. To do this we need to have a feeling for expansion techniques. Suppose we have an equation for a variable x, with one small term of size ϵ, expressed as

$$f(x) - \epsilon g(x) = 0 \tag{2.9}$$

Here, ϵ is a small constant, f and g can represent a general integro-differential operator, and the solution of (2.9) when $\epsilon = 0$ is x_0, so that $f(x_0) = 0$. We can look for a solution x to (2.9) of the form

$$x = x_0 + \epsilon x_1 + \epsilon^2 x_2 + \ldots \tag{2.10}$$

Inserting (2.10) in (2.9) yields

$$f(x_0 + \epsilon x_1 + \epsilon^2 x_2 + \ldots) = \epsilon g(x_0 + \epsilon x_1 + \epsilon^2 x_2 + \ldots) \tag{2.11}$$

After Taylor expanding f and g, one obtains

$$f(x_0) + \frac{df}{dx_0}(\epsilon x_1 + \epsilon^2 x_2 + \ldots) + \frac{1}{2}\frac{d^2 f}{dx_0^{\,2}}(\epsilon x_1 + \epsilon^2 x_2 + \ldots)^2$$
$$+ \ldots = \epsilon\left[g(x_0) + \frac{dg}{dx_0}(\epsilon x_1 + \epsilon^2 x_2 + \ldots) + \ldots\right] \tag{2.12}$$

where $df/dx_0 \equiv df/dx|_{x_0}$, etc. Equating the coefficients of each power of ϵ yields

$$f(x_0) = 0 \tag{2.13}$$

which determines x_0, and

$$x_1 \frac{df}{dx_0} = g(x_0) \tag{2.14}$$

which determines

$$x_1 = \frac{g(x_0)}{df/dx_0}$$

The approximate solution $x = x_0 + \epsilon x_1$ is called the "solution of (2.9) to order ϵ." (Caution: some authors call this "the solution to order ϵ^2.") We must always be careful with what we mean by small in these discussions. Something is "of order ϵ" if it goes to zero as $\epsilon \to 0$; thus, $10^{133}\epsilon$ is of order ϵ while 10^{-133} is of order one.

Consider a particle gyrating in a magnetic field $B_0\hat{z}$ that increases in the $\hat{y}$-direction, as shown in Fig. 2.3. Let us guess what happens. The gyroradius will be smaller at large y than at small y, so the particle will drift as shown in Fig. 2.4 for

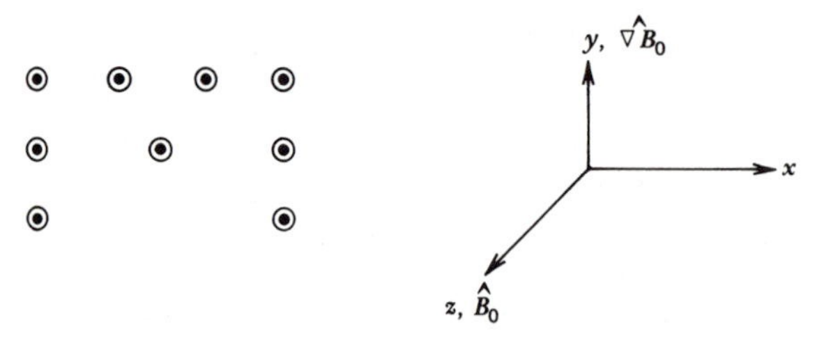

Fig. 2.3 A magnetic field that increases in intensity in the $\hat{y}$-direction.

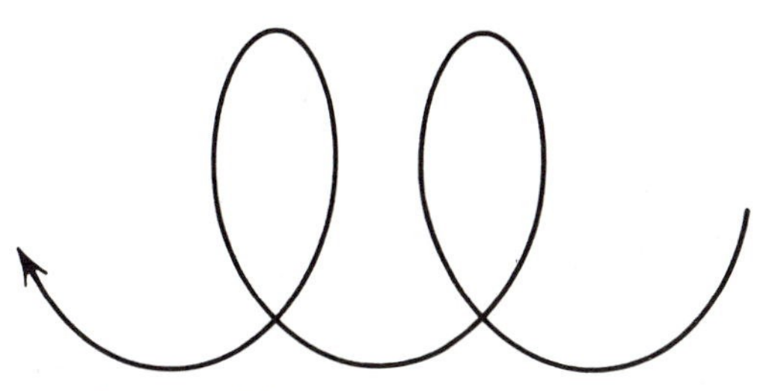

Fig. 2.4 Ion ∇B drift.

ions and in Fig. 2.5 for electrons. Thus, electrons and ions drift in opposite directions and, in a plasma, a net current results.

The force on a charged particle is

$$m_s\dot{\mathbf{v}} = \frac{q_s}{c}(\mathbf{v} \times \mathbf{B}) \tag{2.15}$$

Taylor expanding **B** about the guiding center of the particle,

$$\mathbf{B} = \mathbf{B}_0 + (\mathbf{r} \cdot \nabla)\mathbf{B}_0 \tag{2.16}$$

where $\mathbf{B}_0$ is measured at the guiding center, and where **r** is measured from the guiding center (see Section 1.5), and inserting in (2.15), one obtains

$$m_s\dot{\mathbf{v}} = \frac{q_s}{c}(\mathbf{v} \times \mathbf{B}_0) + \frac{q_s}{c}[\mathbf{v} \times (\mathbf{r} \cdot \nabla)\mathbf{B}_0] \tag{2.17}$$

EXERCISE What assumption is being made in (2.16)?

Expanding

$$\mathbf{v} = \mathbf{v}_0 + \mathbf{v}_1 \tag{2.18}$$

we have

$$m_s\dot{\mathbf{v}}_0 = \frac{q_s}{c}(\mathbf{v}_0 \times \mathbf{B}_0) \tag{2.19}$$

which yields gyromotion, and

$$m_s\dot{\mathbf{v}}_1 = \frac{q_s}{c}(\mathbf{v}_1 \times \mathbf{B}_0) + \frac{q_s}{c}[\mathbf{v}_0 \times (\mathbf{r} \cdot \nabla)\mathbf{B}_0] \tag{2.20}$$

where we are treating $\mathbf{r}\cdot\nabla$ as a small quantity, and to be consistent **r** must be calculated using only $\mathbf{v}_0$.

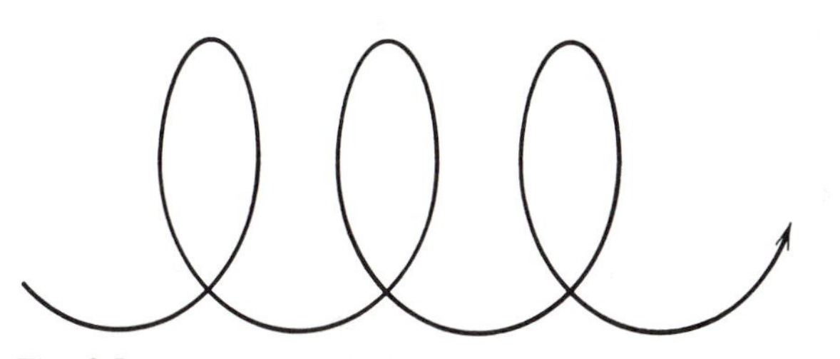

Fig. 2.5 Electron ∇B drift.

Now we are only interested in that part of $\mathbf{v}_1$ that represents steady drift motion; therefore, after averaging both sides of (2.20) over a gyroperiod, upon which the left side vanishes, we have

$$0 = \frac{q_s}{c}\,\overline{\mathbf{v}_1 \times \mathbf{B}_0} + \frac{q_s}{c}\,\overline{[\mathbf{v}_0 \times (\mathbf{r}\cdot\nabla)\mathbf{B}_0]} \tag{2.21}$$

Taking $\mathbf{v}_1 \perp \mathbf{B}_0$, we obtain

$$\mathbf{v}_1 = \frac{1}{B_0^2}\,\overline{[\mathbf{v}_0 \times (\mathbf{r}\cdot\nabla)\mathbf{B}_0] \times \mathbf{B}_0} \tag{2.22}$$

Since the magnetic field $\mathbf{B}$ varies only in the $\hat{y}$-direction,

$$(\mathbf{r}\cdot\nabla)\mathbf{B}_0 = y\,\frac{\partial B_0}{\partial y}\,\hat{z} \tag{2.23}$$

Then

$$\mathbf{v}_0 \times (\mathbf{r}\cdot\nabla)\mathbf{B}_0 = \begin{vmatrix} \hat{i} & \hat{j} & \hat{k} \\ v_{0x} & v_{0y} & 0 \\ 0 & 0 & y\dfrac{\partial B_0}{\partial y} \end{vmatrix} = \hat{i}\,v_{0y}y\,\frac{\partial B_0}{\partial y} - \hat{j}\,v_{0x}y\,\frac{\partial B_0}{\partial y} \tag{2.24}$$

From (1.25) and (1.27),

$$\mathbf{r} = \left(\frac{-v_0}{\Omega_s}\cos\Omega_s t,\ \frac{v_0}{\Omega_s}\sin\Omega_s t, 0\right) \tag{2.25}$$

while (2.6) can be written

$$\mathbf{v}_0 = v_0(\sin\Omega_s t, \cos\Omega_s t, 0) \tag{2.26}$$

When we average the right side of (2.24) over a gyroperiod, the first term vanishes, and the second term yields for (2.22), using $\overline{\sin^2\Omega_s t} = 1/2$,

$$\mathbf{v}_1 = \frac{-1}{2B_0}\,\frac{v_0^2}{\Omega_s}\,\frac{\partial B_0}{\partial y}\,\hat{x} \tag{2.27}$$

or

$$\boxed{\mathbf{v}_1 = \frac{1}{2B_0^2}\,\frac{v_0^2}{\Omega_s}\,(\mathbf{B}_0 \times \nabla B_0)} \tag{2.28}$$

This is the *grad-B* drift. Recalling that Ω_s contains the sign of the charge, we see that the drift is in opposite directions for electrons and ions.

EXERCISE Given an electron and a proton with equal energies, compare the magnitude of the grad-B drifts.

2.4 CURVATURE DRIFTS

Suppose a particle is moving along a field line while gyrating about it. If the field line curves, without changing magnitude, then the particle tries to follow the field line because all motions across field lines are resisted. It therefore feels a centrifugal force $\mathbf{F}_c$ outward (Fig. 2.6), equal to

$$\mathbf{F}_c = \frac{mv_{\parallel}^2}{R_B} \hat{R}_B \tag{2.29}$$

Our general drift equation (2.8) then predicts

$$\mathbf{v}_d = \frac{c}{q_s} \frac{\mathbf{F}_{\perp} \times \mathbf{B}_0}{B_0^2} = \frac{cmv_{\parallel}^2}{R_B q_s B_0^2} \hat{R}_B \times \mathbf{B}_0 \tag{2.30}$$

or

$$\boxed{\mathbf{v}_d = \frac{v_{\parallel}^2}{\Omega_s R_B} (\hat{R}_B \times \mathbf{B}_0)} \tag{2.31}$$

This is the *curvature drift*.

In a cylindrically symmetric vacuum field, it turns out that $\nabla B_0 = (-B_0/R_B)\hat{R}_B$ (see p. 26 of Ref. [1]); thus we may add the grad-B drift to the curvature drift to obtain

$$\boxed{\mathbf{v}_d^{\text{tot}} = \frac{(\hat{R}_B \times \hat{B}_0)}{\Omega_s R_B} (v_{\parallel}^2 + \tfrac{1}{2} v_0^2)} \tag{2.32}$$

where we recall that v_0 is the perpendicular speed. A rigorous derivation of (2.31) and (2.32) can be found in Refs. [2] and [3].

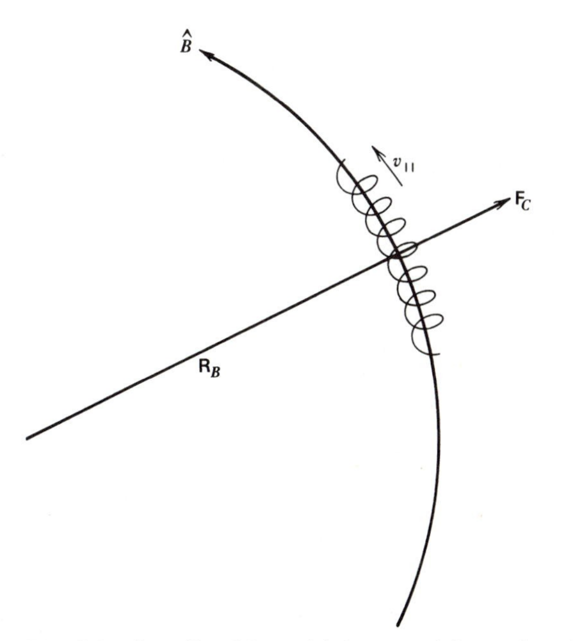

Fig. 2.6 Centrifugal force felt by a particle moving along a curved field line.

2.5 POLARIZATION DRIFT

We discuss next a drift that is the result of an electric field which varies with time. Since the drift is opposite for oppositely charged particles, it leads to a current called the *polarization current*.

Consider a constant magnetic field $B_0\hat{z}$, and an electric field $\mathbf{E}(t) = -\dot{E}t\hat{y}$, where $\dot{E}$ is a constant (Fig. 2.7). The force equation is

$$m_s\dot{\mathbf{v}} = q_s\mathbf{E} + \frac{q_s}{c}\mathbf{v} \times \mathbf{B}_0 \tag{2.33}$$

We expect an $\mathbf{E} \times \mathbf{B}_0$ drift in the $(-)\hat{x}$-direction, which will be increasing with time. Thus, the particle is being accelerated in the $(-)\hat{x}$-direction and, therefore, feels an effective force in the $\hat{x}$-direction. (The effective force is in the direction opposite to the acceleration; when one steps on the gas pedal of a car, one is forced *backward* into the seat.) This effective force should give rise to an $\mathbf{F} \times \mathbf{B}$ drift in the $\hat{y}$-direction. Using this intuition, we consider a solution of (2.33) of the form

$$\mathbf{v} = \mathbf{v}_0 + v_E\hat{x} + v_p\hat{y} \tag{2.34}$$

where $\mathbf{v}_0$ will contain all gyromotion, v_E is the $\mathbf{E} \times \mathbf{B}$ drift, and v_p is the polarization drift, which is assumed to be constant. Substituting (2.34) into (2.33), we obtain

$$m_s(\dot{\mathbf{v}}_0 + \dot{\mathbf{v}}_E) = -q_s\dot{E}t\hat{y} + \frac{q_s}{c}\mathbf{v}_0 \times \mathbf{B}_0 - \frac{q_s}{c}v_E B_0\hat{y} + \frac{q_s}{c}v_p B_0\hat{x} \tag{2.35}$$

The assumed nature of the solution indicates the separation of this equation into pieces, with

$$m_s\dot{\mathbf{v}}_0 = \frac{q_s}{c}\mathbf{v}_0 \times \mathbf{B}_0 \tag{2.36}$$

representing gyromotion,

$$m_s\dot{\mathbf{v}}_E = \frac{q_s}{c}v_p B_0\hat{x} \tag{2.37}$$

giving the polarization drift, and

$$0 = -q_s\dot{E}t\hat{y} - \frac{q_s}{c}v_E B_0\hat{y} \tag{2.38}$$

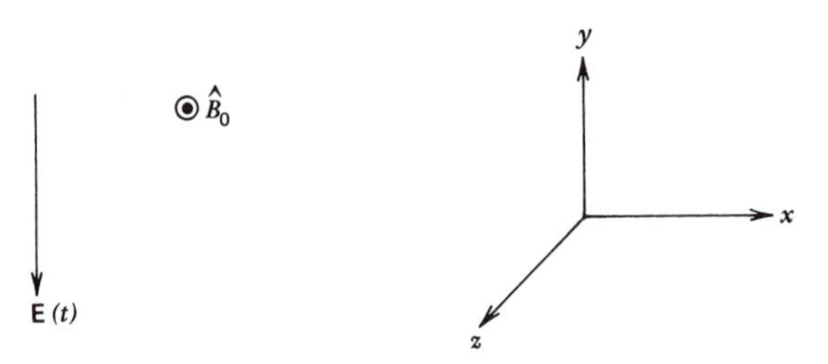

Fig. 2.7 Configuration that leads to a polarization drift.

giving the $\mathbf{E} \times \mathbf{B}$ drift, namely,

$$v_E\hat{x} = -\frac{c\dot{E}t}{B_0}\hat{x} = c\,\frac{\mathbf{E} \times \mathbf{B}_0}{B_0^2} \tag{2.39}$$

Then (2.37) yields the polarization drift

$$v_p = \frac{cm_s}{q_sB_0}(-)\frac{c\dot{E}}{B_0} = -\frac{c}{\Omega_s}\frac{\dot{E}}{B_0} \tag{2.40}$$

which in vector form is

$$\boxed{\mathbf{v}_p = \frac{c}{\Omega_sB_0}\frac{d}{dt}\mathbf{E}} \tag{2.41}$$

We see that (2.34) is an exact solution to (2.33).

EXERCISE If $E(t) = E_0 \cos \omega t$, can you invent a criterion for the validity of (2.41) at each instant of time?

The polarization drift leads to a polarization current J_p, of electrons and protons, given by

$$\mathbf{J}_p = n_0e(\mathbf{v}_{pi} - \mathbf{v}_{pe}) = \frac{n_0c^2}{B_0^2}\frac{d\mathbf{E}}{dt}(m_e + m_i) \tag{2.42}$$

or

$$\boxed{\mathbf{J}_p = \frac{\rho_m c^2}{B_0^2}\frac{d\mathbf{E}}{dt}} \tag{2.43}$$

where ρ_m is the mass density.

2.6 MAGNETIC MOMENT

The preceding sections have discussed drifts due to "forces" perpendicular to the magnetic field. There are also forces parallel to the magnetic field that are very important, leading to the concepts of *magnetic moment* and *adiabatic invariants*.

Recall that the *magnetic moment* of a current loop with current I, area A, in c.g.s. units, is

$$\mu = \frac{IA}{c} \tag{2.44}$$

A charged particle gyrating in a magnetic field is such a current loop, with current $q_s\Omega_s/2\pi$, area $\pi\rho_s^2 = \pi v_\perp^2/\Omega_s^2$ (ρ_s is the gyroradius), and magnetic moment

$$\mu = \frac{q_s\Omega_s}{2\pi c}\pi\frac{v_\perp^2}{\Omega_s^2} = \frac{q_sv_\perp^2}{2c\Omega_s} = \frac{\tfrac{1}{2}m_sv_\perp^2}{B} \tag{2.45}$$

or

$$\boxed{\mu = \frac{W_\perp}{B}} \tag{2.46}$$

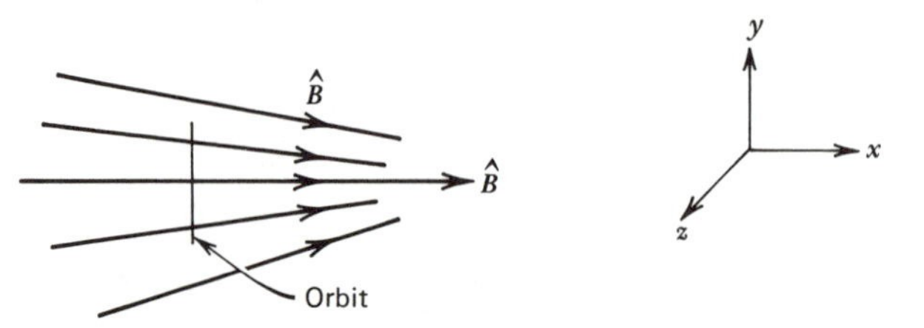

Fig. 2.8 Magnetic field with nonzero grad-B. The orbit shown is a circle in a plane perpendicular to the paper. For the discussion following (2.51), the orbit is actually a spiral with a velocity component $v_{\parallel}$ in the $\hat{B}$-direction.

where $W_{\perp} = ½ m_s v_{\perp}^2$ is that portion of a particle's kinetic energy which is perpendicular to the magnetic field.

We know that magnetic moments feel a force $-\mu \nabla B$ in an inhomogeneous magnetic field. How does this work out for a charged particle? Consider a particle gyrating about the axis of a cylindrically symmetric magnetic field, whose magnitude is changing along the axis, as shown in Fig. 2.8.

EXERCISE Does the field in Fig. 2.8 satisfy Maxwell's equations?

In the figure, the vertical line is a side view of the gyrating particle. Notice that at the top of the orbit (Fig. 2.9), for a positively charged particle, the $\mathbf{v} \times \mathbf{B}$ force has one component giving gyromotion about the field, and another component pointing in the $(-)\hat{x}$-direction. This latter component is constant around the gyro-orbit, and the particle is steadily accelerated away from regions of strong field.

EXERCISE Show that this works the same for either sign of the charge.

The force in the $(-)\hat{x}$-direction, evaluated anywhere on the orbit, is

$$F = \frac{q_s}{c} (\mathbf{v} \times \mathbf{B})_x = \frac{|q_s|}{c} v_{\perp} B_r \tag{2.47}$$

where r is the distance from the x-axis, and B_r is the component of the magnetic field in the y-z plane in Fig. 2.8. In cylindrical coordinates,

$$\nabla \cdot \mathbf{B} = 0 = \frac{\partial B_x}{\partial x} + \frac{1}{r} \frac{\partial}{\partial r} (r B_r)$$

Solving this equation with $B_r = 0$ at $r = 0$, and $B_r \ll B_x$ everywhere, one ob-

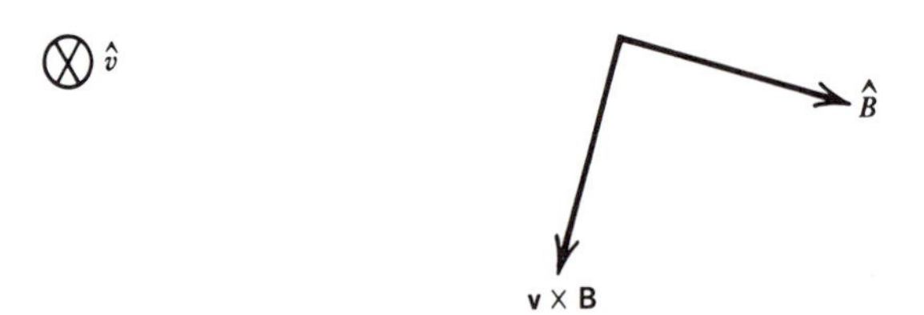

Fig. 2.9 The vectors $\hat{v}$, $\hat{B}$, and $\mathbf{v} \times \mathbf{B}$ for a particle with $q_s > 0$ at the top of the orbit in Fig. 2.8.

tains

$$B_r = -\frac{r}{2}\frac{\partial B_x}{\partial x} = -\frac{r}{2}|\nabla B| \tag{2.48}$$

Inserting (2.48) in (2.47), with r equal to the gyroradius ρ_s, we obtain

$$\mathbf{F} = -\frac{|q_s|}{c}\frac{\rho_s}{2}v_\perp \nabla B = -\frac{m_s v_\perp^2}{2B}\nabla B \tag{2.49}$$

or

$$\boxed{\mathbf{F} = -\mu \nabla B} \tag{2.50}$$

as expected.

Knowing the force on the particle allows the calculation of its orbit. First one needs to know how the magnetic moment μ changes along the orbit. The remarkable fact is that *the magnetic moment is constant* along the orbit, provided the field does not change much in one gyroperiod. Let us prove this.

The Lorentz force on a charged particle (with $\mathbf{E} = 0$) is $\mathbf{F} = (q_s/c)\,\mathbf{v} \times \mathbf{B}$. A small component $\mathbf{F}_\perp$ of this force acts to accelerate the particle in the direction perpendicular to the local magnetic field *and* parallel to the component of the particle velocity $v_\perp$ used in the definition of μ. (In Fig. 2.8, let the particle shown have a positive velocity component $v_\parallel$ along $\hat{x}$. Then at the top of the orbit, $\mathbf{v} \times \mathbf{B}$ has a component of magnitude $|v_\parallel B_r|$ pointing into the paper.) This force is into the paper at the top of the orbit of the particle in Fig. 2.8, and is given by

$$F_\perp = -\frac{q_s}{c}v_\parallel B_r \tag{2.51}$$

where $v_\parallel$ is the component of velocity in the $\hat{x}$-direction and B_r is negative. The perpendicular energy of the particle then changes with time according to

$$\frac{d}{dt}\left(\frac{1}{2}m_s v_\perp^2\right) = v_\perp F_\perp = \frac{-q_s}{c}v_\perp v_\parallel B_r \tag{2.52}$$

When we use (2.48), this becomes (with $r = \rho_s$ at the location of the particle)

$$\frac{d}{dt}W_\perp = \frac{q_s}{c}v_\perp v_\parallel \frac{\rho_s}{2}\frac{\partial B_x}{\partial x} = \frac{1}{2}m_s v_\perp^2 v_\parallel \frac{1}{B}\frac{\partial B}{\partial x} \tag{2.53}$$

where $B_x \approx B$, or

$$\frac{d}{dt}W_\perp = W_\perp v_\parallel \frac{1}{B}\frac{\partial B}{\partial x} \tag{2.54}$$

EXERCISE Combine (2.50) and (2.53) to show that total energy is conserved.

The time rate of change of the magnetic moment is

$$\begin{aligned}\frac{d\mu}{dt} &= \frac{d}{dt}\left(\frac{W_\perp}{B}\right) = \frac{1}{B}\frac{dW_\perp}{dt} - \frac{1}{B^2}W_\perp\frac{dB}{dt}\\ &= W_\perp v_\parallel \frac{1}{B^2}\frac{\partial B}{\partial x} - \frac{1}{B^2}W_\perp v_\parallel \frac{\partial B}{\partial x}\\ &= 0\end{aligned} \tag{2.55}$$

where the rate of change of **B** along the particle orbit $dB/dt = v_\parallel \, \partial B/\partial x$ has been used. Thus,

$$\boxed{\mu = \text{constant}} \tag{2.56}$$

along the particle orbit, to within the accuracy of this calculation. This is an example of an *adiabatic invariant*, a quantity that is constant under slow changes in an external parameter. Although our derivation treated spatially varying magnetic fields, the same result holds for magnetic fields with slow time variation.

We are now in the position to understand the principle of *mirror confinement*, which is the basis for one of the two major approaches to magnetic fusion, and is also the reason for the existence of the earth's magnetosphere. Consider a magnetic field created by two coils, as shown in Fig. 2.10. A particle that starts at $x = 0$ with energy W_0 and magnetic moment μ conserves both of these quantities. Suppose the particle initially has $v_\parallel > 0$; it moves to the right and feels a force to the left, $F = -\mu \nabla B$. Does the particle get reflected by the force, or does it go past $x = x_0$ to be lost from the machine? This will depend on its initial $v_\parallel^0$ and $v_\perp^0$. We have

$$\mu = \frac{\tfrac{1}{2} m v_\perp^2}{B} = \frac{\tfrac{1}{2} m v_\perp^{0^2}}{B_{\min}} = \text{const} \tag{2.57}$$

and

$$W = \tfrac{1}{2} m (v_\perp^2 + v_\parallel^2) = W_0 = \text{const} \tag{2.58}$$

As the particle moves to the right, B increases and $W_\perp = \tfrac{1}{2} m v_\perp^2$ must increase to satisfy (2.57). If $W_\perp$ ever reaches W_0, then all the energy will be in perpendicular motion, $v_\parallel$ will vanish, and the particle will be reflected back toward the mirror machine. This happens if

$$\mu = \frac{\tfrac{1}{2} m v_\perp^{0^2}}{B_{\min}} > \frac{W_0}{B_{\max}} = \frac{\tfrac{1}{2} m (v_\perp^{0^2} + v_\parallel^{0^2})}{B_{\max}} \tag{2.59}$$

or

$$\frac{v_\perp^{0^2}}{v_\perp^{0^2} + v_\parallel^{0^2}} > \frac{B_{\min}}{B_{\max}} \tag{2.60}$$

Defining the *pitch angle* $\theta = \tan^{-1}(v_\perp^0 / v_\parallel^0)$, we have from (2.60)

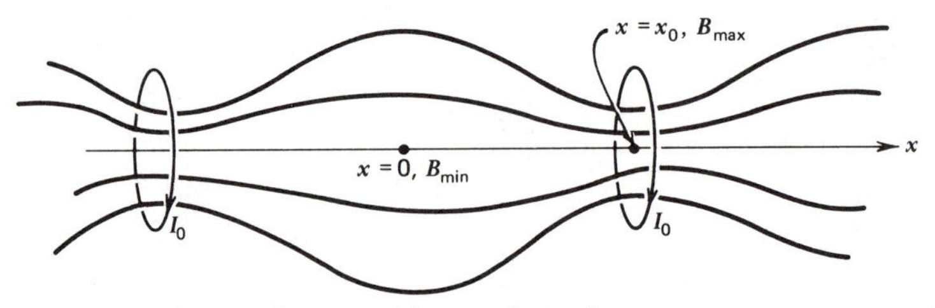

Fig. 2.10 Simple mirror machine configuration.

$$\boxed{\sin\theta > \left(\frac{B_{\min}}{B_{\max}}\right)^{1/2} = \frac{1}{R^{1/2}}} \tag{2.61}$$

where we have introduced the mirror ratio $R \equiv B_{\max}/B_{\min}$. Particles whose pitch angles satisfy (2.61) at the center of the machine are confined. Those that do not are lost out from the ends. While our derivation applies only to particles circling the central magnetic field line, a similar statement is true for off-axis particles.

2.7 ADIABATIC INVARIANTS

The magnetic moment, constant under slow spatial or temporal changes in the magnetic field, is one example of an *adiabatic invariant*. It often turns out that in a system with a coordinate q, and its conjugate momentum p, the action, defined by

$$J = \oint p\, dq \tag{2.62}$$

is a constant under a slow change in an external parameter. Here, we have assumed that when there is no change in the external parameter, the motion is periodic, and $\oint$ represents an integral over one period of the motion. In the case of a charged particle in a magnetic field, we could take, for example, measuring x from the guiding center: $p = mv_x$, $q = x$, and

$$\begin{aligned} J &= \oint m_s v_x\, dx = \oint_0^{2\pi/\Omega_s} m_s v_\perp^2 \sin^2(\Omega_s t)\, dt \\ &= \frac{\pi m_s v_\perp{}^2}{\Omega_s} = \frac{2m_s \pi c}{q_s}\left(\frac{W_\perp}{B}\right) = \frac{2m_s\pi c}{q_s}\mu \end{aligned} \tag{2.63}$$

which is the magnetic moment to within a constant.

One famous example of the constancy of action was derived at the 1911 Solvay conference, where Lorentz asked: "What happens if we slowly shorten the string of a swinging pendulum?" The next morning, Einstein answered: "Action [= energy/frequency] is conserved."

Let us now demonstrate the invariance of the action in the general case. We have in mind the picture of a particle bouncing in a potential well, with the shape of the well changing slowly with time, as shown in Fig. 2.11. In Hamiltonian mechanics, we have a Hamiltonian $H(p,q,\lambda)$ where p is the momentum ($m\dot{x}$ in the

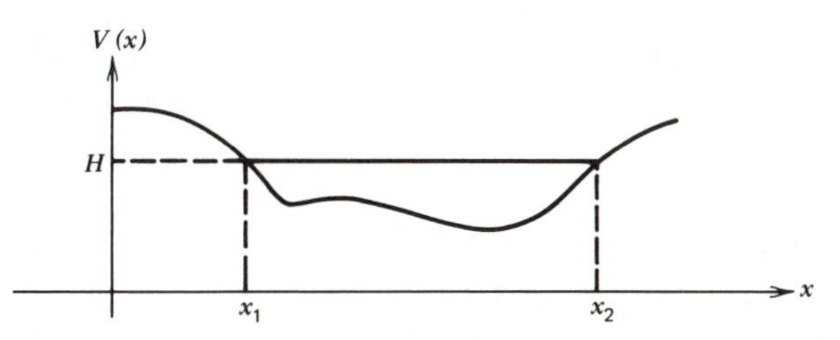

Fig. 2.11 Periodic motion in a slowly changing potential well.

figure), q is the coordinate (x in the figure), and λ is the parameter that determines the shape of the well. When λ is constant, the Hamiltonian is constant; when λ changes,

$$\frac{dH}{dt} = \frac{\partial H}{\partial \lambda}\frac{d\lambda}{dt} \tag{2.64}$$

The Hamiltonian equations of motion are

$$\dot{q} = \frac{\partial H}{\partial p}, \qquad \dot{p} = -\frac{\partial H}{\partial q} \tag{2.65}$$

The time derivative of the action

$$J = \oint p\,dq = 2\int_{q_1}^{q_2} p\,dq \tag{2.66}$$

is

$$\dot{J} = 2\int_{q_1}^{q_2} \dot{p}\,dq + 2p(q_2)\dot{q}_2 - 2p(q_1)\dot{q}_1 \tag{2.67}$$

The last two terms vanish since the momentum is zero at the turning points; then

$$\dot{J} = -2\int_{q_1}^{q_2} \frac{\partial H}{\partial q}\,dq = -2[H(q_2) - H(q_1)] = 0 \tag{2.68}$$

since $\partial H/\partial q$ is to be taken at fixed λ, implying $H(q_2)$ and $H(q_1)$ are to be taken at fixed λ, and H is a constant except for its variation with λ. We have thus shown the approximate constancy of any action variable in the presence of periodic motion and slow changes of external parameters. Note that the present derivation is heuristic, since it has been assumed that the variation is so slow that the turning points $q_2(t)$ and $q_1(t)$ can be treated as continuous functions of time and can be differentiated as in (2.67) [4]. A rigorous treatment of this problem can be found in the fundamental paper of Kruskal [5]; see also Goldstein [6].

The knowledge of an adiabatic invariant can be very useful in predicting particle behavior. Let us return to the mirror machine, where the constancy of the magnetic moment adiabatic invariant has already enabled us to determine which particles will be confined and which will be lost. Consider a confined particle. The confined particle executes periodic motion between x_1 and x_2. Thus,

$$J_2 = \oint p\,dq = \oint mv_x\,dx \tag{2.69}$$

must be a constant of the motion, even when the entire mirror field undergoes slow temporal changes, or when the mirror field is not axisymmetric.

There is yet a third adiabatic invariant. As the charged particle bounces from x_1 to x_2, it drifts perpendicular to the field lines, with speed v_d. Eventually, it comes all the way around the mirror machine. Because this is like a huge gyro-orbit about the axis of the magnetic field, we define a new adiabatic invariant

$$J_3 = \oint v_d\,dl \tag{2.70}$$

where l is the distance around the mirror machine measured at some fixed x, for example x_1. It turns out that J_3 is proportional to the total magnetic flux enclosed by the drifting motion. This invariant is useful when the mirror field is not axisymmetric, or when it undergoes slow temporal changes.

EXERCISE Sketch the earth's magnetosphere, and discuss the motion leading to μ, J_2, and J_3.

2.8 PONDEROMOTIVE FORCE

All of the examples of single particle motion considered in previous sections involve motion in a magnetic field, with or without an electric field. There is one very important single particle effect, the *ponderomotive force* or Miller force, that occurs in spatially varying high frequency electric fields, with or without an accompanying magnetic field. Consider a charged particle oscillating in a high frequency electric field $E(t) = E_0 \cos(\omega t)$. The motion is then a sinusoidal variation of distance with time, as shown in Fig. 2.12. Now suppose the electric field has an amplitude that varies smoothly in space, $E(x,t) = E_0(x) \cos(\omega t)$, being stronger to the right and weaker to the left. Then the first oscillation brings the particle into regions of strong field, where it can be given a strong push to the left (see Fig. 2.13). When the field turns around, the particle is in a region of weaker field, and the push to the right is not as strong. The net result is a displacement to the left, which continues in succeeding cycles as an *acceleration* away from the region of strong field [7].

Mathematically, the force equation is

$$m_s\ddot{x} = q_s E = q_s E_0(x) \cos \omega t \tag{2.71}$$

It is convenient to decompose x into a slowly varying component x_0, called the *oscillation center* (compare this concept to the guiding center in a magnetic field) and a rapidly varying component x_1, $x = x_0 + x_1$. Here, $x_0 = \overline{x}$ where $\overline{(\quad)}$ indicates a time average over the short time $2\pi/\omega$. Making a Taylor expansion of

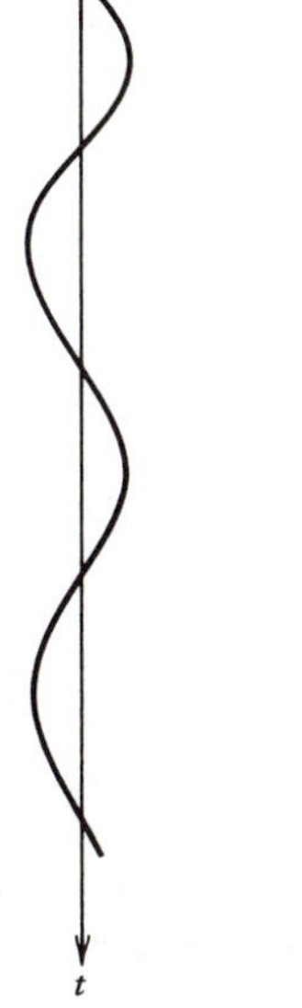

Fig. 2.12 Sinusoidal motion of a charged particle in a high-frequency electric field.

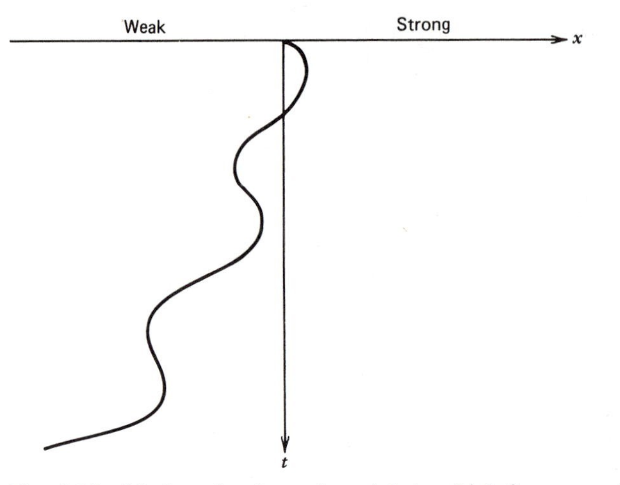

Fig. 2.13 Motion of a charged particle in a high-frequency electric field that is weaker to the left and stronger to the right.

$E_0(x)$ about the oscillation center x_0, (2.71) becomes

$$m_s(\ddot{x}_0 + \ddot{x}_1) = q_s \left(E_0 + x_1 \frac{dE_0}{dx}\right) \cos \omega t \tag{2.72}$$

where dE_0/dx is to be evaluated at x_0. Averaging (2.72) over time, we get

$$m_s\ddot{x}_0 = q_s \left.\frac{dE_0}{dx}\right|_{x_0} \overline{x_1 \cos \omega t} \tag{2.73}$$

To obtain an equation for x_1, we note that $\ddot{x}_1 >> \ddot{x}_0$ since x_1 is high frequency; moreover, in the spirit of the Taylor expansion we have $E_0 >> x_1(dE_0/dx)$; therefore (2.72) is approximately

$$m_s\ddot{x}_1 = q_sE_0 \cos \omega t \tag{2.74}$$

with solution $x_1 = -(q_sE_0/m_s\omega^2)(\cos \omega t)$; inserting this in (2.73) and performing the time average one obtains

$$\ddot{x}_0 = -\frac{q_s^2E_0}{2m_s^2\omega^2}\frac{dE_0}{dx} \tag{2.75}$$

so that the ponderomotive force $F_p = m_s\ddot{x}_0$ is

$$\boxed{F_p = -\frac{q_s^2}{4m_s\omega^2}\frac{d}{dx}(E_0^2)} \tag{2.76}$$

This formula will be easier to remember if we introduce the jitter speed $\tilde{v} \equiv (\dot{x}_1)_{\max} = q_sE_0/m_s\omega$; then

$$F_p = -\frac{m_s}{4}\frac{d}{dx}(\tilde{v}^2) \tag{2.77}$$

This force is very important in such applications as laser fusion, electron beam fusion, radio frequency heating of tokamaks, radio frequency plugging of mirrors, radio frequency modification of the ionosphere, and solar radio bursts. The study of the effects of ponderomotive force is one of the areas of current basic plasma physics research. Notice that the overall mass dependence is as given in (2.76), so that the ponderomotive force acts much more strongly on electrons than on ions. A more complete derivation of the ponderomotive force, including the magnetic field in an electromagnetic wave, can be found in Schmidt [4].

2.9 DIFFUSION

We conclude this chapter with a brief discussion of the effects of collisions on the location of the guiding center of a particle in a magnetic field. The discussion of Chapter 1 shows that the effects of many small angle collisions in a plasma are more important than the effects of rare large angle collisions. However, it is simplest to consider here a single large angle collision between two charged particles; the results can then be qualitatively applied to determine the effects of many small angle collisions.

Consider first the head-on collision between two electrons at $\mathbf{x} = 0$, as shown in Fig. 2.14. The last gyro-orbit, and the guiding center, of each particle before the collision are indicated in the upper half of the figure. After the collision, electron

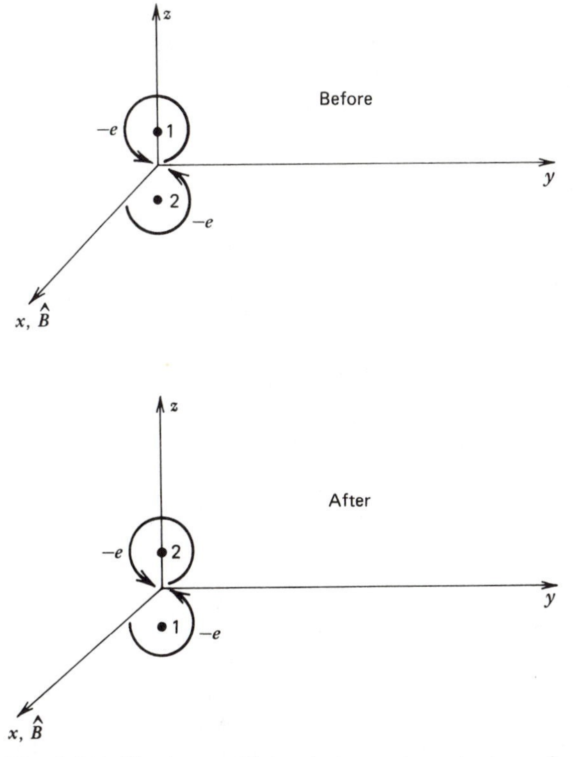

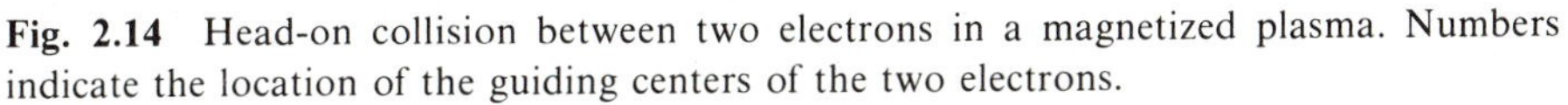
Fig. 2.14 Head-on collision between two electrons in a magnetized plasma. Numbers indicate the location of the guiding centers of the two electrons.

number 2 has the same orbit that electron number 1 had before the collision, and vice versa. Thus, the locations of the two guiding centers have been interchanged, and there is no net motion of the electrons. We conclude that collisions between like particles do not lead to diffusion of those particles across magnetic field lines.

Next, consider the (almost) head-on collision between an electron and a slightly more energetic positron at $\mathbf{x} = 0$, as shown in Fig. 2.15. The last gyro-orbit and the guiding center of each particle before the collision are indicated in the upper half of the figure. After the collision, the electron has slightly more energy than the positron, and both guiding centers have moved by two gyro-radii in the $(-)\hat{z}$-direction. Thus, the center-of-mass of the system has moved a substantial distance in the $(-)\hat{z}$-direction. We conclude that collisions between unlike particles can cause significant diffusion of particles across magnetic field lines. Further discussion of diffusion can be found in Ref. [1].

This completes our discussion of single-particle motion in prescribed electric and magnetic fields. In the next chapter, we begin a systematic treatment of plasma physics in which the electromagnetic fields and the particle orbits are determined self-consistently.

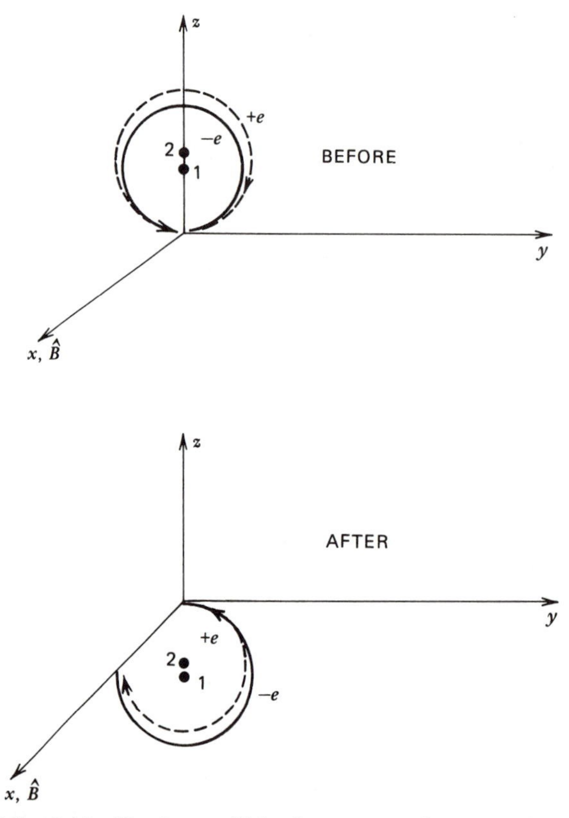

Fig. 2.15 Head-on collision between an electron and a slightly more energetic positron in a magnetized plasma.

REFERENCES

[1] F. F. Chen, *Introduction to Plasma Physics*, Plenum, New York, 1974.

[2] A. Baños, Jr., *J. Plasma Phys., 1*, 305 (1967).

[3] T. G. Northrop, *The Adiabatic Motion of Charged Particles*, Wiley, New York, 1963.

[4] G. Schmidt, *Physics of High Temperature Plasmas*, Academic, New York, 1966.

[5] M. Kruskal, *J. Math. Phys.*, *3*, 806 (1962).

[6] H. Goldstein, *Classical Mechanics*, 2nd ed., Addison-Wesley, Reading, Mass., 1980, Sect. 11-7.

[7] L. D. Landau and E. M. Lifshitz, *Mechanics*, Addison-Wesley, Reading, Mass., 1960, Sect. 30.

PROBLEMS

2.1 Example of a Drift

(a) Consider a particle of charge q and mass m, initially at rest at $(x,y,z) = (0,0,0)$, in the presence of a static magnetic field $\mathbf{B} = B_0\hat{z}$ and $\mathbf{E} = E_0\hat{y}$. Taking $E_0, B_0 > 0$, sketch the orbit of the particle when $q > 0$.

(b) Derive an exact expression for the orbit $[x(t), y(t), z(t)]$ of the particle in part (a). Does this result agree with the sketch of part (a)?

(c) Show that the orbit in (b) can be separated into an oscillatory term and a constant drift term. After averaging in time over the oscillatory motion, is there any net acceleration? If not, how are the forces in the problem balanced?

(d) In what direction is the drift for $q > 0$? For $q < 0$? If there were many particles of various charges and masses present, would there be any net current?

(e) Suppose the electric field were replaced by a force F_0 in the $\hat{y}$-direction. What would be the drift velocity? [*Hint:* Guess the answer using the last part of (c).]

2.2 Grad-B Drift

Let us derive the grad-B drift in a different way. With the force equation $\mathbf{F} = (q/c)\mathbf{v} \times \mathbf{B}$, insert the zero order orbit and the Taylor expanded magnetic field. Average $\mathbf{F}$ over one gyroperiod to obtain an average force. Insert this average force into the general drift equation (2.8), and compare the resulting drift to the grad-B drift (2.28).

2.3 Polarization Drift

Let us get the polarization drift (2.41) in a faster but less rigorous manner. With the given electric field $\mathbf{E} = -\dot{E}t\hat{y}$, calculate an $\mathbf{E} \times \mathbf{B}$ drift. Relate the resulting accelerated drift to a force (being careful with signs), plug in the **F** × **B** formula (2.8), and compare the result to (2.41).

2.4 Mirror Machines

(a) A mirror machine has mirror ratio 2. A Maxwellian group of electrons is released at the center of the machine. In the absence of collisions, what fraction of these electrons is confined?

(b) Suppose the mirror machine has initially equal densities $n \approx 10^{13}$ cm^{-3} of electrons and protons, each Maxwellian with a temperature 1 keV $\approx 10^7$ °C. The machine is roughly one meter in size in both directions. Recalling our discussion of collisions from Chapter 1, estimate very roughly the time for
   (1) loss of the unconfined electrons;
   (2) loss of the unconfined ions;
   (3) loss of many of the initially confined electrons (due primarily to which kind of collision?); why do not all of the electrons leave?;
   (4) loss of the initially confined ions (due primarily to which kind of collision?).

   For fusion purposes (supposing the protons were replaced by deuterium or tritium) which of these numbers is the most relevant?

2.5 Drift Energy

A particle of mass m and charge q in a uniform magnetic field $\mathbf{B} = B_0\hat{z}$ is set into motion in the $\hat{x}$-direction by an electric field $E(t)\hat{y}$ that varies slowly from zero to a final value E_0. Thus, at the final time the particle has an **E** × **B** drift $\mathbf{v}_0$.

(a) Use energy arguments to show that the particle's guiding center must have been displaced a distance v_0/Ω in the direction of the electric field.

(b) Integrate the polarization drift velocity from time zero to time infinity to obtain a displacement. Does the answer agree with (a)?

CHAPTER 3

Plasma Kinetic Theory I: Klimontovich Equation

3.1 INTRODUCTION

In this chapter, we begin a study of the basic equations of plasma physics. The word "kinetic" means "pertaining to motion," so that plasma kinetic theory is the theory of plasma taking into account the motions of all of the particles. This can be done in an exact way, using the *Klimontovich equation* of the present chapter or the *Liouville equation* of the next chapter. However, we are usually not interested in the exact motion of all of the particles in a plasma, but rather in certain average or approximate characteristics. Thus, the greatest usefulness of the exact Klimontovich and Liouville equations is as starting points for the derivation of approximate equations that describe the average properties of a plasma.

In classical plasma physics, we think of the particles as point particles, each with a given charge and mass. Suppose we have a gas consisting of only one particle. This particle has an orbit $\mathbf{X}_1(t)$ in three-dimensional configuration space $\mathbf{x}$. The orbit $\mathbf{X}_1(t)$ is the set of positions $\mathbf{x}$ occupied by the particle at successive times t. Likewise, the particle has an orbit $\mathbf{V}_1(t)$ in three-dimensional velocity space $\mathbf{v}$. We combine three-dimensional configuration space $\mathbf{x}$ and three-dimensional velocity space $\mathbf{v}$ into six-dimensional phase space $(\mathbf{x},\mathbf{v})$. The density of one particle in this phase space is

$$N(\mathbf{x},\mathbf{v},t) = \delta[\mathbf{x} - \mathbf{X}_1(t)]\delta[\mathbf{v} - \mathbf{V}_1(t)] \tag{3.1}$$

where $\delta[\mathbf{x} - \mathbf{X}_1] \equiv \delta(x - X_1)\delta(y - Y_1)\delta(z - Z_1)$, etc. (The properties of the Dirac delta function are reviewed in Ref. [1], p. 29, and in Ref. [2], pp. 53–54.) Note that $\mathbf{X}_1, \mathbf{V}_1$ are the Lagrangian coordinates of the particle itself, whereas $\mathbf{x}, \mathbf{v}$ are the Eulerian coordinates of the phase space.

EXERCISE At any time t, the density of particles integrated over all phase space must yield the total number of particles in the system. Verify this for the density (3.1).

Next, suppose we have a system with two point particles, with respective orbits $[\mathbf{X}_1(t), \mathbf{V}_1(t)]$ and $[\mathbf{X}_2(t), \mathbf{V}_2(t)]$ in phase space $(\mathbf{x},\mathbf{v})$. By analogy to (3.1), the particle density is

$$N(\mathbf{x},\mathbf{v},t) = \sum_{i=1}^{2} \delta[\mathbf{x} - \mathbf{X}_i(t)]\delta[\mathbf{v} - \mathbf{V}_i(t)] \tag{3.2}$$

EXERCISE Repeat the previous exercise for (3.2).

Now suppose that a system contains two species of particles, electrons and ions, and each species has N_0 particles. Then the density N_s of species s is

$$N_s(\mathbf{x},\mathbf{v},t) = \sum_{i=1}^{N_0} \delta[\mathbf{x} - \mathbf{X}_i(t)]\delta[\mathbf{v} - \mathbf{V}_i(t)] \tag{3.3}$$

and the total density N is

$$N(\mathbf{x},\mathbf{v},t) = \sum_{e,i} N_s(\mathbf{x},\mathbf{v},t) \tag{3.4}$$

EXERCISE Repeat the previous exercise for (3.4).

If we know the exact positions and velocities of the particles at one time, then we know them at all later times. This can be seen as follows. The position $\mathbf{X}_i(t)$ of particle i satisfies the equation

$$\dot{\mathbf{X}}_i(t) = \mathbf{V}_i(t) \tag{3.5}$$

where an overdot means a time derivative. Likewise, the velocity $\mathbf{V}_i(t)$ of particle i satisfies the Lorentz force equation

$$m_s\dot{\mathbf{V}}_i(t) = q_s\mathbf{E}^m[\mathbf{X}_i(t),t] + \frac{q_s}{c}\,\mathbf{V}_i(t) \times \mathbf{B}^m[\mathbf{X}_i(t),t] \tag{3.6}$$

where the superscript m indicates that the electric and magnetic fields are the *m*icroscopic fields self-consistently produced by the point particles themselves, together with externally applied fields. [On the right of (3.6), the portion of $\mathbf{E}^m$ and $\mathbf{B}^m$ produced by particle i itself is deleted.] The microscopic fields satisfy Maxwell's equations

$$\nabla \cdot \mathbf{E}^m(\mathbf{x},t) = 4\pi\rho^m(\mathbf{x},t) \tag{3.7}$$

$$\nabla \cdot \mathbf{B}^m(\mathbf{x},t) = 0 \tag{3.8}$$

$$\nabla \times \mathbf{E}^m(\mathbf{x},t) = -\frac{1}{c}\frac{\partial \mathbf{B}^m(\mathbf{x},t)}{\partial t} \tag{3.9}$$

and

$$\nabla \times \mathbf{B}^m(\mathbf{x},t) = \frac{4\pi}{c}\,\mathbf{J}^m(\mathbf{x},t) + \frac{1}{c}\frac{\partial \mathbf{E}^m(\mathbf{x},t)}{\partial t} \tag{3.10}$$

The microscopic charge density is

$$\rho^m(\mathbf{x},t) = \sum_{e,i} q_s \int d\mathbf{v}\, N_s(\mathbf{x},\mathbf{v},t) \tag{3.11}$$

while the microscopic current is

$$\mathbf{J}^m(\mathbf{x},t) = \sum_{e,i} q_s \int d\mathbf{v}\mathbf{v} N_s(\mathbf{x},\mathbf{v},t) \tag{3.12}$$

EXERCISE Convince yourself that (3.11) and (3.12) yield the correct charge density and current.

Equations 3.7 to 3.12 determine the exact fields in terms of the exact particle orbits, while (3.5) and (3.6) determine the exact particle orbits in terms of the exact fields. The entire set of equations is closed, so that if the positions and velocities of all particles, and the fields, are known exactly at one time, then they are known exactly at all later times.

3.2 KLIMONTOVICH EQUATION

An exact equation for the evolution of a plasma is obtained by taking the time derivative of the density N_s. From (3.3), this is

$$\frac{\partial N_s(\mathbf{x},\mathbf{v},t)}{\partial t} = -\sum_{i=1}^{N_0} \dot{\mathbf{X}}_i \cdot \nabla_{\mathbf{x}}\, \delta[\mathbf{x} - \mathbf{X}_i(t)]\delta[\mathbf{v} - \mathbf{V}_i(t)]$$
$$- \sum_{i=1}^{N_0} \dot{\mathbf{V}}_i \cdot \nabla_{\mathbf{v}}\, \delta[\mathbf{x} - \mathbf{X}_i(t)]\delta[\mathbf{v} - \mathbf{V}_i(t)] \tag{3.13}$$

where we have used the relations

$$\frac{\partial}{\partial a} f(a - b) = -\frac{\partial}{\partial b} f(a - b)$$

and

$$\frac{d}{dt} f[g(t)] = \frac{df}{dg}\, \dot{g}$$

and where $\nabla_{\mathbf{x}} \equiv (\partial_x, \partial_y, \partial_z)$ and $\nabla_{\mathbf{v}} \equiv (\partial_{v_x}, \partial_{v_y}, \partial_{v_z})$. Using (3.5) and (3.6), we can write $\dot{\mathbf{X}}_i$ and $\dot{\mathbf{V}}_i$ in terms of $\mathbf{V}_i$ and the fields $\mathbf{E}^m$ and $\mathbf{B}^m$, whereupon (3.13) becomes

$$\frac{\partial N_s(\mathbf{x},\mathbf{v},t)}{\partial t} = -\sum_{i=1}^{N_0} \mathbf{V}_i \cdot \nabla_{\mathbf{x}}\, \delta[\mathbf{x} - \mathbf{X}_i]\delta[\mathbf{v} - \mathbf{V}_i]$$
$$- \sum_{i=1}^{N_0} \left\{ \frac{q_s}{m_s} \mathbf{E}^m[\mathbf{X}_i(t),t] + \frac{q_s}{m_s c} \mathbf{V}_i \times \mathbf{B}^m[\mathbf{X}_i(t),t] \right\}$$
$$\cdot \nabla_{\mathbf{v}}\, \delta[\mathbf{x} - \mathbf{X}_i]\delta[\mathbf{v} - \mathbf{V}_i] \tag{3.14}$$

An important property of the Dirac delta function is

$$a\delta(a - b) = b\delta(a - b)$$

EXERCISE How would one prove this relation?

This relation allows us to replace $\mathbf{V}_i(t)$ with $\mathbf{v}$, and $\mathbf{X}_i(t)$ with $\mathbf{x}$, on the right of (3.14) (but not in the arguments of the delta functions) so that (3.14) becomes

$$\frac{\partial N_s(\mathbf{x},\mathbf{v},t)}{\partial t} = - \mathbf{v} \cdot \nabla_\mathbf{x} \sum_{i=1}^{N_0} \delta[\mathbf{x} - \mathbf{X}_i]\delta[\mathbf{v} - \mathbf{V}_i]$$
$$- \left[\frac{q_s}{m_s} \mathbf{E}^m(\mathbf{x},t) + \frac{q_s}{m_s c} \mathbf{v} \times \mathbf{B}^m(\mathbf{x},t)\right] \cdot \nabla_\mathbf{v} \sum_{i=1}^{N_0} \delta[\mathbf{x} - \mathbf{X}_i]\delta[\mathbf{v} - \mathbf{V}_i] \tag{3.15}$$

But the two summations on the right of (3.15) are just the density (3.3); therefore

$$\boxed{\frac{\partial N_s(\mathbf{x},\mathbf{v},t)}{\partial t} + \mathbf{v} \cdot \nabla_\mathbf{x} N_s + \frac{q_s}{m_s}\left(\mathbf{E}^m + \frac{\mathbf{v}}{c} \times \mathbf{B}^m\right) \cdot \nabla_\mathbf{v} N_s = 0} \tag{3.16}$$

This is the exact *Klimontovich equation* (Klimontovich [3]; Dupree [4]).

The Klimontovich equation, together with Maxwell's equations, constitute an exact description of a plasma. Given the initial positions and velocities of the particles, the initial densities $N_e(\mathbf{x},\mathbf{v},t = 0)$ and $N_i(\mathbf{x},\mathbf{v},t = 0)$ are given exactly by (3.3). The initial fields are then chosen to be consistent with Maxwell's equations (3.7) to (3.12). With these initial conditions the problem is completely deterministic, and the densities and fields are exactly determined for all time.

In practice, we never carry out this procedure. The Klimontovich equation contains every one of the exact single particle orbits. This is far more information than we want or need. What we really want is information about certain average properties of the plasma. We do not really care about all of the individual electromagnetic fields contributed by the individual charges. What we do care about is the average long-range electric field, which might exist over many thousands or millions of interparticle spacings. The usefulness of the Klimontovich equation comes from its role as a starting point in the derivation of equations that describe the average properties of a plasma.

The Klimontovich equation can be thought of as expressing the incompressibility of the "substance" $N_s(\mathbf{x},\mathbf{v},t)$ as it moves about in the $(\mathbf{x},\mathbf{v})$ phase space. (Is it any wonder that a point particle is incompressible?) This can be seen as follows. Imagine a hypothetical particle with charge q_s, mass m_s, which at time t finds itself at the position $(\mathbf{x},\mathbf{v})$. This hypothetical particle has an orbit in phase space determined by the fields in the system. Imagine taking a time derivative of any quantity along this orbit (such a time derivative is called a *convective derivative*). This derivative must include the time variation produced by the changing position in $(\mathbf{x},\mathbf{v})$ space as well as the explicit time variation of the quantity. Thus, it must be given by

$$\frac{D}{Dt} \equiv \frac{\partial}{\partial t} + \left.\frac{d\mathbf{x}}{dt}\right|_{\text{orbit}} \cdot \nabla_\mathbf{x} + \left.\frac{d\mathbf{v}}{dt}\right|_{\text{orbit}} \cdot \nabla_\mathbf{v} \tag{3.17}$$

where by $d\mathbf{x}/dt|_{\text{orbit}}$ we mean the change in position $\mathbf{x}$ of the hypothetical particle with time; likewise for $d\mathbf{v}/dt|_{\text{orbit}}$. But for our hypothetical particle at position $(\mathbf{x},\mathbf{v})$

in phase space we know that

$$\left.\frac{d\mathbf{x}}{dt}\right|_{\text{orbit}} = \mathbf{v} \tag{3.18}$$

and

$$\left.\frac{d\mathbf{v}}{dt}\right|_{\text{orbit}} = \frac{q_s}{m_s}\left[\mathbf{E}^m(\mathbf{x},t) + \frac{\mathbf{v}}{c} \times \mathbf{B}^m(\mathbf{x},t)\right] \tag{3.19}$$

Thus,

$$\frac{D}{Dt} = \frac{\partial}{\partial t} + \mathbf{v}\cdot\nabla_{\mathbf{x}} + \frac{q_s}{m_s}\left[\mathbf{E}^m(\mathbf{x},t) + \frac{\mathbf{v}}{c} \times \mathbf{B}^m(\mathbf{x},t)\right]\cdot\nabla_{\mathbf{v}} \tag{3.20}$$

and the Klimontovich equation (3.16) simply says

$$\boxed{\frac{D}{Dt} N_s(\mathbf{x},\mathbf{v},t) = 0} \tag{3.21}$$

The density of particles of species s is a constant in time, as measured along the orbit of a hypothetical particle of species s. This is true whether we are moving along the orbit of an actual particle, in which case the density is infinite, or whether we are moving along a hypothetical orbit that is not occupied by an actual particle, in which case the density is zero. Note that the density is only constant as measured along orbits of hypothetical particles; in $(\mathbf{x},\mathbf{v})$ space at a given time it is not constant but is zero or infinite.

There is yet a third way to think of the Klimontovich equation. Any fluid in which the fluid density $f(\mathbf{r},t)$ is neither created nor destroyed satisfies a continuity equation

$$\partial_t f(\mathbf{r},t) + \nabla_{\mathbf{r}}\cdot(f\mathbf{V}) = 0 \tag{3.22}$$

where $\nabla_{\mathbf{r}}$ is the divergence vector in the phase space under consideration, and $\mathbf{V}$ is a vector that gives the time rate of change of a fluid element at a point in phase space. (See, for example, Symon [5], p. 317.) In the present case, $\nabla_{\mathbf{r}} = (\nabla_{\mathbf{x}},\nabla_{\mathbf{v}})$ and $\mathbf{V} = (d\mathbf{x}/dt|_{\text{orbit}}, d\mathbf{v}/dt|_{\text{orbit}})$. Since the particle density is neither created nor destroyed, it must satisfy a continuity equation of the form

$$\partial_t N_s(\mathbf{x},\mathbf{v},t) + \nabla_{\mathbf{x}}\cdot(\mathbf{v}N_s) + \nabla_{\mathbf{v}}\cdot\left\{\frac{q_s}{m_s}\left[\mathbf{E}^m + \frac{\mathbf{v}}{c}\times\mathbf{B}^m\right]N_s\right\} = 0 \tag{3.23}$$

It is left as a problem to demonstrate that the continuity equation (3.23) is equivalent to the Klimontovich equation (3.16).

3.3 PLASMA KINETIC EQUATION

Although the Klimontovich equation is exact, we are really not interested in exact solutions of it. These would contain all of the particle orbits, and would thus be far too detailed for any practical purpose. What we really would like to know are the average properties of a plasma. The Klimontovich equation tells us whether or not a particle with infinite density is to be found at a given point $(\mathbf{x},\mathbf{v})$ in phase space.

What we really want to know is how many particles are likely to be found in a small volume $\Delta\mathbf{x}\,\Delta\mathbf{v}$ of phase space whose center is at $(\mathbf{x},\mathbf{v})$. Thus, we really are not interested in the spikey function $N_s(\mathbf{x},\mathbf{v},t)$, but rather in the smooth function

$$f_s(\mathbf{x},\mathbf{v},t) \equiv \langle N_s(\mathbf{x},\mathbf{v},t)\rangle \tag{3.24}$$

The most rigorous way to interpret $\langle\ \ \rangle$ is as an ensemble average [6] over an infinite number of realizations of the plasma, prepared according to some prescription. For example, we could prepare an ensemble of equal temperature plasmas, each in thermal equilibrium, and each with a test charge q_T at the origin of configuration space. The resulting f_e and f_i would then be consistent with the discussion of Debye shielding in Section 1.2.

There is another useful interpretation of the *distribution function* $f_s(\mathbf{x},\mathbf{v},t)$, the number of particles of species s per unit configuration space per unit velocity space. Suppose we are interested in long range electric and magnetic fields that extend over distances much larger than a Debye length. Then we can imagine a box, centered around the point $\mathbf{x}$ in configuration space, of a size much greater than a mean interparticle spacing, but much smaller than a Debye length (this is easy to do in a plasma; why?) We can now count the number of particles of species s in the box at time t with velocities in the range $\mathbf{v}$ to $\mathbf{v} + \Delta\mathbf{v}$, divide by (the size of the box multiplied by $\Delta v_x\,\Delta v_y\,\Delta v_z$), and call the result $f_s(\mathbf{x},\mathbf{v},t)$. This number will of course fluctuate with time but, if there are very many particles in the box, the fluctuations will be tiny and the $f_s(\mathbf{x},\mathbf{v},t)$ obtained in this manner will agree very well with that obtained in the more rigorous ensemble averaging procedure.

An equation for the time evolution of the distribution function $f_s(\mathbf{x},\mathbf{v},t)$ can be obtained from the Klimontovich equation (3.16) by ensemble averaging. We define δN_s, $\delta\mathbf{E}$, and $\delta\mathbf{B}$ by

$$\begin{aligned} N_s(\mathbf{x},\mathbf{v},t) &= f_s(\mathbf{x},\mathbf{v},t) + \delta N_s(\mathbf{x},\mathbf{v},t) \\ \mathbf{E}^m(\mathbf{x},\mathbf{v},t) &= \mathbf{E}(\mathbf{x},\mathbf{v},t) + \delta\mathbf{E}(\mathbf{x},\mathbf{v},t) \end{aligned} \tag{3.25}$$

and

$$\mathbf{B}^m(\mathbf{x},\mathbf{v},t) = \mathbf{B}(\mathbf{x},\mathbf{v},t) + \delta\mathbf{B}(\mathbf{x},\mathbf{v},t)$$

where $\mathbf{B} \equiv \langle\mathbf{B}^m\rangle$, $\mathbf{E} \equiv \langle\mathbf{E}^m\rangle$, and $\langle\delta N_s\rangle = \langle\delta\mathbf{E}\rangle = \langle\delta\mathbf{B}\rangle = 0$. Inserting these definitions into (3.16) and ensemble averaging, we obtain

$$\begin{aligned} \frac{\partial f_s(\mathbf{x},\mathbf{v},t)}{\partial t} &+ \mathbf{v}\cdot\nabla_\mathbf{x} f_s + \frac{q_s}{m_s}\left(\mathbf{E} + \frac{\mathbf{v}}{c}\times\mathbf{B}\right)\cdot\nabla_\mathbf{v} f_s \\ &= -\frac{q_s}{m_s}\left\langle\left(\delta\mathbf{E} + \frac{\mathbf{v}}{c}\times\delta\mathbf{B}\right)\cdot\nabla_\mathbf{v}\,\delta N_s\right\rangle \end{aligned} \tag{3.26}$$

Equation (3.26) is the exact form of the *plasma kinetic equation*. We shall meet other forms of this equation in the next chapter.

The left side of (3.26) consists only of terms that vary smoothly in $(\mathbf{x},\mathbf{v})$ space. The right side is the ensemble average of the products of very spikey quantities like $\delta\mathbf{E} = \mathbf{E}^m - \langle\mathbf{E}^m\rangle$ and δN_s. Thus, the left side of (3.26) contains terms that are insensitive to the discrete-particle nature of the plasma, while the right side of (3.26) is very sensitive to the discrete-particle nature of the plasma. But the discrete-particle nature of a plasma is what gives rise to collisional effects, so that the left side of (3.26) contains smoothly varying functions representing collective effects, while the right side represents the collisional effects. We have seen in Section 1.6 that the ratio of the importance of collisional effects to the importance of collective effects is sometimes given by $1/\Lambda$, which is a very small number. We might guess that for many phenomena in a plasma, the right side of (3.26) has a size $1/\Lambda$ compared to each of the terms on the left side; thus the right side can be neglected for the study of such phenomena. This indeed is the case, as shown in the next two chapters.

This important point can be illustrated by a hypothetical exercise. Imagine that we break each electron into an infinite number of pieces, so that $n_0 \to \infty$, $m_e \to 0$, and $e \to 0$, while $n_0 e$ = constant, e/m_e = constant, and v_e = constant.

EXERCISE Show that in this hypothetical exercise, ω_e = constant, λ_e = constant, but $T_e \to 0$ and $\Lambda_e \to \infty$.

Then any volume, no matter how small, would contain an infinite number of point particles, each represented by a delta function with infinitesimal charge. Statistical mechanics tells us that the relative fluctuations in such a plasma would vanish, since the fluctuations in the number of particles N_0 in a certain volume is proportional to the square root of that number. Thus, on the right side of (3.26) we have $\delta N_s \sim N_0^{1/2} \sim \Lambda_e^{1/2}$, and $\delta\mathbf{E}$ and $\delta\mathbf{B}$, which are produced by δN_s behaving like (from Poisson's equation) $\sim e\delta N_s \sim N_0^{-1}N_0^{1/2} \sim N_0^{-1/2} \sim \Lambda_e^{-1/2}$, so that the right side becomes constant. On the left, however, each term becomes infinite as $f_s \to \infty$. Thus, the relative importance of the right side vanishes $\sim N_0^{-1} \sim \Lambda_e^{-1}$, and we have

$$\boxed{\frac{\partial f_s(\mathbf{x},\mathbf{v},t)}{\partial t} + \mathbf{v}\cdot\nabla f_s + \frac{q_s}{m_s}\left(\mathbf{E} + \frac{\mathbf{v}}{c}\times\mathbf{B}\right)\cdot\nabla_{\mathbf{v}} f_s = 0} \tag{3.27}$$

which is the *Vlasov [7] equation* (sometimes referred to as the collisionless Boltzmann equation). This approximate equation, which neglects collisional effects, is often called the most important equation in plasma physics. Its properties will be explored in detail in Chapter 6.

The fields **E** and **B** of (3.27) are the ensemble averaged fields of (3.25). They must satisfy the ensemble averaged versions of Maxwell's equations (3.7) to (3.12), which are

$$\nabla\cdot\mathbf{E}(\mathbf{x},t) = 4\pi\rho$$

$$\nabla\cdot\mathbf{B}(\mathbf{x},t) = 0$$

$$\nabla\times\mathbf{E}(\mathbf{x},t) = -\frac{1}{c}\frac{\partial\mathbf{B}}{\partial t}$$

$$\nabla\times\mathbf{B}(\mathbf{x},t) = \frac{4\pi}{c}\mathbf{J} + \frac{1}{c}\frac{\partial\mathbf{E}}{\partial t}$$

$$\rho(\mathbf{x},t) \equiv \langle\rho^m\rangle = \sum_{e,i} q_s\int d\mathbf{v}\, f_s(\mathbf{x},\mathbf{v},t)$$

and

$$\mathbf{J}(\mathbf{x},t) \equiv \langle\mathbf{J}^m\rangle = \sum_{e,i} q_s\int d\mathbf{v}\,\mathbf{v} f_s(\mathbf{x},\mathbf{v},t) \tag{3.28}$$

In the next two chapters we shall approach the plasma kinetic equation (3.26) from another direction, and shall use approximate methods to evaluate the collisional right side. In Chapter 6 we shall take up the study of the Vlasov equation (3.27).

REFERENCES

[1] J. D. Jackson, *Classical Electrodynamics*, 2nd ed., Wiley, New York, 1975.

[2] K. Gottfried, *Quantum Mechanics*, Benjamin, New York, 1966.

[3] Yu. L. Klimontovich, *The Statistical Theory of Non-equilibrium Processes in a Plasma*, M.I.T. Press, Cambridge, Mass., 1967.

[4] T. H. Dupree, *Phys. Fluids*, *6*, 1714 (1963).

[5] K. R. Symon, *Mechanics*, 3rd ed., Addison-Wesley, Reading, Mass., 1971.

[6] F. Reif, *Fundamentals of Statistical and Thermal Physics*, McGraw-Hill, New York, 1965.

[7] A. A. Vlasov, *J. Phys. (U.S.S.R.)*, *9*, 25 (1945).

PROBLEM

3.1 Klimontovich as Continuity

Prove that the continuity equation (3.23) is equivalent to the Klimontovich equation (3.16).

CHAPTER 4

Plasma Kinetic Theory II: Liouville Equation

4.1 INTRODUCTION

In addition to the Klimontovich equation, there is another equation, the *Liouville equation*, which provides an exact description of a plasma. Like the Klimontovich equation, the Liouville equation is of no direct use, but provides a starting point for the construction of approximate statistical theories. One of the most useful practical results of this approach is to provide us with an approximate form for the right side of the plasma kinetic equation (3.26), which tells us how the distribution function changes in time due to collisions.

The Klimontovich equation describes the behavior of individual particles. By contrast, the Liouville equation describes the behavior of *systems*. Consider first a "system" consisting of one charged particle. Suppose we measure this particle's position in a coordinate system $\mathbf{x}_1$; then the orbit of the particle $\mathbf{X}_1(t)$ is the set of positions $\mathbf{x}_1$ occupied by the particle at consecutive times t. Likewise, in velocity space we denote the orbit of the particle $\mathbf{V}_1(t)$; this is the set of velocities taken by the particle at consecutive times t; these velocities are measured in a coordinate system $\mathbf{v}_1$. We thus have a phase space $(\mathbf{x}_1,\mathbf{v}_1) = (x_1,y_1,z_1,v_{x_1},v_{y_1},v_{z_1})$. In this six-dimensional phase space there is one "system" consisting of one particle. The *density of systems* in this phase space is

$$N(\mathbf{x}_1,\mathbf{v}_{1,}t) = \delta[\mathbf{x}_1 - \mathbf{X}_1(t)]\delta[\mathbf{v}_1 - \mathbf{V}_1(t)] \tag{4.1}$$

Next, consider a system of two particles. We introduce a set of coordinate axes for each particle. Particle 1 has $(\mathbf{x}_1,\mathbf{v}_1)$ coordinate axes as before. Particle 2 has $(\mathbf{x}_2,\mathbf{v}_2)$ coordinate axes that lay right on top of the $(\mathbf{x}_1,\mathbf{v}_1)$ coordinate axes. The orbit $\mathbf{X}_1(t)$, $\mathbf{V}_1(t)$ of particle 1 is measured with respect to the $(\mathbf{x}_1,\mathbf{v}_1)$ coordinate axes, while the orbit $\mathbf{X}_2(t)$, $\mathbf{V}_2(t)$ of particle 2 is measured with respect to the $(\mathbf{x}_2,\mathbf{v}_2)$ coordinate axes. We now introduce an entirely new phase space, having twelve dimensions. The phase space is

$$(\mathbf{x}_1,\mathbf{v}_1,\mathbf{x}_2,\mathbf{v}_2) = (x_1,y_1,z_1,v_{x_1},v_{y_1},v_{z_1},x_2,y_2,z_2,v_{x_2},v_{y_2},v_{z_2}) \tag{4.2}$$

In this twelve-dimensional phase space, there is one system that is occupying the point $[\mathbf{x}_1 = \mathbf{X}_1(t), \mathbf{v}_1 = \mathbf{V}_1(t), \mathbf{x}_2 = \mathbf{X}_2(t), \mathbf{v}_2 = \mathbf{V}_2(t)]$ at time t. The density of systems in this phase space is

$$N(\mathbf{x}_1,\mathbf{v}_1,\mathbf{x}_2,\mathbf{v}_2,t) = \delta[\mathbf{x}_1 - \mathbf{X}_1(t)]\delta[\mathbf{v}_1 - \mathbf{V}_1(t)]\delta[\mathbf{x}_2 - \mathbf{X}_2(t)]\delta[\mathbf{v}_2 - \mathbf{V}_2(t)] \tag{4.3}$$

EXERCISE Show that there is indeed one system in the phase space by integrating the density (4.3) over all phase space.

Note that the density N in (4.3) is completely different from the density N_s used in the previous chapter in the discussion of the Klimontovich equation. The density N_s in Ch. 3 is the density of particles in six-dimensional phase space. The density N in (4.3) is the density of systems (each having two particles) in twelve-dimensional phase space.

Finally, suppose that we have a system of N_0 particles. With each particle i, $i = 1,2, \ldots N_0$, we associate a six-dimensional coordinate system $(\mathbf{x}_i,\mathbf{v}_i)$. Using these $6N_0$ coordinate axes, we construct a $6N_0$-dimensional phase space, analogous to the twelve-dimensional phase space in (4.2). There is one system in $6N_0$-dimensional phase space; therefore the density of systems, by analogy with the density of systems (4.3), is

$$N(\mathbf{x}_1,\mathbf{v}_1,\mathbf{x}_2,\mathbf{v}_2 \ldots \mathbf{x}_{N_0},\mathbf{v}_{N_0},t) = \prod_{i=1}^{N_0} \delta[\mathbf{x}_i - \mathbf{X}_i(t)]\delta[\mathbf{v}_i - \mathbf{V}_i(t)] \tag{4.4}$$

where $\Pi_{j=1}^{n} f_j \equiv f_1 f_2 \ldots f_n$.

EXERCISE Use (4.4) to prove that there is one system in all of phase space.

4.2 LIOUVILLE EQUATION

As with the Klimontovich equation in Chapter 3, the Liouville equation is obtained by taking the time derivative of the appropriate density. In this case, we take the time derivative of the density of systems (4.4). Because the density of systems (4.4) is the product of $6N_0$ terms, its time derivative involves the sum of $6N_0$ terms. Using the relation

$$\frac{\partial}{\partial t}\,\delta[\mathbf{x}_i - \mathbf{X}_i(t)] = -\frac{\partial \mathbf{X}_i}{\partial t}\cdot\nabla_{\mathbf{x}_i}\,\delta[\mathbf{x}_i - \mathbf{X}_i(t)] \tag{4.5}$$

and similar relations encountered in the previous chapter, the time derivative of (4.4) is

$$\frac{\partial N}{\partial t} + \sum_{i=1}^{N_0} \mathbf{V}_i(t)\cdot\nabla_{\mathbf{x}_i} \prod_{j=1}^{N_0} \delta(\mathbf{x}_j - \mathbf{X}_j)\delta(\mathbf{v}_j - \mathbf{V}_j)$$

$$+ \sum_{i=1}^{N_0} \dot{\mathbf{V}}_i\cdot\nabla_{\mathbf{v}_i} \prod_{j=1}^{N_0} \delta(\mathbf{x}_j - \mathbf{X}_j)\delta(\mathbf{v}_j - \mathbf{V}_j) = 0 \tag{4.6}$$

Using $a\delta(a - b) = b\delta(a - b)$ to replace $\mathbf{V}_i$ by $\mathbf{v}_i$, and similarly for $\dot{\mathbf{V}}_i$ so that for the remainder of this chapter

$$\dot{\mathbf{V}}_i(t) = \frac{q_{s_i}}{m_{s_i}}\left[\mathbf{E}^m(\mathbf{x}_i,t) + \frac{\mathbf{v}_i}{c} \times \mathbf{B}^m(\mathbf{x}_i,t) \right] \tag{4.7}$$

and noting that the products are just the density of systems N, (4.6) becomes

$$\boxed{\frac{\partial N}{\partial t} + \sum_{i=1}^{N_0} \mathbf{v}_i \cdot \nabla_{\mathbf{x}_i} N + \sum_{i=1}^{N_0} \dot{\mathbf{V}}_i(t) \cdot \nabla_{\mathbf{v}_i} N = 0} \tag{4.8}$$

which is the *Liouville equation*. When combined with Maxwell's equations and the Lorentz force equation, the Liouville equation is an exact description of a plasma. For a two-component plasma with $N_0/2$ electrons and $N_0/2$ ions, the expression for $\mathbf{V}_i(t)$ will depend upon whether the ith particle is an electron or a proton. The Liouville equation has all of the advantages and all of the disadvantages of the Klimontovich equation. Because it contains all of the exact six-dimensional orbits of the individual particles in a single system orbit in $6N_0$-dimensional space, it contains far more information than we want or need. Its usefulness is as a starting point in deriving a reduced statistical description, which with appropriate approximations can yield practical information.

Equation (4.8) has the form of a convective time derivative in the $6N_0$-dimensional phase space,

$$\frac{D}{Dt} N(\mathbf{x}_1,\mathbf{v}_1,\mathbf{x}_2,\mathbf{v}_2, \ldots, \mathbf{x}_{N_0},\mathbf{v}_{N_0},t) = 0 \tag{4.9}$$

where

$$\frac{D}{Dt} \equiv \frac{\partial}{\partial t} + \sum_{i=1}^{N_0} \mathbf{v}_i \cdot \nabla_{\mathbf{x}_i} + \sum_{i=1}^{N_0} \dot{\mathbf{V}}_i(t) \cdot \nabla_{\mathbf{v}_i} \tag{4.10}$$

Here, $\dot{\mathbf{V}}_i(t)$ is expressed in terms of the position $(\mathbf{x}_1,\mathbf{v}_1,\mathbf{x}_2,\mathbf{v}_2, \ldots, \mathbf{x}_{N_0},\mathbf{v}_{N_0})$ of the system in $6N_0$-dimensional phase space, since that position determines the positions of the particles in six-dimensional space and thus the fields at all points in six-dimensional space through Maxwell's equations. Thus, the convective time derivative, taken along the system orbit in $6N_0$-dimensional phase space, is zero. The density of systems is incompressible.

The Liouville equation (4.8) can also be put in the form of a continuity equation. Recall the vector identity $\nabla \cdot (a\mathbf{b}) = \mathbf{b} \cdot \nabla a + a\nabla \cdot \mathbf{b}$. Then

$$\mathbf{v}_i \cdot \nabla_{\mathbf{x}_i} N = \nabla_{\mathbf{x}_i} \cdot (\mathbf{v}_i N) \tag{4.11}$$

since $\mathbf{v}_i$ and $\mathbf{x}_i$ are independent variables. Similarly,

$$\dot{\mathbf{V}}_i \cdot \nabla_{\mathbf{v}_i} N = \nabla_{\mathbf{v}_i} \cdot (\dot{\mathbf{V}}_i N) \tag{4.12}$$

since

$$\nabla_{\mathbf{v}_i} \cdot \dot{\mathbf{V}}_i = \nabla_{\mathbf{v}_i} \cdot \left\{ \frac{q_{s_i}}{m_{s_i}} \left[\mathbf{E}^m(\mathbf{x}_i,t) + \frac{\mathbf{v}_i}{c} \times \mathbf{B}^m(\mathbf{x}_i,t) \right] \right\} = 0 \tag{4.13}$$

EXERCISE Prove (4.13).

Then the Liouville equation (4.8) becomes

$$\frac{\partial N}{\partial t} + \sum_{i=1}^{N_0} \nabla_{\mathbf{x}_i} \cdot (\mathbf{v}_i N) + \sum_{i=1}^{N_0} \nabla_{\mathbf{v}_i} \cdot (\dot{\mathbf{V}}_i N) = 0 \tag{4.14}$$

In the form of a continuity equation, the Liouville equation expresses the conservation of systems in $6N_0$-dimensional phase space.

As we have introduced it, the Liouville equation describes the exact orbit of a single point in $6N_0$-dimensional phase space. An example is shown in Fig. 4.1, which is a projection of the orbit onto three of the $6N_0$ dimensions. As the individual particles of the system move about in six-dimensional space, the system itself moves along a continuous orbit in $6N_0$-dimensional phase space.

Suppose that we have an ensemble of such systems, prepared at time t_0. At any later time $t \geq t_0$, we define

$$f_{N_0}(\mathbf{x}_1, \mathbf{v}_1, \mathbf{x}_2, \mathbf{v}_2, \ldots, \mathbf{x}_{N_0}, \mathbf{v}_{N_0}, t)\, d\mathbf{x}_1 d\mathbf{v}_1 d\mathbf{x}_2 d\mathbf{v}_2 \ldots d\mathbf{x}_{N_0} d\mathbf{v}_{N_0}$$

to be the probability that a particular system is at the point $(\mathbf{x}_1, \mathbf{v}_1, \ldots, \mathbf{x}_{N_0}, \mathbf{v}_{N_0})$ in $6N_0$-dimensional phase space, that is, the probability that $\mathbf{X}_1(t)$ lies between $\mathbf{x}_1$ and $\mathbf{x}_1 + d\mathbf{x}_1$, *and* $\mathbf{V}_1(t)$ lies between $\mathbf{v}_1$ and $\mathbf{v}_1 + d\mathbf{v}_1$ *and* $\mathbf{X}_2(t)$ lies between $\mathbf{x}_2$ and $\mathbf{x}_2 + d\mathbf{x}_2$, *and* etc. Since f_{N_0} is a probability density, its integral over all $6N_0$ dimensions must be unity.

Each system in the ensemble moves along an orbit like that shown in Fig. 4.1. We can think of this orbit as carrying its "piece" of probability along with it. A large probability for point A in Fig. 4.1 at time t_0 implies a large probability for point B at time t. In other words, we can think of the probability density as a fluid moving in the $6N_0$-dimensional phase space. Each element in the probability fluid moves along an exact orbit as given by the solution of the Liouville equation (4.8). Since each element of probability fluid moves along a continuous orbit, and since probability is neither created nor destroyed, the probability fluid must satisfy a

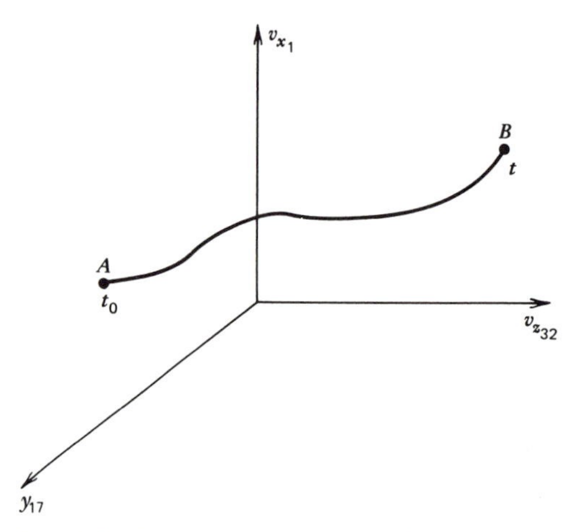

Fig. 4.1 A projection onto three dimensions of a typical system orbit in $6\,N_0$-dimensional phase space.

continuity equation in $6N_0$-dimensional phase space of the form (4.14). Thus, f_{N_0} must satisfy

$$\frac{\partial f_{N_0}}{\partial t} + \sum_{i=1}^{N_0} \nabla_{\mathbf{x}_i}\cdot(\mathbf{v}_i f_{N_0}) + \sum_{i=1}^{N_0} \nabla_{\mathbf{v}_i}\cdot(\dot{\mathbf{V}}_i f_{N_0}) = 0 \tag{4.15}$$

where $\dot{\mathbf{V}}_i(t)$ is, as usual, calculated from the Lorentz force equation (4.7) and the fields $\mathbf{E}^m$ and $\mathbf{B}^m$ are the exact fields appropriate to the system that occupies this particular point in $6N_0$-dimensional phase space.

We shall only be concerned with smooth functions f_{N_0}. Thus, we might think of a drop of ink placed in a glass of water. The initial drop contains all those systems that have a finite probability of being represented in the ensemble of systems at time t_0. Ignoring diffusion, the drop may lengthen, contract, distort, squeeze, break into pieces, deform, etc., as time progresses. However, the total volume of ink is always constant; the total probability is always unity. The convection of the probability ink is expressed mathematically by reversing the steps that led from the Liouville equation (4.8) to the continuity equation (4.14). (See Problem 4.1.) Equation (4.15) becomes

$$\boxed{\frac{\partial f_{N_0}}{\partial t} + \sum_{i=1}^{N_0} \mathbf{v}_i\cdot\nabla_{\mathbf{x}_i} f_{N_0} + \sum_{i=1}^{N_0} \dot{\mathbf{V}}_i\cdot\nabla_{\mathbf{v}_i} f_{N_0} = 0} \tag{4.16}$$

which by (4.10) is

$$\frac{Df_{N_0}}{Dt} = 0 \tag{4.17}$$

Equation (4.16) is the Liouville equation for the probability density f_{N_0}. Thus, the density of the probability ink is a constant provided that we move with the ink. The probability density f_{N_0} is incompressible.

4.3 BBGKY HIERARCHY

As discussed above, the density f_{N_0} represents the joint probability density that particle 1 has coordinates between $(\mathbf{x}_1,\mathbf{v}_1)$ and $(\mathbf{x}_1 + d\mathbf{x}_1,\mathbf{v}_1 + d\mathbf{v}_1)$ *and* particle 2 has coordinates between $(\mathbf{x}_2,\mathbf{v}_2)$ and $(\mathbf{x}_2 + d\mathbf{x}_2,\mathbf{v}_2 + d\mathbf{v}_2)$, *and* etc. We may also consider reduced probability distributions

$$f_k(\mathbf{x}_1,\mathbf{v}_1,\mathbf{x}_2,\mathbf{v}_2,\ldots,\mathbf{x}_k,\mathbf{v}_k,t) \equiv V^k \int d\mathbf{x}_{k+1}d\mathbf{v}_{k+1}\ldots d\mathbf{x}_{N_0}d\mathbf{v}_{N_0}f_{N_0} \tag{4.18}$$

which give the joint probability of particles 1 through k having the coordinates $(\mathbf{x}_1,\mathbf{v}_1)$ to $(\mathbf{x}_1 + d\mathbf{x}_1,\mathbf{v}_1 + d\mathbf{v}_1)$ *and . . . and* $(\mathbf{x}_k,\mathbf{v}_k)$ to $(\mathbf{x}_k + d\mathbf{x}_k,\mathbf{v}_k + d\mathbf{v}_k)$, irrespective of the coordinates of particles $k + 1, k + 2, \ldots, N_0$. The factor V^k on the right of (4.18) is a normalization factor, where V is the finite spatial volume in which f_{N_0} is nonzero for all $\mathbf{x}_1,\mathbf{x}_2,\ldots,\mathbf{x}_{N_0}$ (Fig. 4.2). At the end of our theoretical development, we will take the limit $N_0 \to \infty$, $V \to \infty$, in such a way that $n_0 = N_0/V$ is a constant giving the average number of particles per unit real space. For the present, we assume that $f_{N_0} \to 0$ as $x_i \to \pm\infty$ or $y_i \to \pm\infty$ or $z_i \to \pm\infty$ for any

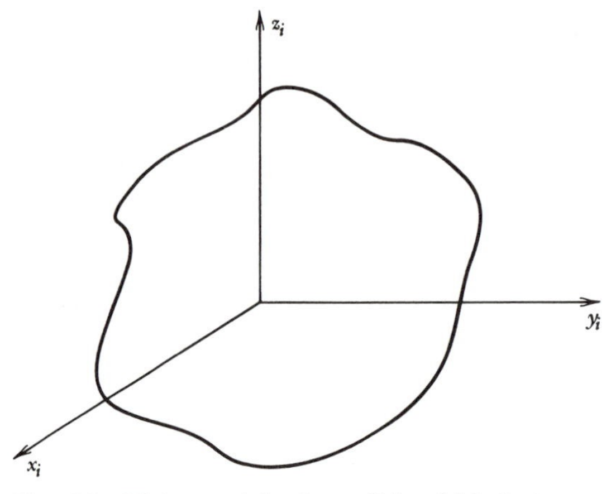

Fig. 4.2 Finite spatial volume V in which f_{N_0} is nonzero for any $\mathbf{x}_i$, $i = 1, \ldots, N_0$.

i. Likewise, because there are no particles with infinite speed, $f_{N_0} \to 0$ as $v_{x_i} \to \pm\infty$ or $v_{y_i} \to \pm\infty$ or $v_{z_i} \to \pm\infty$ for any i.

In this development, we do not care which one of the N_0 particles is called particle number 1, etc. Thus, we always choose probability densities f_{N_0} that are completely symmetric with respect to the particle labels. For example,

$$\begin{aligned} &f_{N_0}(\ldots z_7 = 2 \text{ cm} \ldots z_{13} = 5 \text{ cm} \ldots t) \\ &= f_{N_0}(\ldots z_7 = 5 \text{ cm} \ldots z_{13} = 2 \text{ cm} \ldots t) \end{aligned} \tag{4.19}$$

provided all of the other independent variables are the same. Here, we must interchange all of the $i = 7$ variables with all of the $i = 13$ variables. This means that when we set $k = 1$ in (4.18), the function $f_1(\mathbf{x}_1, \mathbf{v}_1, t)$ is (to within a normalization constant) the number of particles per unit real space per unit velocity space. Thus, this function $f_1(\mathbf{x}_1, \mathbf{v}_1, t)$ has the same meaning (to within a normalization constant) as the function $f_s(\mathbf{x}, \mathbf{v}, t)$ introduced in the previous chapter in connection with the plasma kinetic equation.

To keep the theory as simple as possible, we shall ignore any external electric and magnetic fields. We shall deal with only one species of N_0 particles; it is easy enough to generalize the results to a plasma with two species of $N_0/2$ particles each at the end of the development. For some purposes, such as calculating electron-electron collisional effects, the second species can be introduced as a smeared-out ion background of density n_0, which simply neutralizes the total electron charge. Finally, we adopt the *Coulomb model*, which ignores the magnetic fields produced by the charged particle motion. In this model, the acceleration

$$\dot{\mathbf{V}}_i(t) = \sum_{j=1}^{N_0} \mathbf{a}_{ij} \tag{4.20}$$

where

$$\mathbf{a}_{ij} = \frac{q_s^2}{m_s |\mathbf{x}_i - \mathbf{x}_j|^3} (\mathbf{x}_i - \mathbf{x}_j) \tag{4.21}$$

is the acceleration of particle i *due to* the Coulomb electric field of particle j. Since a particle exerts no force on itself, we use (4.21) only if $i \neq j$; if $i = j$, we use $\mathbf{a}_{ii} = 0$. Equation (4.21) replaces Maxwell's equations and the Lorentz force law. The Liouville equation (4.16) becomes

$$\frac{\partial f_{N_0}}{\partial t} + \sum_{i=1}^{N_0} \mathbf{v}_i \cdot \nabla_{\mathbf{x}_i} f_{N_0} + \sum_{i=1}^{N_0} \sum_{j=1}^{N_0} \mathbf{a}_{ij} \cdot \nabla_{\mathbf{v}_i} f_{N_0} = 0 \tag{4.22}$$

Equations for the reduced distributions f_k are obtained by integrating the Liouville equation (4.22) over all $\mathbf{x}_{k+1}, \mathbf{v}_{k+1}, \mathbf{x}_{k+2}, \mathbf{v}_{k+2}, \ldots \mathbf{x}_{N_0}, \mathbf{v}_{N_0}$. For example, to obtain the equation for f_{N_0-1} we integrate (4.22) over all $\mathbf{x}_{N_0}$ and $\mathbf{v}_{N_0}$, obtaining

$$\overset{①}{\int d\mathbf{x}_{N_0}\, d\mathbf{v}_{N_0} \frac{\partial f_{N_0}}{\partial t}} + \overset{②}{\int d\mathbf{x}_{N_0}\, d\mathbf{v}_{N_0} \sum_{i=1}^{N_0} \mathbf{v}_i \cdot \nabla_{\mathbf{x}_i} f_{N_0}}$$

$$+ \overset{③}{\int d\mathbf{x}_{N_0}\, d\mathbf{v}_{N_0} \sum_{i=1}^{N_0} \sum_{j=1}^{N_0} \mathbf{a}_{ij} \cdot \nabla_{\mathbf{v}_i} f_{N_0}} = 0 \tag{4.23}$$

Term ① is easy, since we can move the time derivative outside the integral to obtain

$$① = \frac{\partial}{\partial t} \int d\mathbf{x}_{N_0}\, d\mathbf{v}_{N_0} f_{N_0} = V^{1-N_0} \frac{\partial}{\partial t} f_{N_0-1} \tag{4.24}$$

where the definition (4.18) has been used. Term ② is also easy. In the first $N_0 - 1$ terms in the sum, the integration variables are independent of the operator $\mathbf{v}_i \cdot \nabla_{\mathbf{x}_i}$; this operator can then be moved outside the integration and we again obtain a term proportional to f_{N_0-1}. The last term in the sum, with $i = N_0$, is

$$\begin{aligned}
&\int d\mathbf{x}_{N_0}\, d\mathbf{v}_{N_0} (v_{x_{N_0}} \partial_{x_{N_0}} + v_{y_{N_0}} \partial_{y_{N_0}} + v_{z_{N_0}} \partial_{z_{N_0}}) f_{N_0} \\
&= \int d\mathbf{v}_{N_0}\, dy_{N_0}\, dz_{N_0}\, v_{x_{N_0}} f_{N_0} \Big|_{x_{N_0} = -\infty}^{x_{N_0} = +\infty} + 2 \text{ similar terms} \\
&= 0
\end{aligned} \tag{4.25}$$

since f_{N_0} vanishes at the boundaries of the system that have been placed at $x_{N_0} = \pm\infty$, etc. Thus,

$$② = V^{1-N_0} \sum_{i=1}^{N_0-1} \mathbf{v}_i \cdot \nabla_{\mathbf{x}_i} f_{N_0-1} \tag{4.26}$$

Term ③ is not much harder. Splitting the double sum

$$\sum_{i=1}^{N_0} \sum_{j=1}^{N_0} g_{ij} = \sum_{i=1}^{N_0-1} \sum_{j=1}^{N_0-1} g_{ij} + \sum_{j=1}^{N_0-1} g_{N_0 j} + \sum_{i=1}^{N_0-1} g_{iN_0} + g_{N_0 N_0}$$

we get

$$③ = V^{1-N_0} \sum_{i=1}^{N_0-1} \sum_{j=1}^{N_0-1} \mathbf{a}_{ij} \cdot \nabla_{\mathbf{v}_i} f_{N_0-1}$$

$$+ \int d\mathbf{x}_{N_0}\, d\mathbf{v}_{N_0} \sum_{j=1}^{N_0-1} \mathbf{a}_{N_0 j} \cdot \nabla_{\mathbf{v}_{N_0}} f_{N_0}$$

$$+ \int d\mathbf{x}_{N_0}\, d\mathbf{v}_{N_0} \sum_{i=1}^{N_0-1} \mathbf{a}_{iN_0} \cdot \nabla_{\mathbf{v}_i} f_{N_0} \tag{4.27}$$

where the $i = N_0, j = N_0$ term has been discarded because $\mathbf{a}_{N_0 N_0} = 0$. The second term on the right vanishes after direct integration with respect to $dv_{x_{N_0}}$ and evaluation at $v_{x_{N_0}} = \pm\infty$, etc. The remaining terms in ①, ②, and ③, after multiplication by V^{N_0-1}, are

$$\overset{①}{\frac{\partial}{\partial t} f_{N_0-1}} + \overset{②}{\sum_{i=1}^{N_0-1} \mathbf{v}_i \cdot \nabla_{\mathbf{x}_i} f_{N_0-1}} + \overset{③}{\sum_{i=1}^{N_0-1} \sum_{j=1}^{N_0-1} \mathbf{a}_{ij} \cdot \nabla_{\mathbf{v}_i} f_{N_0-1}}$$

$$+ \overset{④}{V^{N_0-1} \sum_{i=1}^{N_0-1} \int d\mathbf{x}_{N_0}\, d\mathbf{v}_{N_0}\, \mathbf{a}_{iN_0} \cdot \nabla_{\mathbf{v}_i} f_{N_0}} = 0 \tag{4.28}$$

This is the desired equation for f_{N_0-1}. Notice that it does not depend only on f_{N_0-1}; the last term ④ depends on f_{N_0}. We have made no approximations in deriving (4.28); within the Coulomb model, it is exact.

Having succeeded in deriving the equation for f_{N_0-1}, let us proceed to derive the equation for f_{N_0-2}. To do this, we integrate (4.28) over all $\mathbf{x}_{N_0-1}$ and over all $\mathbf{v}_{N_0-1}$. As in (4.24), term ① yields $V\, \partial_t f_{N_0-2}$.

EXERCISE Use the definition (4.18) to explain the difference between the power of V encountered here and that encountered in (4.24).

As in (4.26), term ② yields one term that vanishes upon integration, leaving a sum from 1 to $N_0 - 2$. In term ③, we do as in (4.27); we split the double $(N_0 - 1)$ sum into a double $(N_0 - 2)$ sum plus two single $(N_0 - 2)$ sums, the $i = N_0 - 1$, $j = N_0 - 1$ term vanishing since $\mathbf{a}_{N_0-1, N_0-1} = 0$. Term ③ becomes

$$③ = V \sum_{i=1}^{N_0-2} \sum_{j=1}^{N_0-2} \mathbf{a}_{ij} \cdot \nabla_{\mathbf{v}_i} f_{N_0-2}$$

$$+ \int d\mathbf{x}_{N_0-1}\, d\mathbf{v}_{N_0-1} \sum_{i=1}^{N_0-2} \mathbf{a}_{i, N_0-1} \cdot \nabla_{\mathbf{v}_i} f_{N_0-1}$$

$$+ \int d\mathbf{x}_{N_0-1}\, d\mathbf{v}_{N_0-1} \sum_{j=1}^{N_0-2} \mathbf{a}_{N_0-1, j} \cdot \nabla_{\mathbf{v}_{N_0-1}} f_{N_0-1} \tag{4.29}$$

The last term on the right vanishes upon direct integration with respect to $\mathbf{v}_{N_0-1}$.

For term ④ we have

$$④ = V^{N_0-1} \sum_{i=1}^{N_0-1} \int d\mathbf{x}_{N_0-1}\, d\mathbf{v}_{N_0-1}\, d\mathbf{x}_{N_0}\, d\mathbf{v}_{N_0}\, \mathbf{a}_{iN_0} \cdot \nabla_{\mathbf{v}_i} f_{N_0}$$

$$= V^{N_0-1} \sum_{i=1}^{N_0-2} \int d\mathbf{x}_{N_0}\, d\mathbf{v}_{N_0}\, \mathbf{a}_{iN_0} \cdot \nabla_{\mathbf{v}_i} \int d\mathbf{x}_{N_0-1}\, d\mathbf{v}_{N_0-1}\, f_{N_0} \tag{4.30}$$

where the $N_0 - 1$ term in the sum vanishes upon doing the $d\mathbf{v}_{N_0-1}$ integration. The variables $(\mathbf{x}_{N_0}, \mathbf{v}_{N_0})$ and $(\mathbf{x}_{N_0-1}, \mathbf{v}_{N_0-1})$ are simply dummy variables of integration on the far right of (4.30). Therefore, we can switch the labels N_0 and $N_0 - 1$, so that $\mathbf{a}_{iN_0}$ becomes $\mathbf{a}_{i,N_0-1}$. The density f_{N_0} can stay the same, however, because it has the symmetry property (4.19). Equation (4.30) becomes

$$④ = V^{N_0-1} \sum_{i=1}^{N_0-2} \int d\mathbf{x}_{N_0-1}\, d\mathbf{v}_{N_0-1}\, \mathbf{a}_{i,N_0-1} \cdot \nabla_{\mathbf{v}_i} \underbrace{\int d\mathbf{x}_{N_0}\, d\mathbf{v}_{N_0} f_{N_0}}_{V^{1-N_0} f_{N_0-1}}$$

$$= \sum_{i=1}^{N_0-2} \int d\mathbf{x}_{N_0-1}\, d\mathbf{v}_{N_0-1}\, \mathbf{a}_{i,N_0-1} \cdot \nabla_{\mathbf{v}_i} f_{N_0-1} \tag{4.31}$$

which is identical with the middle term on the right of term ③ in (4.29). Collecting all of the remaining terms in ①, ②, ③, and ④ and dividing by V, we obtain

$$\frac{\partial}{\partial t} f_{N_0-2} + \sum_{i=1}^{N_0-2} \mathbf{v}_i \cdot \nabla_{\mathbf{x}_i} f_{N_0-2} + \sum_{i=1}^{N_0-2} \sum_{j=1}^{N_0-2} \mathbf{a}_{ij} \cdot \nabla_{\mathbf{v}_i} f_{N_0-2}$$
$$+ \frac{2}{V} \sum_{i=1}^{N_0-2} \int d\mathbf{x}_{N_0-1}\, d\mathbf{v}_{N_0-1}\, \mathbf{a}_{i,N_0-1} \cdot \nabla_{\mathbf{v}_i} f_{N_0-1} = 0 \tag{4.32}$$

This equation for f_{N_0-2} is quite similar in structure to (4.28) for f_{N_0-1}. Notice again that this equation does not involve only f_{N_0-2}, but also involves f_{N_0-1} in the last term on the left.

By comparing (4.28) and (4.32), we see a pattern emerging. Using the same manipulations that we have been using (see Problem 4.2), we can generate an equation similar to (4.28) and (4.32) for arbitrary k. This equation is

$$\boxed{\begin{aligned} &\frac{\partial}{\partial t} f_k + \sum_{i=1}^{k} \mathbf{v}_i \cdot \nabla_{\mathbf{x}_i} f_k + \sum_{i=1}^{k} \sum_{j=1}^{k} \mathbf{a}_{ij} \cdot \nabla_{\mathbf{v}_i} f_k \\ &+ \frac{(N_0 - k)}{V} \sum_{i=1}^{k} \int d\mathbf{x}_{k+1}\, d\mathbf{v}_{k+1}\, \mathbf{a}_{i,k+1} \cdot \nabla_{\mathbf{v}_i} f_{k+1} = 0 \end{aligned}} \tag{4.33}$$

for $k = 1, 2, \ldots, N_0 - 2$. This is the *BBGKY hierarchy* (*B*ogoliubov [1]; *B*orn and *G*reen [2]; *K*irkwood [3, 4]; and *Y*von [5]). Each equation for f_k is coupled to the next higher equation through the f_{k+1} term.

EXERCISE Verify that (4.22) for f_{N_0} and (4.32) for f_{N_0-2} are in agreement with (4.33). Verify that (4.28) for f_{N_0-1} is in agreement with (4.33), provided that f_{N_0} is replaced by $V^{N_0} f_{N_0}$ in (4.33) [see (4.18)].

As it stands, the BBGKY hierarchy (4.33) is still exact (within the Coulomb model) and is just as hard to solve as the original Liouville equation (4.22). It consists of N_0 coupled integro-differential equations. Progress will come only when we take just the first few equations, for $k = 1, k = 2$, etc., and then use an approximation to close the set and cut off the dependence on higher equations.

From (4.33) the $k = 1$ equation is

$$\partial_t f_1(\mathbf{x}_1,\mathbf{v}_1,t) + \mathbf{v}_1 \cdot \nabla_{\mathbf{x}_1} f_1 + \frac{N_0 - 1}{V} \int d\mathbf{x}_2\, d\mathbf{v}_2\, \mathbf{a}_{12} \cdot \nabla_{\mathbf{v}_1} f_2(\mathbf{x}_1,\mathbf{v}_1,\mathbf{x}_2,\mathbf{v}_2,t) = 0 \tag{4.34}$$

This is coupled to the $k = 2$ equation through f_2. One way to proceed is to find some approximation for f_2 in terms of f_1. If we can do this, then (4.34) will be written entirely in terms of f_1, and we will have a complete description of the time evolution of $f_1(\mathbf{x}_1,\mathbf{v}_1,t)$ given the initial value $f_1(\mathbf{x}_1,\mathbf{v}_1,t = 0)$.

This is a good point at which to repeat our interpretation of the functions $f_1(\mathbf{x}_1,\mathbf{v}_1,t)$ and $f_2(\mathbf{x}_1,\mathbf{v}_1,\mathbf{x}_2,\mathbf{v}_2,t)$. We have said before that f_1 is equivalent to f_s in the plasma kinetic equation; when multiplied by $n_0 \equiv N_0/V$, it is the ensemble averaged number of particles per unit real space per unit velocity space at the point $(\mathbf{x}_1,\mathbf{v}_1)$ in six-dimensional phase space.

EXERCISE Use the definition to show that $\int d\mathbf{v}_1 f_1(\mathbf{x}_1,\mathbf{v}_1,t) = 1$ provided that none of the functions f_k, $k = 1,2, \ldots, N_0$ depend upon the positions $\mathbf{x}_1,\mathbf{x}_2, \ldots, \mathbf{x}_{N_0}$.

We may also say that $f_1(\mathbf{x}_1,\mathbf{v}_1,t)d\mathbf{x}_1\, d\mathbf{v}_1$ is the probability that a given particle finds itself in the region of phase space between $(\mathbf{x}_1,\mathbf{v}_1)$ and $(\mathbf{x}_1 + d\mathbf{x}_1,\mathbf{v}_1 + d\mathbf{v}_1)$. The interpretation of f_2 is similar to the interpretation of f_1. The function f_2 is the ensemble averaged number of particles per unit $\mathbf{x}_1$ real space per unit $\mathbf{x}_2$ real space per unit $\mathbf{v}_1$ velocity space per unit $\mathbf{v}_2$ velocity space. We may also say that $f_2(\mathbf{x}_1,\mathbf{v}_1,\mathbf{x}_2,\mathbf{v}_2,t)$ is proportional to the joint probability that particle 1 finds itself at $(\mathbf{x}_1,\mathbf{v}_1)$ *and* particle 2 finds itself at $(\mathbf{x}_2,\mathbf{v}_2)$. Since in this discussion all particles are of the same species, we know that an exact expression for f_2 would include the fact that no two particles (electrons, for example) can occupy the same spatial location. Thus, an exact expression for f_2 must have the property that $f_2 \rightarrow 0$ as $\mathbf{x}_1 \rightarrow \mathbf{x}_2$, regardless of the values of $\mathbf{v}_1$ and $\mathbf{v}_2$. In developing an approximate expression for f_2, we could of course lose this property. Another property that f_2 should have is symmetry with respect to the particle labels: $f_2(\mathbf{x}_1,\mathbf{v}_1,\mathbf{x}_2,\mathbf{v}_2,t) = f_2(\mathbf{x}_2,\mathbf{v}_2,\mathbf{x}_1,\mathbf{v}_1,t)$. This symmetry occurs because the original f_{N_0} has such symmetry, by assumption.

It turns out that f_2 has an intimate relation to f_1, which can be seen by an elementary example from probability theory. Suppose we have two loaded dice, each of which always rolls a five. Then the probability distribution for the value of the throws of either die is

$$P_1(x) = \delta(x - 5) \tag{4.35}$$

The joint probability that the value of the first die will be x and the value of the second die will be y is

$$P_2(x,y) = \delta(x - 5)\delta(y - 5) \tag{4.36}$$

But by (4.35) this is just

$$P_2(x,y) = P_1(x)P_1(y) \tag{4.37}$$

This separation always occurs when two quantities are *statistically independent*; that is, the value of one quantity does not depend on the value of the other quantity. Thus, it is always useful in considering joint probability distributions to factor out the piece that would be there if the two quantities were uncorrelated. Thus, for the dice we have

$$P_2(x,y) = P_1(x)P_1(y) + \delta P(x,y) \tag{4.38}$$

where $\delta P(x,y) = 0$ by (4.37). For a plasma, we define the *correlation function* $g(\mathbf{x}_1,\mathbf{v}_1,\mathbf{x}_2,\mathbf{v}_2,t)$ by

$$\begin{aligned} f_2(\mathbf{x}_1,\mathbf{v}_1,\mathbf{x}_2,\mathbf{v}_2,t) &= f_1(\mathbf{x}_1,\mathbf{v}_1,t)f_1(\mathbf{x}_2,\mathbf{v}_2,t) \\ &\quad + g(\mathbf{x}_1,\mathbf{v}_1,\mathbf{x}_2,\mathbf{v}_2,t) \end{aligned} \tag{4.39}$$

This is the first step in the *Mayer* [6] *cluster expansion*.

EXERCISE From the definitions of f_{N_0} and f_k, convince yourself that f_2 has the same units as $f_1 f_1$.

We are ready to insert the form (4.39) into the equation (4.34) for f_1, which becomes

$$\begin{aligned} &\partial_t f_1(\mathbf{x}_1,\mathbf{v}_1,t) + \mathbf{v}_1 \cdot \nabla_{\mathbf{x}_1} f_1 \\ &+ n_0 \int d\mathbf{x}_2\, d\mathbf{v}_2\, \mathbf{a}_{12} \cdot \nabla_{\mathbf{v}_1} [f_1(\mathbf{x}_1,\mathbf{v}_1,t)f_1(\mathbf{x}_2,\mathbf{v}_2,t) \\ &\qquad + g(\mathbf{x}_1,\mathbf{v}_1,\mathbf{x}_2,\mathbf{v}_2,t)] = 0 \end{aligned} \tag{4.40}$$

where we have replaced $(N_0 - 1)/V$ by n_0 because we are interested only in systems with $N_0 >> 1$.

Suppose one assumes that the correlation function vanishes. That is, we assume that the particles in the plasma behave as if they were completely independent of the particular positions and velocities of the other particles. This assumption would be exactly valid if we performed the pulverization procedure discussed in the previous chapter, in which $n_0 \to \infty$, $e \to 0$, $m_e \to 0$, $\Lambda \to \infty$, $n_0 e$ = constant, e/m_e = constant, v_e = constant, ω_e = constant, and λ_e = constant. Then each particle would have zero charge, and its presence would not affect any other particle. Collective effects could of course still happen, as these involve only f_1 and not g. When we set g equal to zero, (4.40) becomes

$$\begin{aligned} &\partial_t f_1 + \mathbf{v}_1 \cdot \nabla_{\mathbf{x}_1} f_1 \\ &+ [n_0 \int d\mathbf{x}_2\, d\mathbf{v}_2\, \mathbf{a}_{12} f_1(\mathbf{x}_2,\mathbf{v}_2,t)] \cdot \nabla_{\mathbf{v}_1} f_1(\mathbf{x}_1,\mathbf{v}_1,t) = 0 \end{aligned} \tag{4.41}$$

But the quantity in brackets is just the acceleration $\mathbf{a}_{12}$ produced on particle 1 by particle 2, integrated over the probability distribution $f_1(\mathbf{x}_2,\mathbf{v}_2,t)$ of particle 2. This is the ensemble averaged acceleration experienced by particle 1 due to all other particles,

$$\mathbf{a}(\mathbf{x}_1,t) \equiv n_0 \int d\mathbf{x}_2\, d\mathbf{v}_2\, \mathbf{a}_{12} f_1(\mathbf{x}_2,\mathbf{v}_2,t) \tag{4.42}$$

EXERCISE Convince yourself that $\mathbf{a}$ is normalized correctly.

Then (4.41) becomes

$$\boxed{\partial_t f_1 + \mathbf{v}_1 \cdot \nabla_{\mathbf{x}_1} f_1 + \mathbf{a} \cdot \nabla_{\mathbf{v}_1} f_1 = 0} \tag{4.43}$$

which we recognize as our old friend the Vlasov equation.

The Vlasov equation is probably the most useful equation in plasma physics, and a large portion of this book is devoted to its study. For our present purposes, however, it is not enough. It does not include the collisional effects that are represented by the two-particle correlation function g. We would like to have at least an approximate equation that does include collisional effects and that, therefore, predicts the temporal evolution of f_1 due to collisions. We must therefore return to the exact $k = 1$ equation (4.40) and find some method to evaluate g.

Since g is defined through (4.39) as $g = f_2 - f_1 f_1$, we must go back to the $k = 2$ equation in the BBGKY hierarchy in order to obtain an equation for f_2 and, hence, for g. Setting $k = 2$ in (4.33) and using $(N_0 - 2)/V \approx n_0$, one has

$$\overset{①}{\partial_t f_2} + \overset{②}{(\mathbf{v}_1 \cdot \nabla_{\mathbf{x}_1} + \mathbf{v}_2 \cdot \nabla_{\mathbf{x}_2}) f_2} + \overset{③}{(\mathbf{a}_{12} \cdot \nabla_{\mathbf{v}_1} + \mathbf{a}_{21} \cdot \nabla_{\mathbf{v}_2}) f_2}$$

$$+ \overset{④}{n_0 \int d\mathbf{x}_3\, d\mathbf{v}_3 (\mathbf{a}_{13} \cdot \nabla_{\mathbf{v}_1} + \mathbf{a}_{23} \cdot \nabla_{\mathbf{v}_2}) f_3} = 0 \tag{4.44}$$

We have seen that it is useful to factor out the part $f_1 f_1$ of $f_2 = f_1 f_1 + g$, which exists when the particles are uncorrelated. Likewise, it is useful to factor from f_3 the part that would exist when the particles are uncorrelated, plus those parts that result from two-particle correlations. This leads to the next step in the Mayer cluster expansion, which is

$$\begin{aligned} f_3(123) &= f_1(1) f_1(2) f_1(3) + f_1(1) g(23) \\ &\quad + f_1(2) g(13) + f_1(3) g(12) + h(123) \end{aligned} \tag{4.45}$$

where we have introduced a simplified notation: $(1) \equiv (\mathbf{x}_1, \mathbf{v}_1)$, $(2) \equiv (\mathbf{x}_2, \mathbf{v}_2)$, and $(3) \equiv (\mathbf{x}_3, \mathbf{v}_3)$. Equation (4.45) will be explored further in Problems 4.4 and 4.5.

Our procedure is to insert (4.45) into (4.44) and neglect $h(123)$. This means that we neglect three-particle correlations, or three-body collisions. It turns out that these correlations are of higher order in the plasma parameter Λ; therefore their neglect is quite well justified for many purposes. The resulting set of equations constitute two equations in two unknowns f_1 and g. Thus, we have truncated the BBGKY hierarchy while retaining the effects of collisions to a good approximation.

Inserting (4.45) for f_3 and $f_2 = f_1 f_1 + g$ into the $k = 2$ BBGKY equation (4.44), we find for the numbered terms:

$$① = \overset{ⓐ}{\dot{f}_1(1) f_1(2)} + \overset{ⓔ}{\dot{f}_1(2) f_1(1)} + \dot{g}(12)$$

$$② = \overset{ⓑ}{\mathbf{v}_1 \cdot \nabla_{\mathbf{x}_1} f_1(1) f_1(2)} + \mathbf{v}_1 \cdot \nabla_{\mathbf{x}_1} g(12) + \{1 \leftrightarrow 2\}$$

$$\textcircled{3} = \mathbf{a}_{12} \cdot \nabla_{\mathbf{v}_1} f_1(1) f_1(2) + \underbrace{\mathbf{a}_{12} \cdot \nabla_{\mathbf{v}_1} g(12)}_{\textcircled{c}} + \{1 \leftrightarrow 2\}$$

$$\textcircled{4} = n_0 \int d3\, \mathbf{a}_{13} \cdot \nabla_{\mathbf{v}_1} [f_1(1) f_1(2) f_1(3) + f_1(1) g(23)$$

$$+ \overset{\textcircled{d}}{f_1(2) g(13)} + f_1(3) g(12)] + \{1 \leftrightarrow 2\} \tag{4.46}$$

where $d3 \equiv d\mathbf{x}_3\, d\mathbf{v}_3$ and $\{1 \leftrightarrow 2\}$ means that all of the preceding terms on the right side are repeated with the symbols 1 and 2 interchanged. Recall that $g(12) = g(21)$ by the symmetry of f_2. Many of the terms in (4.46) can be eliminated using the $k = 1$ BBGKY equation (4.40). For example,

$$\textcircled{a} + \textcircled{b} + \textcircled{c} + \textcircled{d} = \{\dot{f}_1(1) + \mathbf{v}_1 \cdot \nabla_{\mathbf{x}_1} f_1(1)$$

$$+ n_0 \int d3\, \mathbf{a}_{13} \cdot \nabla_{\mathbf{v}_1} [f_1(1) f_1(3) + g(13)]\} f_1(2)$$

$$= [\text{left side of (4.40)}] f_1(2) = 0 \tag{4.47}$$

Term $\textcircled{e}$ likewise combines with three of the $\{1 \leftrightarrow 2\}$ terms to vanish, leaving

$$\boxed{\begin{aligned} &\dot{g}(12) + (\mathbf{v}_1 \cdot \nabla_{\mathbf{x}_1} + \mathbf{v}_2 \cdot \nabla_{\mathbf{x}_2}) g(12) = \\ &- (\mathbf{a}_{12} \cdot \nabla_{\mathbf{v}_1} + \mathbf{a}_{21} \cdot \nabla_{\mathbf{v}_2})[f_1(1) f_1(2) + g(12)] \\ &- \{n_0 \int d3\, \mathbf{a}_{13} \cdot \nabla_{\mathbf{v}_1} [f_1(1) g(23) + f_1(3) g(12)] + \{1 \leftrightarrow 2\}\} \end{aligned}} \tag{4.48}$$

Together with (4.40) which in the condensed notation reads

$$\boxed{\begin{aligned} &\dot{f}_1(1) + \mathbf{v}_1 \cdot \nabla_{\mathbf{x}_1} f_1(1) + n_0 \int d2\, \mathbf{a}_{12} \\ &\cdot \nabla_{\mathbf{v}_1} [f_1(1) f_1(2) + g(12)] = 0 \end{aligned}} \tag{4.49}$$

we have two equations in the two unknowns f_1 and g. We have truncated the BBGKY hierarchy by ignoring three-particle correlations.

In practice, (4.48) and (4.49) are impossibly difficult to solve, either analytically or numerically. They are two coupled nonlinear integro-differential equations in a twelve-dimensional phase space. The present thrust of plasma kinetic theory consists in finding certain approximations to $g(12)$ that are then inserted in (4.49). Using the definition of the acceleration $\mathbf{a}$ in (4.42), we rewrite (4.49) as

$$\boxed{\dot{f}_1(1) + \mathbf{v}_1 \cdot \nabla_{\mathbf{x}_1} f_1 + \mathbf{a} \cdot \nabla_{\mathbf{v}_1} f_1 = - n_0 \int d2\, \mathbf{a}_{12} \cdot \nabla_{\mathbf{v}_1} g(12)} \tag{4.50}$$

which is in exactly the same form as the plasma kinetic equation (3.26).

Most of the discussion in this chapter has been exact, in particular, the derivation of the Liouville equation and the BBGKY hierarchy. Even the approximations that lead to (4.48) and (4.49) are extremely good ones, for example, $1 \ll N_0$ and the neglect of three-particle collisions. By contrast, the approximations needed to convert (4.48) and (4.49) into manageable form are sometimes quite drastic and less justifiable, as will be seen in the next chapter. Further discussion of the Liouville equation and the BBGKY hierarchy can be found in the books of Montgomery and Tidman [7], Montgomery [8], Clemmow and Dougherty [9], Krall and Trivelpiece [10], and Klimontovich [11].

REFERENCES

[1] N. N. Bogoliubov, *Problems of a Dynamical Theory in Statistical Physics*, State Technical Press, Moscow, 1946.

[2] M. Born and H. S. Green, *A General Kinetic Theory of Liquids*, Cambridge University Press, Cambridge, England, 1949.

[3] J. G. Kirkwood, *J. Chem. Phys.*, *14*, 180 (1946).

[4] J. G. Kirkwood, *J. Chem. Phys.*, *15*, 72 (1947).

[5] J. Yvon, *La Theorie des Fluides et l'Equation d'Etat*, Hermann et Cie, Paris, 1935.

[6] J. E. Mayer and M. G. Mayer, *Statistical Mechanics*, Wiley, New York, 1940.

[7] D. C. Montgomery and D. A. Tidman, *Plasma Kinetic Theory*, McGraw-Hill, New York, 1964.

[8] D. C. Montgomery, *Theory of the Unmagnetized Plasma*, Gordon and Breach, New York, 1971.

[9] P. C. Clemmow and J. P. Dougherty, *Electrodynamics of Particles and Plasmas*, Addison-Wesley, Reading, Mass., 1969.

[10] N. A. Krall and A. W. Trivelpiece, *Principles of Plasma Physics*, McGraw-Hill, New York, 1973.

[11] Yu. L. Klimontovich, *The Statistical Theory of Non-equilibrium Processes in a Plasma*, M.I.T. Press, Cambridge, Mass., 1967.

PROBLEMS

4.1 Continuity vs. Convective

Demonstrate the equivalence between the convective derivative form of the Liouville equation (4.16) and the continuity equation (4.15).

4.2 BBGKY Hierarchy

Integrate (4.32) over all $\mathbf{x}_{N_0-2}$ and $\mathbf{v}_{N_0-2}$ to obtain the $k = N_0 - 3$ equation of the BBGKY hierarchy, and compare your result to (4.33).

4.3 Normalization

Explain in detail the normalization of (4.42).

4.4 Three-Point Correlations (Coins)

In (4.45) we define a three-point joint probability function f_3 in terms of the one-point probability f_1, the two-point correlation function g, and the three-point correlation function h. Suppose we apply this kind of thinking to the case of three coins, each of which can come up heads (+) or tails (−). What is the meaning of f_3 in this case? Write out f_3 in the form (4.45), and evaluate f_3, f_1, g, and h in each of the following cases.

(a) All three coins are "honest," that is, each coin is equally likely to come up heads or tails, and each coin is unaffected by any other coin.
(b) Because the coins are mysteriously locked together, in any one throw all three are heads or tails, the result changing randomly from throw to throw.
(c) All three coins always come up tails.
(d) The first two coins always come up heads, while the third is honest. Note that here the probability functions are not symmetric, so that, for example, $f_1(1)$ is not the same function as $f_1(3)$.

4.5 Three-Point Correlations (Dice)

In (4.45) we define a three-point joint probability function f_3 in terms of the one-point probability f_1, the two-point correlation function g, and the three-point correlation function h. Suppose we apply this kind of thinking to the case of three dice, each of which can take on integer values from one through six. What is the meaning of f_3 in this case? Write out f_3 in the form (4.45), and evaluate f_3, f_1, g, and h in each of the following cases.

(a) All three dice are "honest," that is, the value of each die is equally likely one through six and is independent of the value of any other die.
(b) Because the dice are mysteriously locked together, in one throw all three always show the same value, the value changing randomly from throw to throw with all six values equally likely.
(c) All of the dice always come up "five."
(d) The first two dice always come up "two"; the other one is "honest."

4.6 BBGKY Hierarchy

In this chapter, we derive the BBGKY hierarchy from the Liouville equation. This can be done in a completely different way [10], starting with the Klimontovich equation. Explain, by using words and writing equations only for illustration, how the $k = 1$ and $k = 2$ equations of the BBGKY hierarchy can be obtained from the Klimontovich equation.

CHAPTER 5

Plasma Kinetic Theory III: Lenard–Balescu Equation

5.1 BOGOLIUBOV'S HYPOTHESIS

In the preceding chapter, the BBGKY hierarchy is truncated by neglecting three-particle correlations (three-body collisions). For a good plasma, this is probably a very good approximation, although no rigorous proof exists. The spirit of the approximation is the same as that of Section 1.6, where the collision frequency is calculated as a series of two-body collisions, even though the particle is interacting with Λ particles simultaneously. Since the collision of particle A with particle B is usually a small angle collision, its effect on the orbit of particle A is small, thus making a negligible effect on the simultaneous collision of particle A with particle C.

The result of our truncation of the BBGKY hierarchy is the set of coupled equations (4.48) and (4.50) in the two unknowns $f_1(\mathbf{x}_1,\mathbf{v}_1,t)$ and $g(\mathbf{x}_1,\mathbf{v}_1,\mathbf{x}_2,\mathbf{v}_2,t)$. These equations are quite intractable in general. However, there is one set of simplifying assumptions that is both physically very important and allows the exact (almost) solution of (4.48) and (4.50).

Consider a spatially homogeneous ensemble of plasmas. This means that any function of one spatial variable must be independent of that variable; so $f_1(\mathbf{x}_1,\mathbf{v}_1,t) = f_1(\mathbf{v}_1,t)$ and $\mathbf{a}(\mathbf{x}_1,t) = \mathbf{a}(t) = 0$ by (4.21) and (4.42). Any ensemble averaged function of two spatial variables can only be a function of the difference between those variables; therefore we write $g = g(\mathbf{x}_1 - \mathbf{x}_2,\mathbf{v}_1,\mathbf{v}_2,t)$. With these assumptions, (4.50) simplifies considerably and becomes

$$\partial_t f_1(\mathbf{v}_1,t) = -n_0 \int d\mathbf{x}_2\, d\mathbf{v}_2\, \mathbf{a}_{12} \cdot \nabla_{\mathbf{v}_1} g(\mathbf{x}_1 - \mathbf{x}_2,\mathbf{v}_1,\mathbf{v}_2,t) \tag{5.1}$$

Equation 4.48 simplifies since two terms are of the form

$$[n_0 \int d3\ \mathbf{a}_{13} f_1(3)] \cdot \nabla_{\mathbf{v}_1} g(12) = \mathbf{a} \cdot \nabla_{\mathbf{v}_1} g(12) = 0 \tag{5.2}$$

leaving

$$\begin{aligned} \partial_t g(\mathbf{x}_1 - \mathbf{x}_2, \mathbf{v}_1, \mathbf{v}_2, t) + \mathbf{v}_1 \cdot \nabla_{\mathbf{x}_1} g(12) + \mathbf{v}_2 \cdot \nabla_{\mathbf{x}_2} g(12) \\ + (\mathbf{a}_{12} \cdot \nabla_{\mathbf{v}_1} + \mathbf{a}_{21} \cdot \nabla_{\mathbf{v}_2}) g(12) \\ + n_0 \int d3\ \mathbf{a}_{13} \cdot \nabla_{\mathbf{v}_1} f_1(1) g(23) + n_0 \int d3\ \mathbf{a}_{23} \cdot \nabla_{\mathbf{v}_2} f_1(2) g(13) \\ = -(\mathbf{a}_{12} \cdot \nabla_{\mathbf{v}_1} + \mathbf{a}_{21} \cdot \nabla_{\mathbf{v}_2}) f_1(1) f_1(2) \end{aligned} \tag{5.3}$$

We now wish to argue that the fourth term on the left is smaller than all the other terms and can be discarded. Recall the pulverization procedure of the previous chapter. By that argument, as well as the discussion of collisions in Section 1.6, we argue that the two-point correlation function g is higher order in the plasma parameter Λ than f_1; thus $g/f_1 \sim \Lambda^{-1}$. The acceleration $\mathbf{a}_{12} \sim e^2/m_e \sim \Lambda^{-1}$ since e/m_e is constant and $e \sim n_0^{-1} \sim \Lambda^{-1}$; here, we phrase our discussion in terms of electrons. Thus, all terms in (5.3) are $\sim \Lambda^{-1}$ except for the fourth term on the left, which is $\sim \Lambda^{-2}$. We discard this term, leaving

$$\boxed{\frac{\partial g(12)}{\partial t} + V_1 g + V_2 g = S} \tag{5.4}$$

where V_1 and V_2 are operators defined by

$$\begin{aligned} V_1 g(12) &= \mathbf{v}_1 \cdot \nabla_{\mathbf{x}_1} g(12) \\ &+ [n_0 \int d3 \mathbf{a}_{13}\ g(23)] \cdot \nabla_{\mathbf{v}_1} f_1(1) \end{aligned} \tag{5.6}$$

$$\begin{aligned} V_2 g(12) &= \mathbf{v}_2 \cdot \nabla_{\mathbf{x}_2} g(12) \\ &+ [n_0 \int d3 \mathbf{a}_{23}\ g(13)] \cdot \nabla_{\mathbf{v}_2} f_1(2) \end{aligned} \tag{5.7}$$

and the source function S is

$$S(\mathbf{x}_1 - \mathbf{x}_2, \mathbf{v}_1, \mathbf{v}_2) = -(\mathbf{a}_{12} \cdot \nabla_{\mathbf{v}_1} + \mathbf{a}_{21} \cdot \nabla_{\mathbf{v}_2}) f_1(1) f_1(2) \tag{5.8}$$

In this chapter we alternate between the notations (1) and $(\mathbf{x}_1, \mathbf{v}_1)$ depending on convenience. For simplicity, we suppose that we are dealing with an electron plasma. A neutralizing ion background can be thought to be present; it is considered to be smoothed out so that it does not contribute explicitly to the acceleration $\mathbf{a}_{ij}$, which by (4.21) is

$$\mathbf{a}_{ij} = \frac{e^2}{m_e |\mathbf{x}_i - \mathbf{x}_j|^3} (\mathbf{x}_i - \mathbf{x}_j) \tag{5.9}$$

The important physical situation to which this discussion applies is as follows. Imagine a beam of electrons incident on a Maxwellian electron plasma in the

$\hat{x}$-direction. Then the function

$$F(v_x) \equiv \int dv_y \, dv_z \, f_1(\mathbf{v}) \tag{5.10}$$

has the form shown in Fig. 5.1. Ignoring questions of stability (see Chapter 6), we recognize that the beam of electrons represented by the bump at large positive v_x will experience collisions that will eventually ($t \to \infty$) produce a new Maxwellian at a higher temperature. By the discussion of Section 1.6 we can predict the time scale for this process to be $\sim v_{ee} \sim \omega_e/\Lambda$. The solution of (5.1) and (5.4) which we are about to obtain should yield a very good theoretical description for this important process. This evolution is encountered in such applications as electron beam-pellet fusion and (when generalized to ions) ohmic heating of tokamaks.

The further assumption that allows us to solve the (still very complicated) set of equations (5.1) and (5.4) is *Bogoliubov's hypothesis*. The assumption is that the two-point correlation function g relaxes on a time scale very short compared to the time scale on which f_1 relaxes [1]. Imagine introducing a test electron into a plasma. The other electrons will adjust to the presence of the test electron in roughly the time it takes for them to have a collision with it. With a typical speed v_e and a typical length λ_e, the time for a collision is $\sim \lambda_e/v_e \sim \omega_e^{-1}$. By contrast, the time for f_1 to change because of collisions is $\sim \Lambda\omega_e^{-1}$; thus it is indeed quite reasonable to assume that g relaxes quickly compared to f_1. Mathematically, we incorporate this assumption by ignoring the time dependence of $f_1(\mathbf{v}_1,t)$ and $f_1(\mathbf{v}_2,t)$ in the source function S on the right of (5.4). Equation (5.4) is then a linear equation for g with a known, constant (in time) source function on the right. We can solve such a linear equation for $g(\mathbf{x}_1 - \mathbf{x}_2,\mathbf{v}_1,\mathbf{v}_2,t \to \infty)$ where $t \to \infty$ is understood to refer to the short time scale on which g relaxes. The solution for g will then depend on the factors $f_1(\mathbf{v}_1,t)$ and $f_1(\mathbf{v}_2,t)$ in the source function (5.8). When this solution for g is substituted into the right side of (5.1), there results a single nonlinear integro-differential equation in the one unknown function f_1. We

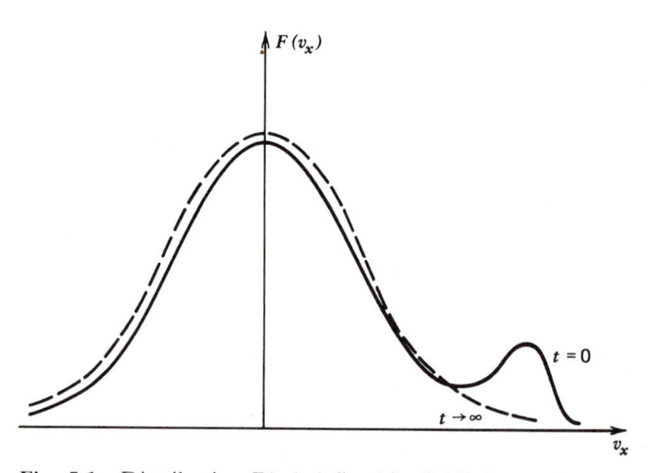

Fig. 5.1 Distribution $F(v_x)$ defined in (5.10) for an electron beam incident on a plasma.

have finally achieved our goal of truncating the BBGKY hierarchy and have expressed the entire plasma kinetic equation (5.1) in terms of the one unknown function $f_1(\mathbf{v}_1,t)$.

The implementation of this procedure is straightforward but complicated. In order to understand it, it is useful to have first studied the material in Chapter 6 on the Vlasov equation. Thus, we will not perform the derivation here; it is included in Appendix A. The reader who is studying plasma physics for the first time may wish to accept the results as given here, and proceed to read Appendix A after a thorough study of Chapter 6.

The solution of (5.1) and (5.4) uses the techniques of Fourier transformation in space, Laplace transformation in time, and their inverses. The conventions used in this book are as follows:

$$f(\mathbf{k}) = \int \frac{d\mathbf{x}}{(2\pi)^3}\, e^{-i\mathbf{k}\cdot\mathbf{x}} f(\mathbf{x}) \tag{5.11}$$

$$f(\mathbf{x}) = \int d\mathbf{k}\, e^{i\mathbf{k}\cdot\mathbf{x}}\, f(\mathbf{k}) \tag{5.12}$$

$$f(\omega) = \int_0^\infty dt\, e^{i\omega t} f(t) \tag{5.13}$$

$$f(t) = \int_L \frac{d\omega}{2\pi}\, e^{-i\omega t} f(\omega) \tag{5.14}$$

where the integrals over $\mathbf{x}$, $\mathbf{k}$, and t are usually along the real axes while the integral over ω is along the Laplace inversion contour to be discussed later.

Expressed in terms of the difference variable $\mathbf{x} = \mathbf{x}_1 - \mathbf{x}_2$, the acceleration $\mathbf{a}_{12}$ in (5.9) is

$$\mathbf{a}_{12}(\mathbf{x}) = \frac{e^2}{m_e|\mathbf{x}|^3}\,\mathbf{x} \tag{5.15}$$

with Fourier transform

$$\mathbf{a}_{12}(\mathbf{k}) = \frac{-i\mathbf{k}}{m_e}\,\varphi(k) \tag{5.16}$$

where

$$\varphi(k) = \frac{e^2}{2\pi^2 k^2} \tag{5.17}$$

is the Fourier transform of the Coulomb potential

$$\varphi(x) = \frac{e^2}{|\mathbf{x}|} \tag{5.18}$$

(See Problem 5.1.) Then, as shown in Appendix A, the solution of (5.1) and (5.4), under the Bogoliubov hypothesis, is

$$\boxed{\begin{aligned}\frac{\partial f(\mathbf{v},t)}{\partial t} &= -\frac{8\pi^4 n_0}{m_e^2}\,\nabla_{\mathbf{v}}\cdot\int d\mathbf{k}\, d\mathbf{v}'\, \mathbf{k}\mathbf{k}\cdot\frac{\varphi^2(k)}{|\epsilon(\mathbf{k},\mathbf{k}\cdot\mathbf{v})|^2}\\ &\quad\times \delta[\mathbf{k}\cdot(\mathbf{v}-\mathbf{v}')][f(\mathbf{v})\nabla_{\mathbf{v}'} f(\mathbf{v}') - f(\mathbf{v}')\nabla_{\mathbf{v}} f(\mathbf{v})]\end{aligned}} \tag{5.19}$$

which is the *Lenard–Balescu equation* (Refs. [2] to [6]). In this equation, we have dropped the subscript 1 from $\mathbf{v}_1$, and the subscript 1 from f_1, and have used the *dielectric function*

$$\epsilon(\mathbf{k},\omega) = 1 + \frac{\omega_e^2}{k^2}\int d\mathbf{v}\,\frac{\mathbf{k}\cdot\nabla_{\mathbf{v}} f(\mathbf{v})}{\omega - \mathbf{k}\cdot\mathbf{v}} \tag{5.20}$$

which will be studied in detail in the next chapter. The velocity integral must be performed along the Landau contour, as discussed in the next chapter. The interpretation of the Lenard–Balescu equation (5.19), and several alternate forms, will be discussed in the next section.

5.2 LENARD-BALESCU EQUATION

The Lenard–Balescu equation (5.19) is obtained from the BBGKY hierarchy after several assumptions: three-particle correlations are negligible, the ensemble of plasmas is spatially homogeneous, and the two-particle correlation function g relaxes much faster than the one-particle distribution function f_1. Thus, the Lenard–Balescu equation is applicable to situations such as the collisional relaxation of a beam in a plasma, but is not applicable in general to the collisional damping of spatially inhomogeneous wave motion or any phenomena that involve high frequencies like ω_e.

The right side of (5.19) represents the physics of two-particle collisions, since the right side of (5.1) is proportional to the two-particle correlation function g. This is indicated by the factor $\varphi(k)/\epsilon(\mathbf{k},\mathbf{k}\cdot\mathbf{v})$, which appears squared. It will be shown in the next chapter that the dielectric function $\epsilon(\mathbf{k},\omega)$ represents the plasma shielding of the field of a test charge. Thus, this term in (5.19) represents the interaction of one particle (together with its shielding cloud) with the potential field of another particle (together with its shielding cloud); that is, the collision of two shielded particles.

There is a problem with the Lenard–Balescu equation (5.19) as it stands. If one converts the k integration into spherical coordinates, and takes into account the forms (5.17) of $\varphi(k)$ and (5.20) of $\epsilon(\mathbf{k},\omega)$, one finds that at large k the integral diverges like $\int dk/k \sim \ln k$. Thus, just as in the derivation of the collision frequency in Section 1.6, we find a logarithmic divergence at large k, or small distances. In Section 1.6 we cut off the spatial integral at the lower limit p_0, where p_0 is the impact parameter for large angle collisions. It is argued in Section 1.6 that the physical formulation is not valid for large angle collisions, thus producing an unphysical divergence at short distances. The same thing is going on here. The derivation of the Lenard–Balescu equation is based on the assumption that in the expression

$$f_2(12) = f_1(1)\,f_1(2) + g(12) \tag{5.21}$$

we have $|g| << |f_1 f_1|$. This assumption led us to discard a term in (5.3) to obtain (5.4). However, this assumption is not always valid. It is not possible for two electrons to get very close to each other; therefore, we must have $f_2 \to 0$ as $\mathbf{x}_1 \to \mathbf{x}_2$, which implies $g = -f_1 f_1$. Thus, for small values of $|\mathbf{x}_1 - \mathbf{x}_2|$ (large k), it is not correct to assume $|g| << |f_1 f_1|$. In practice, since the divergence is logarithmic, we can simply cut off the integral in (5.19) at some upper limit wave number corresponding to some lower limit spatial scale. For this purpose, the impact parameter (Landau length) p_0 for large angle collisions (see Section 1.6) would be a reasonable choice.

The Lenard–Balescu equation (5.19) has several desirable features [4–5]. These are:

(a) If $f \geq 0$ at $t = 0$, $f \geq 0$ at all t.
(b) Particles are conserved: $d/dt \int d\mathbf{v}\, f(\mathbf{v},t) = 0$.
(c) Momentum is conserved: $d/dt \int d\mathbf{v}\mathbf{v}\, f(\mathbf{v},t) = 0$.
(d) Kinetic energy is conserved: $d/dt \int d\mathbf{v}\, v^2 f(\mathbf{v},t) = 0$.
(e) Any Maxwellian is a time-independent solution.
(f) As $t \to \infty$, any f satisfying (a) approaches a Maxwellian.

A simplified but fairly accurate form of the Lenard–Balescu equation (5.19) can be obtained as follows. We rewrite (5.19) in the form

$$\frac{\partial f(\mathbf{v},t)}{\partial t} = -\nabla_{\mathbf{v}} \cdot \int d\mathbf{v}'\, \overleftrightarrow{\mathbf{Q}}(\mathbf{v},\mathbf{v}') \cdot (\nabla_{\mathbf{v}} - \nabla_{\mathbf{v}'}) f(\mathbf{v}) f(\mathbf{v}') \tag{5.22}$$

with the tensor

$$\overleftrightarrow{\mathbf{Q}}(\mathbf{v},\mathbf{v}') = -\frac{8\pi^4 n_0}{m_e^2} \int d\mathbf{k}\, \frac{\mathbf{k}\mathbf{k}\varphi^2(k)}{|\epsilon(\mathbf{k},\mathbf{k}\cdot\mathbf{v})|^2}\, \delta[\mathbf{k}\cdot(\mathbf{v}-\mathbf{v}')]$$

$$= -\frac{2n_0 e^4}{m_e^2} \int d\mathbf{k}\, \frac{\mathbf{k}\mathbf{k}}{k^4} \frac{\delta[\mathbf{k}\cdot(\mathbf{v}-\mathbf{v}')]}{\left|1 + \dfrac{\psi}{k^2\lambda_e^2}\right|^2} \tag{5.23}$$

where the definition (5.17) has been used, and where the dimensionless function ψ is found from (5.20) to be

$$\psi(\mathbf{k},\mathbf{k}\cdot\mathbf{v}) = v_e^2 \int d\mathbf{v}'\, \frac{\mathbf{k}\cdot\nabla_{\mathbf{v}'} f(\mathbf{v}')}{\mathbf{k}\cdot(\mathbf{v}-\mathbf{v}')} \tag{5.24}$$

Again, the velocity integral must be performed along the Landau contour, as discussed in the next chapter. The wave number integral in (5.23) is performed as follows. When we orient the $\hat{k}_1$ axis in the $\mathbf{v} - \mathbf{v}'$ direction, the Q_{ij} component of the tensor $\overleftrightarrow{\mathbf{Q}}$ is

$$Q_{ij}(\mathbf{v},\mathbf{v}') = -\frac{2n_0 e^4}{m_e^2} \int dk_1\, dk_2\, dk_3\, \frac{k_i k_j}{k^4} \frac{1}{|\mathbf{v}-\mathbf{v}'|} \frac{\delta(k_1)}{|1 + (\psi/k^2\lambda_e^2)|^2} \tag{5.25}$$

The factor $\delta(k_1)$ implies $Q_{ij} = 0$ if either $i = 1$ or $j = 1$. The k_1 integration is trivially performed using this factor. In cylindrical coordinates with $k_2 = k\cos\theta$, $k_3 = k\sin\theta$, and cutting off the integration at an upper wave number $k_0 = p_0^{-1}$, we find, using Q_{33} as an example,

$$Q_{33}(\mathbf{v},\mathbf{v}') = -\frac{2n_0 e^4}{m_e^2|\mathbf{v}-\mathbf{v}'|} \int_0^{2\pi} d\theta \sin^2\theta \int_0^{k_0} \frac{dk}{k} \frac{1}{|1 + (\psi/k^2\lambda_e^2)|^2} \tag{5.26}$$

Since ψ is a function of θ but not of k [see (5.24)], the wave number integration can be performed (Problem 5.3). The result is

$$Q_{33}(\mathbf{v},\mathbf{v}') = -\frac{n_0 e^4}{m_e^2|\mathbf{v}-\mathbf{v}'|}\int_0^{2\pi} d\theta \, \frac{\mathrm{Im}[\psi \ln(1 + k_0^2\lambda_e^2/\psi)]}{\mathrm{Im}(\psi)} \sin^2\theta \quad (5.27)$$

It turns out (as can be seen more clearly after a study of the following chapter) that the dimensionless function ψ is of order unity. In addition, we recognize the factor $k_0\lambda_e = \lambda_e/p_0$ from Section 1.6 to be (within factors of order unity) the plasma parameter Λ. Thus, we neglect unity compared to $k_0^2\lambda_e^2/\psi$, and $\ln(\psi)$ compared to $\ln(k_0^2\lambda_e^2) \approx \ln(\Lambda^2) = 2\ln\Lambda$, to obtain

$$Q_{33}(\mathbf{v},\mathbf{v}') = Q_{22}(\mathbf{v},\mathbf{v}') = -\frac{2\pi n_0 e^4}{m_e^2|\mathbf{v}-\mathbf{v}'|}\ln\Lambda \quad (5.28)$$

Similar arguments yield $Q_{23} = Q_{32} = 0$. A tensor with only the Q_{22} and Q_{33} components nonzero can be conveniently expressed in terms of the unit tensor $\overleftrightarrow{\mathbf{I}} \equiv \hat{k}_1\hat{k}_1 + \hat{k}_2\hat{k}_2 + \hat{k}_3\hat{k}_3$; with $\mathbf{g} \equiv \mathbf{v} - \mathbf{v}'$ and recalling that $\hat{k}_1 = \hat{g}$, we have

$$\overleftrightarrow{\mathbf{Q}}(\mathbf{v},\mathbf{v}') = -\frac{2\pi n_0 e^4 \ln\Lambda}{m_e^2}\,\frac{g^2\,\overleftrightarrow{\mathbf{I}} - \mathbf{g}\mathbf{g}}{g^3} \quad (5.29)$$

This expression is known as the *Landau form* for $\overleftrightarrow{\mathbf{Q}}$.

With $\overleftrightarrow{\mathbf{Q}}$ in the form (5.29), it is possible to put the Lenard–Balescu equation (5.22) in the form of a *Fokker–Planck equation*. The general Fokker–Planck equation is a very important equation in all aspects of statistical physics, and is derived from first principles in Appendix B. Following Montgomery and Tidman [5], we notice that

$$\nabla_{\mathbf{v}}\nabla_{\mathbf{v}} g = \frac{g^2\,\overleftrightarrow{\mathbf{I}} - \mathbf{g}\mathbf{g}}{g^3} \quad (5.30)$$

so that with an integration by parts (5.22) becomes

$$\begin{aligned}
\partial_t f(\mathbf{v},t) &= \frac{2\pi n_0 e^4 \ln\Lambda}{m_e^2}\,\nabla_{\mathbf{v}}\cdot\Big[(\nabla_{\mathbf{v}} f)\cdot\nabla_{\mathbf{v}}\nabla_{\mathbf{v}}\int d\mathbf{v}'\, g f(\mathbf{v}') \\
&\quad - f(\mathbf{v})\int d\mathbf{v}'\,\nabla_{\mathbf{v}}(\nabla_{\mathbf{v}}\cdot\nabla_{\mathbf{v}}) g\, f(\mathbf{v}')\Big] \\
&= \frac{2\pi n_0 e^4 \ln\Lambda}{m_e^2}\Big\{\nabla_{\mathbf{v}}\nabla_{\mathbf{v}} : \Big[f(\mathbf{v})\nabla_{\mathbf{v}}\nabla_{\mathbf{v}}\int d\mathbf{v}'\, g f(\mathbf{v}')\Big] \\
&\quad - 2\,\nabla_{\mathbf{v}}\cdot\Big[f(\mathbf{v})\int d\mathbf{v}'\,\nabla_{\mathbf{v}}(\nabla_{\mathbf{v}}\cdot\nabla_{\mathbf{v}}) g\, f(\mathbf{v}')\Big]\Big\} \\
&= \frac{2\pi n_0 e^4 \ln\Lambda}{m_e^2}\Big\{-4\nabla_{\mathbf{v}}\cdot\Big[f(\mathbf{v})\nabla_{\mathbf{v}}\int d\mathbf{v}'\,\frac{f(\mathbf{v}')}{g}\Big] \\
&\quad + \nabla_{\mathbf{v}}\nabla_{\mathbf{v}} : \Big[f(\mathbf{v})\nabla_{\mathbf{v}}\nabla_{\mathbf{v}}\int d\mathbf{v}'\, g f(\mathbf{v}')\Big]\Big\}
\end{aligned} \quad (5.31)$$

where in the first step we have used $\nabla_{\mathbf{v}} g = -\nabla_{\mathbf{v}'} g$, and in the third step we have used $(\nabla_{\mathbf{v}}\cdot\nabla_{\mathbf{v}})g = 2/g$. This is in the standard form of a Fokker–Planck equation (see Appendix B),

$$\boxed{\frac{\partial f(\mathbf{v},t)}{\partial t} = -\nabla_{\mathbf{v}}\cdot[\mathbf{A} f(\mathbf{v})] + \frac{1}{2}\nabla_{\mathbf{v}}\nabla_{\mathbf{v}} : [\overleftrightarrow{\mathbf{B}} f(\mathbf{v})]} \quad (5.32)$$

where the *coefficient of dynamic friction*

$$\mathbf{A}(\mathbf{v},t) = \frac{8\pi n_0 e^4 \ln \Lambda}{m_e^2} \nabla_{\mathbf{v}} \int d\mathbf{v}' \frac{f(\mathbf{v}',t)}{|\mathbf{v} - \mathbf{v}'|} \tag{5.33}$$

and the *diffusion coefficient*

$$\overleftrightarrow{\mathbf{B}}(\mathbf{v},t) = \frac{4\pi n_0 e^4 \ln \Lambda}{m_e^2} \nabla_{\mathbf{v}}\nabla_{\mathbf{v}} \int d\mathbf{v}' \, |\mathbf{v} - \mathbf{v}'| \, f(\mathbf{v}',t) \tag{5.34}$$

With the coefficients (5.33) and (5.34), Eq. (5.32) is known as the *Landau form* of the Fokker–Planck equation.

The meaning of the terms in the Fokker–Planck equation is discussed in Appendix B. The coefficient of dynamic friction **A** represents the slowing down of a typical particle because of many small angle collisions. The diffusion coefficient represents the increase of a typical particle's velocity (in the direction perpendicular to its instantaneous velocity) because of many small angle collisions. Thus, the two terms on the right side of the Fokker–Planck equation (5.32) tend to balance each other. They are in perfect balance when f is a Maxwellian, as shown in Problem 5.5.

The Landau form of the Fokker–Planck equation (5.32) has been solved numerically by MacDonald et al. [7] (Fig. 5.2). The initial distribution function $f(\mathbf{v},t = 0) = f(|\mathbf{v}|,t = 0)$ is spherically symmetric in velocity space. Figure 5.2 shows the steady progression of the distribution, as time increases, toward a Maxwellian. At late times, there is an overshoot at low speeds, which indicates that it

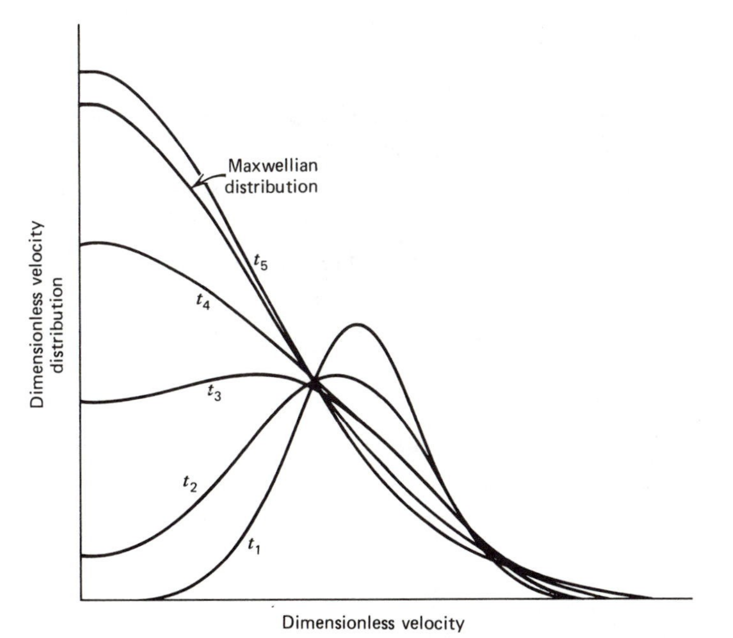

Fig. 5.2 Time evolution of a spherically symmetric electron distribution function as obtained from a numerical solution of the Landau form of the Fokker–Planck equation (5.32) by MacDonald et al. [7].

takes a long time to populate the high speed tail of the Maxwellian. (Remember that Coulomb collisions become quite weak for fast particles.)

There exist even simpler forms of the Fokker–Planck equation [4] but these are not too accurate and are used only to get a rough idea of collisional effects. One is

$$\frac{\partial f}{\partial t} = \nu \nabla_{\mathbf{v}} \cdot [(\mathbf{v} - \mathbf{v}_0) f + v_e^2 \nabla_{\mathbf{v}} f] \tag{5.35}$$

where ν is a collision frequency, and $\mathbf{v}_0$ is a constant velocity. An even cruder model, which is not related to the development of the present chapter, is the *Krook model*,

$$\frac{\partial f}{\partial t} = -\nu(f - f_0) \tag{5.36}$$

where f_0 is the appropriate Maxwellian distribution. Equation (5.36) is also called the BGK equation, after *B*hatnagar, *G*ross, and *K*rook [8].

This brings us to the end of our study of plasma kinetic theory including the effects of two-body collisions. The material in this chapter can be truly appreciated only after a careful study of Appendices A and B. However, Appendix A itself can best be understood after one has mastered the treatment of the Vlasov equation, to which we turn our attention in the next chapter.

REFERENCES

[1] N. N. Bogoliubov, *Problems of a Dynamical Theory in Statistical Physics*, State Technical Press, Moscow, 1946.

[2] A. Lenard, *Ann. Phys. (New York)*, *10*, 390 (1960).

[3] R. Balescu, *Phys. Fluids*, *3*, 52 (1960).

[4] P. C. Clemmow and J. P. Dougherty, *Electrodynamics of Particles and Plasmas*, Addison-Wesley, Reading, Mass., 1969.

[5] D. C. Montgomery and D. A. Tidman, *Plasma Kinetic Theory*, McGraw-Hill, New York, 1964.

[6] D. C. Montgomery, *Theory of the Unmagnetized Plasma*, Gordon and Breach, New York, 1971.

[7] W. M. MacDonald, M. N. Rosenbluth, and W. Chuck, *Phys. Rev.*, *107*, 350 (1957).

[8] P. L. Bhatnagar, E. P. Gross, and M. Krook, *Phys. Rev.*, *94*, 511 (1954).

[9] M. Rosenbluth, W. M. MacDonald, and D. L. Judd, *Phys. Rev.*, *107*, 1 (1957).

PROBLEMS

5.1 Fourier Transforms

Find the Fourier transforms (5.16) and (5.17). (*Hint:* Use spherical polar coordinates with $\mathbf{k} \cdot \mathbf{x} = kr \cos \theta$.)

5.2 Lenard–Balescu Equation

After referring to Clemmow and Dougherty [4], and Montgomery and Tidman [5], sketch the proofs of properties (a) to (f) of the Lenard–Balescu equation as listed below (5.21).

5.3 An Integral

With the help of a table of integrals, perform the integration in (5.26).

5.4 Simpler Derivation of the Landau Form

The development of the Landau form for $\overleftrightarrow{Q}$, from (5.23) to (5.28), is the standard one. However, a simpler one exists. In (5.23), replace ϵ by unity, and cut off the wave number integration at a lower wave number λ_e^{-1} as well as at the upper wave number p_0^{-1}. Show that (5.28) results. The replacement of ϵ by unity is equivalent to ignoring the shielding, as can be seen in (5.20).

5.5 Maxwellian

Show that a Maxwellian is an exact time-independent solution of both the Lenard–Balescu equation (5.19) and the Landau form of the Fokker–Planck equation (5.32).

5.6 Two-Point Correlation Function

Discuss the meaning of $f_2 = f_1 f_1 + g$. Why should g depend on f_1? In particular, how would g change as we turn up the temperature of a Maxwellian?

5.7 Plasmas and Brownian Motion

Discuss the analogy between collisional effects on a particle in a plasma and Brownian motion. Explain why the collisional effects can be described by a Fokker–Planck equation. Thus, using only words, explain how we could use the results of Section 1.6 on collisions to obtain the Fokker–Planck equation directly, without starting from Liouville → BBGKY → Lenard–Balescu → Fokker–Planck. This is actually the technique used by Rosenbluth et al. [9].

5.8 Units

Check all of the units in (5.19) to (5.36). Using crude dimensional arguments, derive the model (5.35) from the Fokker–Planck equation (5.32) and the coefficients (5.33) and (5.34).

CHAPTER 6

Vlasov Equation

6.1 INTRODUCTION

Possibly the single most important equation in plasma physics is the *Vlasov equation*. This equation describes the evolution of the *distribution function* $f_s(\mathbf{x},\mathbf{v},t)$ in six-dimensional phase space. As discussed in Chapter 3, the distribution function $f_s(\mathbf{x},\mathbf{v},t)$ can be thought of as the ensemble averaged number of point particles per unit six-dimensional phase space. It can also be thought of as the number of particles at any given time t, in a small region of the six-dimensional phase space of a single plasma, divided by the volume of the small region of six-dimensional phase space. As discussed in Chapter 3, the Vlasov equation becomes exact in the limit that the number of particles Λ in a Debye cube becomes infinite.

The Vlasov equation arises naturally from the Klimontovich equation (Chapter 3) or from the BBGKY hierarchy (Chapter 4) when the effects of collisions are ignored. For this reason, the Vlasov equation is also called the *collisionless Boltzmann equation*. By ignoring the effects of collisions from the start, we can derive the Vlasov equation as follows.

Consider $f_s(\mathbf{x},\mathbf{v},t)$ as a probability density associated with an ensemble of systems. This probability density can be thought of as a fluid in six-dimensional phase space. Since particles are neither created nor destroyed, this fluid must satisfy a continuity equation with the form

$$\partial_t f_s(\mathbf{x},\mathbf{v},t) + \nabla_{\mathbf{x}} \cdot \left(\left.\frac{d\mathbf{x}}{dt}\right|_{\text{orbit}} f_s \right)$$

$$+ \nabla_{\mathbf{v}} \cdot \left(\left.\frac{d\mathbf{v}}{dt}\right|_{\text{orbit}} f_s \right) = 0 \tag{6.1}$$

where $d/dt|_{\text{orbit}}$ refers to the orbit of the fluid element at the position $(\mathbf{x}, \mathbf{v})$ in phase space. But the fluid represents the probability density of particles; therefore the orbit of the fluid element must be the same as the orbit of a particle of species s at position $\mathbf{x}$ with velocity $\mathbf{v}$. With this identification, we have immediately

$$\left.\frac{d\mathbf{x}}{dt}\right|_{\text{orbit}} = \mathbf{v} \tag{6.2}$$

and

$$\left.\frac{d\mathbf{v}}{dt}\right|_{\text{orbit}} = \frac{q_s}{m_s}\left[\mathbf{E}(\mathbf{x},t) + \frac{\mathbf{v}}{c} \times \mathbf{B}(\mathbf{x},t)\right] \tag{6.3}$$

where because the effects of collisions are being ignored the fields $\mathbf{E}$ and $\mathbf{B}$ are the smooth, ensemble averaged fields satisfying Maxwell's equations (3.28). Equation (6.1) becomes

$$\partial_t f_s(\mathbf{x},\mathbf{v},t) + \nabla_{\mathbf{x}} \cdot (\mathbf{v} f_s) + \frac{q_s}{m_s} \nabla_{\mathbf{v}} \cdot \left[\left(\mathbf{E} + \frac{\mathbf{v}}{c} \times \mathbf{B}\right) f_s\right] = 0 \tag{6.4}$$

With the vector identity $\nabla \cdot (\mathbf{a}b) = b\nabla \cdot \mathbf{a} + \mathbf{a} \cdot \nabla b$, we find

$$\boxed{\partial_t f_s(\mathbf{x},\mathbf{v},t) + \mathbf{v} \cdot \nabla_{\mathbf{x}} f_s + \frac{q_s}{m_s}\left(\mathbf{E} + \frac{\mathbf{v}}{c} \times \mathbf{B}\right) \cdot \nabla_{\mathbf{v}} f_s = 0} \tag{6.5}$$

which is the *Vlasov equation* [1].

EXERCISE Verify that the two terms dropped in going from (6.4) to (6.5) indeed vanish.

When this equation, one for each species, is combined with Maxwell's equations (3.28), we have a complete description of the behavior of a plasma. Although in principle the Vlasov equation only applies to an ensemble of plasmas, in practice we assume that, because of the large number of particles in a single plasma, the fluctuations are so small that the Vlasov equation yields good predictions for a single plasma. Since collisions have been ignored, the Vlasov equation applies only when collisional effects are unimportant. Often, this means that we are limited to phenomena with a characteristic frequency $\omega >> \nu_{ei} \approx \omega_e/\Lambda$.

6.2 EQUILIBRIUM SOLUTIONS

For time scales short compared to a collision time $\nu_{ei}^{-1} \approx \Lambda\, \omega_e^{-1}$, we are interested in finding steady-state solutions to the Vlasov equation (6.5), that is, those for which $\partial_t f_s = 0$. (In this chapter, the words "equilibrium" and "steady-state" are used synonymously.) Of course, there is no guarantee that such steady-state solutions are stable to small perturbations. (A pencil standing on its tip is a steady-state solution, but not a stable one.)

As we look for solutions to the Vlasov equation, it is useful to interpret the left side of (6.5) as the total time derivative of f_s along a particle orbit. Consider a particle of species s whose orbit in six-dimensional phase space is $\mathbf{X}(t)$, $\mathbf{V}(t)$, where $\mathbf{X}(t)$ is the function that gives the position $\mathbf{x}$ in real space of the particle at time t, and $\mathbf{V}(t)$ is the function that gives the position $\mathbf{v}$ in velocity space of the particle at time t. Then the total time derivative of any quantity, measured along the test particle's orbit in phase space, is

$$\begin{aligned} \frac{D}{Dt} &= \partial_t + \frac{d\mathbf{X}(t)}{dt} \cdot \nabla_{\mathbf{x}} + \frac{d\mathbf{V}(t)}{dt} \cdot \nabla_{\mathbf{v}} \\ &= \partial_t + \left.\frac{d\mathbf{x}}{dt}\right|_{\text{orbit}} \cdot \nabla_{\mathbf{x}} + \left.\frac{d\mathbf{v}}{dt}\right|_{\text{orbit}} \cdot \nabla_{\mathbf{v}} \\ &= \partial_t + \mathbf{v} \cdot \nabla_{\mathbf{x}} + \frac{q_s}{m_s}\left(\mathbf{E} + \frac{\mathbf{v}}{c} \times \mathbf{B}\right) \cdot \nabla_{\mathbf{v}} \end{aligned} \tag{6.6}$$

where we have inserted (6.2) and (6.3). Thus, the Vlasov equation (6.5) simply says

$$\frac{D}{Dt} f_s(\mathbf{x},\mathbf{v},t) = 0 \tag{6.7}$$

Knowledge of the form (6.7) gives us one way to solve the Vlasov equation (6.5). Suppose we construct f_s out of functions $C_i(\mathbf{x},\mathbf{v},t)$ that are constants of the motion along the orbit of a particle. Then by (6.7),

$$\frac{D}{Dt} f_s(\{C_i(\mathbf{x},\mathbf{v},t)\}) = \sum_i \frac{\partial f_s}{\partial C_i} \frac{D}{Dt} C_i = 0 \tag{6.8}$$

so that the Vlasov equation (6.5) is satisfied. Thus, any distribution that is a function only of the constants of the motion of the individual particle orbits is a solution of the Vlasov equation.

In the present section we are interested only in equilibrium solutions that do not depend explicitly on time. Noting that the fields $\mathbf{E}$ and $\mathbf{B}$ in (6.5) can be combinations of externally imposed fields and self-consistent fields, we consider the following cases.

CASE A: E = B = 0

In the absence of external fields, the energy $\frac{1}{2}m_s v^2$ and momentum $m_s\mathbf{v} = m_s(v_x, v_y, v_z)$ of a particle are constants of the motion. Thus, any function

$$f_s = f_s(v_x, v_y, v_z) \tag{6.9}$$

is a solution of the time-independent Vlasov equation. This can also be seen by writing the time-independent Vlasov equation with no external fields,

$$\mathbf{v} \cdot \nabla_{\mathbf{x}} f_s = 0 \tag{6.10}$$

to which (6.9) is seen to be a solution.

CASE B: E = 0, B = CONSTANT

In the presence of a uniform background magnetic field, the total particle momentum is no longer a constant. If we choose the magnetic field in the $\hat{z}$-direction, then the constants of the motion are the momentum $m_s v_z$ in the $\hat{z}$-direction and the energy $\frac{1}{2}m_s v_\perp^2 \equiv \frac{1}{2}m_s(v_x^2 + v_y^2)$ in the plane perpendicular to the magnetic field. Thus, any function

$$f_s = f_s(v_\perp, v_z) \tag{6.11}$$

is an equilibrium solution to the Vlasov equation in the presence of a uniform magnetic field.

EXERCISE Show (Chapter 1) that $v_\perp$ is a constant of the motion in a uniform magnetic field. Verify by direct calculation that (6.11) is a solution of the Vlasov equation (6.5) in this case.

CASE C: B = 0, E = E(x) ≠ 0

In the presence of an arbitrary electric field $\mathbf{E}(x) = -\hat{x}\, d/dx\, \varphi(x)$ in the $\hat{x}$-direction, the particle constants of the motion are the momenta $m_s v_y$ and $m_s v_z$, and the energy $\frac{1}{2} m_s v_x^2 + q_s\, \varphi(x)$ associated with motion in the $\hat{x}$-direction. Thus,

$$f_s = f_s(v_x^2 + 2q_s\, \varphi(x)/m_s, v_y, v_z) \tag{6.12}$$

is an equilibrium distribution function. (Note that f_s can also depend upon the sign of v_x.)

EXERCISE Verify by direct substitution that (6.12) is a solution of the Vlasov equation (6.5) in this case.

In addition to these three simple cases, there are other important examples that are used. For example, in Chapter 2 we discussed the adiabatic invariants that are approximate constants of motion. Using these adiabatic invariants, one can construct approximate equilibrium distribution functions. Such solutions find wide applications in the study of magnetic confinement devices such as the tokamak and mirror machine.

6.3 ELECTROSTATIC WAVES

One of the simplest and most instructive predictions of Vlasov theory is the existence of *electrostatic waves*, waves that have only an electric field with no magnetic field, and in the small amplitude limit have a time and spatial dependence

$$\sim \exp\,(i\mathbf{k}\cdot\mathbf{x} - i\omega t) + \text{c.c. (complex conjugate), where } \mathbf{k} \| \mathbf{E}$$

EXERCISE Show that for waves with $\mathbf{k}\|\mathbf{E}$, Maxwell's equations predict no magnetic field.

We begin with the simple situation of a plasma with no applied electric or magnetic fields. Each species has a distribution function

$$f_s = f_{s0} + f_{s1} \tag{6.13}$$

where $f_{s0} = f_{s0}(\mathbf{v})$ is one of the equilibrium solutions discussed in the previous section, and $f_{s1}(\mathbf{x},\mathbf{v},t)$ is a small perturbation associated with the small-amplitude wave. For each species,

$$\int d\mathbf{v}\, f_{s0}(\mathbf{v}) = n_0 \tag{6.14}$$

where n_0 is the average number of particles per unit configuration space. Choosing the electric field in the $\hat{x}$-direction, and treating waves with a spatial variation in the $\hat{x}$-direction only, the Vlasov equation is

$$\partial_t f_s + v_x \partial_x f_s + \frac{q_s}{m_s} E \, \partial_{v_x} f_s = 0 \tag{6.15}$$

With f_{s0} a zero order quantity, and f_{s1} and E small quantities of first order, we look for linearized solutions of (6.15). The zero order terms in (6.15) yield

$$\partial_t f_{s0} + v_x \, \partial_x f_{s0} = 0 \tag{6.16}$$

which is trivially satisfied by our equilibrium solutions $f_{s0} = f_{s0}(\mathbf{v})$. The first order terms in (6.15) are

$$\partial_t f_{s1} + v_x \, \partial_x f_{s1} + \frac{q_s}{m_s} E \, \partial_{v_x} f_{s0} = 0 \tag{6.17}$$

Looking for plane wave solutions $\sim \exp(ikx - i\omega t)$ this is

$$- i\omega f_{s1} + ikv_x f_{s1} = - \frac{q_s}{m_s} E \, \partial_{v_x} f_{s0} \tag{6.18}$$

or

$$f_{s1}(\mathbf{x},\mathbf{v},t) = \frac{-iq_s/m_s}{\omega - kv_x} E \, \partial_{v_x} f_{s0}(\mathbf{v}) \tag{6.19}$$

The only one of Maxwell's equations (3.28) needed for electrostatic waves is Poisson's equation, which in the present case is

$$\begin{aligned} ikE &= 4\pi e(n_i - n_e) \\ &= 4\pi e \int d\mathbf{v} \, (f_i - f_e) \\ &= 4\pi e \int d\mathbf{v} \, (f_{i1} - f_{e1}) \\ &= -i4\pi e^2 E \int d\mathbf{v} \left[\frac{m_i^{-1} \, \partial_{v_x} f_{i0}}{\omega - kv_x} + \frac{m_e^{-1} \, \partial_{v_x} f_{e0}}{\omega - kv_x} \right] \end{aligned} \tag{6.20}$$

Eliminating E from both sides we obtain the *dispersion relation* for electrostatic waves in an unmagnetized plasma,

$$\boxed{1 + \frac{\omega_e^2}{k^2} \int du \, \frac{d_u g(u)}{\omega/k - u} = 0} \tag{6.21}$$

where

$$g(v_x) = \frac{m_e}{n_0 m_i} \int dv_y \, dv_z \, f_{i0}(\mathbf{v}) + \frac{1}{n_0} \int dv_y \, dv_z \, f_{e0}(\mathbf{v}) \tag{6.22}$$

Notice that the ion component of g is reduced by the factor m_e/m_i. For example, if the electrons and ions are Maxwellian, we have

$$f_{s0} = \frac{n_0}{(2\pi)^{3/2} v_s^3} \exp\left[-(v_x^2 + v_y^2 + v_z^2)/2v_s^2\right] \tag{6.23}$$

whereupon

$$g(v_x) = \frac{1}{(2\pi)^{1/2}v_e} \exp(-v_x^2/2v_e^2) + \frac{m_e}{m_i}\frac{1}{(2\pi)^{1/2}v_i} \exp(-v_x^2/2v_i^2) \tag{6.24}$$

where as usual $v_s^2 \equiv T_s/m_s$.

EXERCISE Verify that the Maxwellian (6.23) satisfies the normalization (6.14).

For equal temperatures $T_e = T_i$ we have $v_i << v_e$, and $g(u)$ is as shown in Fig. 6.1. Notice that $g(u)$ has the units (velocity)$^{-1}$. The ion contribution appears tiny when compared to the electron contribution. However, for low frequency motions the ion contribution can be very important, as in the ion-acoustic wave.

Let us use the dispersion relation (6.21) to find the relation between frequency ω and wave number k for high frequency electron plasma waves called *Langmuir waves*. The high frequency of these waves implies that the massive ions do not have time to respond to them, so we ignore the ion contribution to $g(u)$ in (6.22), that is, we let $m_i \to \infty$. This is equivalent to ignoring the ion motion in our derivation of the plasma frequency in Section 1.4. The dispersion relation (6.21) includes an integration over an integrand with a pole at $u = \omega/k$. This pole must be handled with care. For the present, suppose we restrict ourselves to waves such that $\omega/k >> u$ for all u for which $g(u)$ is appreciable, so that $d_u g(u) = 0$ at $u = \omega/k$ (Fig. 6.2). With this assumption, (6.21) can be integrated by parts to obtain

$$1 - \frac{\omega_e^2}{k^2}\int_{-\infty}^{\infty} du\, \frac{g(u)}{(\omega/k - u)^2} = 0 \tag{6.25}$$

where the boundary terms vanish because $g(u \to \pm\infty) = 0$. Expanding the denominator up to and including second order terms in uk/ω, we find

$$1 - \frac{\omega_e^2}{\omega^2}\int du\, g(u)\left(1 + \frac{2uk}{\omega} + \frac{3u^2k^2}{\omega^2}\right) = 0 \tag{6.26}$$

EXERCISE Verify the expansion.

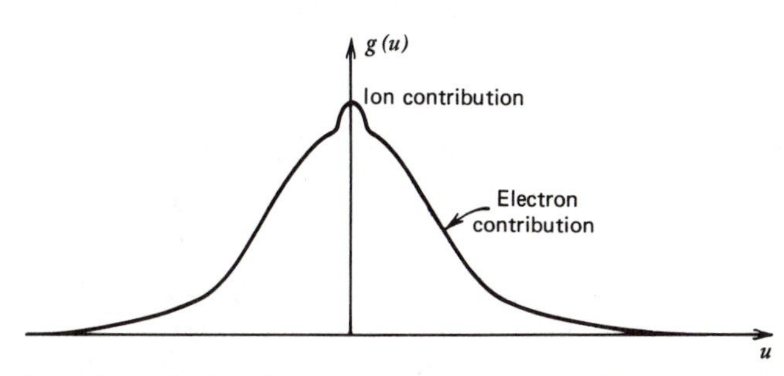

Fig. 6.1 The function $g(u)$ as predicted by (6.22) for an equal temperature Maxwellian plasma.

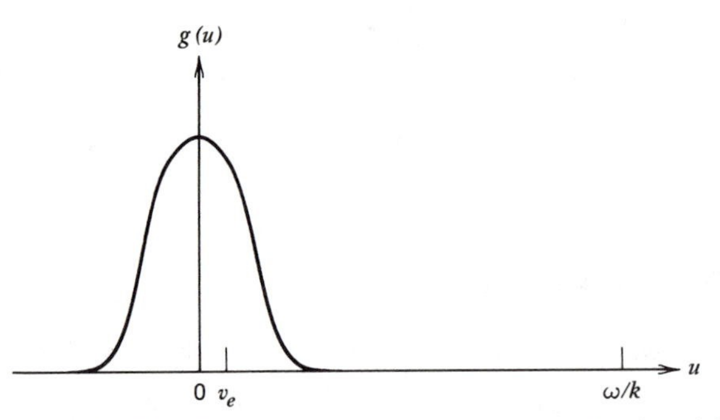

Fig. 6.2 The Langmuir wave calculation of Section 6.3 is appropriate only when the phase speed ω/k of the wave is much larger than the thermal speed v_e.

With $g(u)$ given by the first term on the right of (6.24), we have $\int du\, g(u) = 1$, $\int du\, g(u)u = 0$, $\int du\, g(u)u^2 = v_e^2$.

EXERCISE Verify these statements.

Equation 6.26 then predicts

$$1 - \frac{\omega_e^2}{\omega^2} - \frac{3k^2v_e^2\omega_e^2}{\omega^4} = 0 \tag{6.27}$$

which upon solving for ω^2 and assuming $k^2v_e^2 << \omega^2$ yields

$$\boxed{\omega^2 = \omega_e^2 + 3k^2v_e^2} \tag{6.28}$$

which is the famous *Langmuir wave dispersion relation*; it can easily be committed to memory.

EXERCISE

(a) Obtain (6.28) from (6.27) with the given assumptions.
(b) Verify that (6.28) is consistent with the given assumptions.
(c) Show that (6.28) is equivalent to $\omega = \omega_e(1 + 3k^2\lambda_e^2/2)$.
(d) Use the result of Problem 1.3 to modify (6.28) to include the effects of ion motion.

In the next section we shall return to the dispersion relation (6.21) and show how to properly treat the pole when we do not have $d_u\, g(u) = 0$ at $u = \omega/k$.

6.4 LANDAU CONTOUR

In this section, we present a more complete treatment of electrostatic waves, which includes a careful evaluation of the pole in the dispersion relation (6.21). As shown by Landau [2], the best way to proceed is by solving the Vlasov equation (6.17) and Poisson's equation (6.20) in the context of an initial value problem.

To simplify the discussion, we treat only high frequency Langmuir waves and ignore ion motion. Looking for waves with the spatial dependence $\sim \exp(ikx)$, and denoting the first order electron distribution by $f_1(k,\mathbf{v},t)$, Eq. (6.17) becomes

$$\partial_t f_1 + ikv_x f_1 - (e/m_e)E\, \partial_{v_x} f_{e0} = 0 \tag{6.29}$$

When we use the Laplace transform convention (5.13) to (5.14), and the fact that the Laplace transform of $d_t g(t)$ is $-i\omega g(\omega) - g(t = 0)$, the Laplace transform of (6.29) is

$$\begin{aligned} -i\omega\, f_1(k,\mathbf{v},\omega) &+ ikv_x f_1 \\ - (e/m_e)E(\omega)\partial_{v_x} f_{e0} &= f_1(k,\mathbf{v},t = 0) \end{aligned} \tag{6.30}$$

EXERCISE Demonstrate that the Laplace transform of $d_t g(t)$ is $-i\omega\, g(\omega) - g(t = 0)$.

Poisson's equation (6.20) is in this case

$$ikE(\omega) = -\, 4\pi e \int d\mathbf{v}\; f_1(k,\mathbf{v},\omega) \tag{6.31}$$

Solving (6.30) for $f_1(k,\mathbf{v},\omega)$, we obtain

$$f_1(k,\mathbf{v},\omega) = \frac{(e/m_e)E(\omega)\partial_{v_x} f_{e0} + f_1(k,\mathbf{v},t = 0)}{-i\omega + ikv_x} \tag{6.32}$$

which when substituted in (6.31) yields

$$\begin{aligned} ik\left[1 - \frac{4\pi e^2/m_e}{k^2}\int d\mathbf{v}\; \frac{\partial_{v_x} f_{e0}}{v_x - (\omega/k)}\right] E(\omega) \\ = \frac{-4\pi e}{ik}\int d\mathbf{v}\; \frac{f_1(k,\mathbf{v},t = 0)}{v_x - (\omega/k)} \end{aligned} \tag{6.33}$$

The factor in brackets on the left is the *dielectric function*

$$\epsilon(k,\omega) = 1 - \frac{\omega_e^{\,2}}{k^2}\int du\; \frac{d_u\, g(u)}{u - (\omega/k)} \tag{6.34}$$

where the definition (6.22) of $g(u)$ has been used. Note that $\epsilon(-k,-\omega^*) = \epsilon^*(k,\omega)$. Equation (6.33) then becomes

$$E(\omega) = \frac{4\pi e}{k^2\epsilon(k,\omega)}\int d\mathbf{v}\; \frac{f_1(k,\mathbf{v},t = 0)}{v_x - (\omega/k)} \tag{6.35}$$

As with all Laplace transforms, $E(\omega)$ is defined only for ω_i sufficiently large, and the inverse Laplace transform

$$E(t) = \int_L \frac{d\omega}{2\pi}\; E(\omega)e^{-i\omega t} \tag{6.36}$$

is carried out along the Laplace contour as shown in Fig. 6.3. The Laplace contour must pass above all poles of $E(\omega)$, which by (6.35) includes all zeros of $\epsilon(k,\omega)$. With the Laplace contour as shown in Fig. 6.3, we have $\omega_i > 0$ everywhere on the contour and thus $\omega_i > 0$ in the evaluation of $\epsilon(k,\omega)$ in (6.34). By analytically continuing the function $g(u)$ to the entire complex $u = u_r + iu_i$-plane, we can think of the integration in (6.34) as occurring along the real u-axis, with a pole at $u = \omega/k$ in the upper half u-plane, as shown in Fig. 6.4.

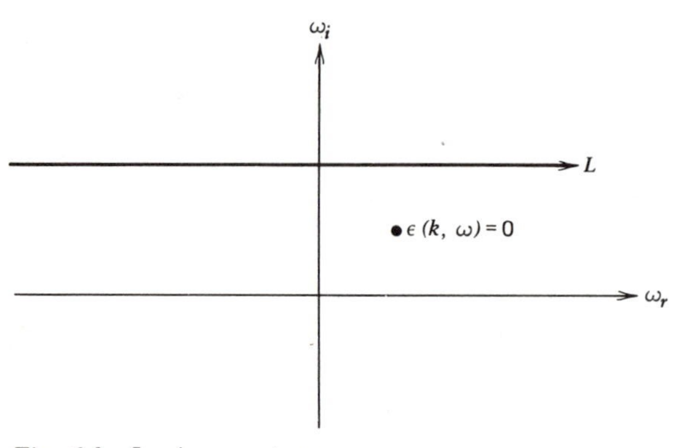

Fig. 6.3 Laplace contour.

The inverse Laplace transform (6.36) is accomplished by analytically continuing $E(\omega)$ to the entire complex ω-plane. By (6.35), when we analytically continue $E(\omega)$, we must analytically continue $\epsilon(k,\omega)$. This means that when ω crosses the real ω-axis from above to below, we cannot allow the pole in Fig. 6.4 to cross the integration contour; if it did, the value of $\epsilon(k,\omega)$ would jump by $-2\pi i$ times the residue at $u = \omega/k$. Thus, we must deform the integration contour in the complex u-plane as shown in Fig. 6.5 when $\omega_i < 0$. The two sets of contours shown in Figs. 6.4 and 6.5 are collectively known as the *Landau contour*. A similar contour must be used to evaluate the other integral in (6.35).

With $E(\omega)$ in (6.35) analytically continued to the entire complex ω-plane, the L contour in (6.36) can be deformed as shown in Fig. 6.6. We do not attempt to close the contour with a semicircle in the lower half ω-plane, since it is not clear from

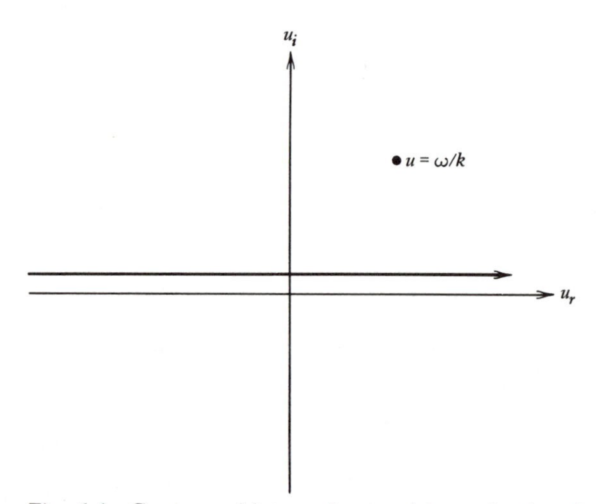

Fig. 6.4 Contour of integration used in evaluating the dielectric function (6.34) when $\omega_i > 0$.

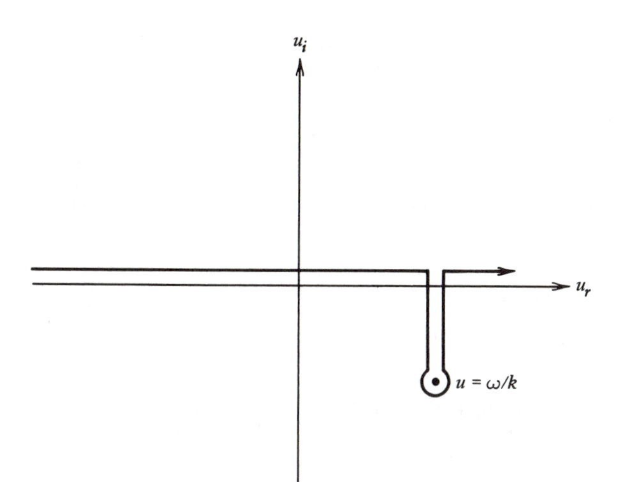

Fig. 6.5 Contour of integration used in evaluating the dielectric function (6.34) when $\omega_i < 0$.

(6.35) that $E(\omega)$ falls off fast enough at large negative ω_i for this to be done. Rather, we treat each part of the contour in Fig. 6.6 separately. There are four types of contributions. First, there are the poles of $E(\omega)$ such as the one at point A in Fig. 6.6. We assume that this pole comes from a zero of $\epsilon(k,\omega)$ rather than from the other factor on the right of (6.35). Denoting the frequency at point A by ω_A, this pole contributes a term to $E(t)$ with the time dependence $\exp(-i\omega_A t)$. We call these contributions the *normal modes*, and we note that since the frequencies of the normal modes are given by $\epsilon(k,\omega) = 0$, they correspond to the waves found by solving the dispersion relation (6.21) in the previous section. After some time has

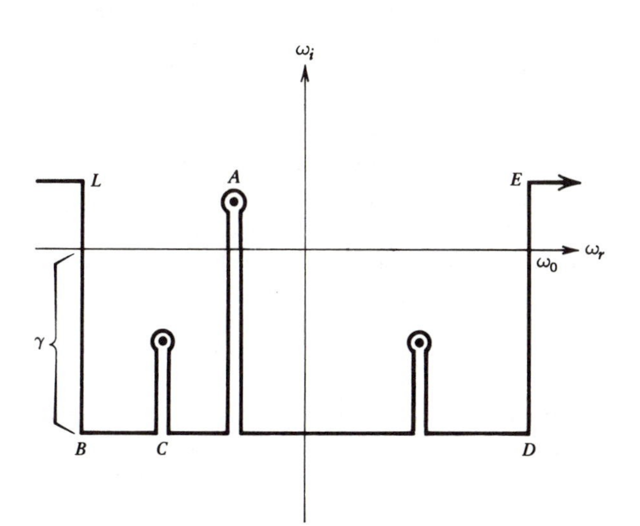

Fig. 6.6 Deformed Laplace contour used in taking the inverse Laplace transform (6.36).

elapsed, the dominant contribution from the normal modes comes from the one with the largest imaginary part of the frequency, as in point A of Fig. 6.6. If this is positive, the normal mode is *unstable* and grows with time; if it is negative, the normal mode is *damped* and decays with time.

Second, there are contributions from segments like the one from point B to point C. With the contour a distance $\gamma > 0$ below the real ω-axis, this contribution is of the form

$$E(t) \sim \int_B^C \frac{d\omega}{2\pi} E(\omega) e^{-i\omega t} = e^{-\gamma t} \int_B^C \frac{d\omega}{2\pi} E(\omega) e^{-i\omega_r t} \tag{6.37}$$

which decays very rapidly with time. Thus, after an initial *transient* period of time, these contributions can be ignored.

Third, there are contributions from the two segments like the one from point D to point E, of the form

$$E(t) \sim \int_D^E \frac{d\omega}{2\pi} E(\omega) e^{-i\omega t}$$

$$= e^{-i\omega_0 t} \int_D^E \frac{d\omega}{2\pi} E(\omega_0 + i\omega_i) e^{\omega_i t} \tag{6.38}$$

These contributions can be ignored since $E(\omega)$ is small for $\omega_0 \rightarrow \infty$; by (6.35) it varies as ω^{-1} for large ω.

Fourth, the segment from point E to infinity gives a contribution

$$E(t) \sim \int_E^\infty \frac{d\omega}{2\pi} e^{-i\omega t} E(\omega) \tag{6.39}$$

which vanishes since for $\omega_0 \rightarrow \infty$ the integrand oscillates infinitely fast.

Thus, the response to an initial perturbation consists of normal modes that oscillate with the normal mode frequencies given by the dispersion relation (6.21), and transients. After some time, the normal mode with the largest imaginary part dominates. This discussion refers to a single wave number k. A spatially localized perturbation in a plasma will have a spectrum of wave numbers, and the response after an initial transient period will consist of a spectrum of normal modes at different wave numbers. If any one of the normal mode frequencies has a positive imaginary part, the plasma is unstable and the perturbation grows with time. If all of the normal mode frequencies have negative imaginary parts, the perturbation eventually damps away. In the next section, we quantitatively evaluate the real and imaginary parts of the normal mode frequencies in terms of the zero-order distribution function $f_{e0}(\mathbf{v})$.

6.5 LANDAU DAMPING

In this section, we return to the Langmuir wave dispersion relation (6.21) and use our knowledge of the integration contours to carefully evaluate the contribution of the pole at $u = \omega/k$. This calculation is especially elegant when we assume that $|\omega_i| << |\omega_r|$, which can be checked after ω is calculated.

Writing the dispersion relation (6.21) in the form

$$\epsilon(k,\omega) = \epsilon_r + i\epsilon_i = 0 \tag{6.40}$$

and Taylor expanding about $\omega = \omega_r$ yields

$$\epsilon_r(k,\omega_r) + i\epsilon_i(k,\omega_r) + i\omega_i \left.\frac{\partial\epsilon_r(k,\omega)}{\partial\omega}\right|_{\omega=\omega_r} = 0 \tag{6.41}$$

Here the term $\sim \omega_i\, \partial\epsilon_i/\partial\omega$ is ignored because it is the product of ω_i, which is small, and $\partial\epsilon_i/\partial\omega$, which can be assumed to be small because $\omega_i \sim \epsilon_i$, which can be seen by equating the real and imaginary parts of (6.41) separately to zero; this yields

$$\boxed{\epsilon_r(k,\omega_r) = 0} \tag{6.42}$$

and

$$\boxed{\omega_i = -\frac{\epsilon_i(k,\omega_r)}{\partial\epsilon_r(k,\omega)/\partial\omega|_{\omega=\omega_r}}} \tag{6.43}$$

When $\omega_i = 0$, the Landau contour is as shown in Fig. 6.7. The Plemelj formula (see Appendix C), as applied to the contour in Fig. 6.7, is

$$\frac{1}{u-a} = P\left(\frac{1}{u-a}\right) + \pi i\delta(u-a) \tag{6.44}$$

where P means principal value; this allows (6.34) to be written

$$\epsilon(k,\omega_r) = 1 - \frac{\omega_e^2}{k^2} P\int_{-\infty}^{\infty} du\, \frac{d_u g(u)}{u - (\omega_r/k)} - \pi i \frac{\omega_e^2}{k^2}\, d_u\, g(u)|_{u=\omega_r/k} \tag{6.45}$$

where for any function $f(u)$,

$$P\int_{-\infty}^{\infty} du\, \frac{f(u)}{u-a} = \lim_{\epsilon\to 0^+}\left[\int_{-\infty}^{a-\epsilon} du\, \frac{f(u)}{u-a} + \int_{a+\epsilon}^{\infty} du\, \frac{f(u)}{u-a}\right] \tag{6.46}$$

The second term on the right of (6.44) comes from integrating around the semicircle in Fig. 6.7, which yields one-half of $2\pi i$ times the residue. The integrand of (6.46) is shown as Fig. 6.8. The two large contributions of opposite sign cancel each other, so that the principal value is not very sensitive to the properties of the function $f(u)$ near the pole at $u = a$. From (6.42) and (6.45) we determine the real part of the frequency ω_r from

$$\epsilon_r(k,\omega_r) = 0 = 1 - \frac{\omega_e^2}{k^2} P\int_{-\infty}^{\infty} du\, \frac{d_u\, g(u)}{u - (\omega/k)} \tag{6.47}$$

For purposes of this integration, we can take $d_u g(u) = 0$ at $u = \omega_r/k$, integrate by parts as in (6.25) to (6.27), and obtain

$$\epsilon_r(k,\omega_r) = 1 - \frac{\omega_e^2}{\omega_r^2} - \frac{3k^2 v_e^2 \omega_e^2}{\omega_r^4} = 0 \tag{6.48}$$

which by (6.28) yields, with the assumption $k^2 v_e^2 << \omega_r^2$,

$$\omega_r^2 = \omega_e^2 + 3k^2 v_e^2 \tag{6.49}$$

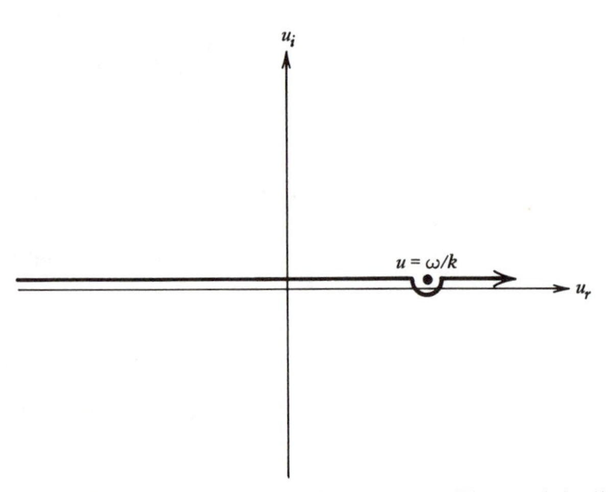

Fig. 6.7 Landau contour when $\omega_i = 0$. The straight line portions of the contour are exactly on the real axis.

[Note that the expressions below (6.26) are valid for an arbitrary $g(u)$, provided that $\int du\; g(u)u = 0$ and provided that one defines v_e by $\int du\; g(u)u^2 = v_e^2$.] In order to calculate ω_i with (6.43) we need

$$\left.\frac{\partial \epsilon_r(k,\omega)}{\partial \omega}\right|_{\omega_r} = \frac{\partial \epsilon_r(k,\omega_r)}{\partial \omega_r} \approx \frac{2\omega_e^2}{\omega_r^3} \approx \frac{2}{\omega_e} \tag{6.50}$$

where terms $\mathcal{O}(k^2\lambda_e^2)$ have been ignored. Then by (6.43) and (6.45),

$$\omega_i = \frac{\pi \omega_e^3}{2k^2}\, d_u g(u)|_{u=\omega_r/k} \tag{6.51}$$

The total normal mode frequency is finally

$$\boxed{\omega = \omega_e \left(1 + \frac{3}{2}\, k^2 \lambda_e^2\right) + i\, \frac{\pi \omega_e^3}{2k^2}\, d_u g(u)|_{u=\omega_r/k}} \tag{6.52}$$

This equation is valid for all Langmuir waves such that $k\lambda_e << 1$. The generalization of (6.52) to encompass all wave numbers has been presented by Jackson [3].

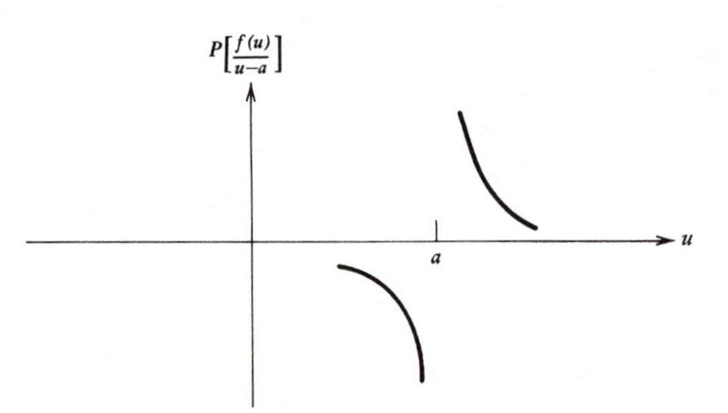

Fig. 6.8 Principal value integrand.

When the slope of the distribution function $d_u g(u)|_{u=\omega_r/k}$ is negative, as with the Maxwellian distribution in Fig. 6.2, Langmuir waves have $\omega_i < 0$ and are *Landau damped.* When the slope is positive at $u = \omega_r/k$, as it is for a range of wave numbers in the "bump-on-tail" distribution in Fig. 5.1 at $t = 0$, then these wave numbers grow exponentially.

With a Maxwellian $g(u)$ given by the first term on the right of (6.24) one can explicitly evaluate ω_i, which is

$$\omega_i = -\frac{\pi\omega_e^3}{2k^2}\frac{u}{(2\pi)^{1/2}v_e^3}\exp(-u^2/2v_e^2)|_{u=\omega_r/k}$$

$$= -\omega_e\left(\frac{\pi}{8}\right)^{1/2}\frac{1}{(k\lambda_e)^3}\exp(-3/2)\exp(-1/2k^2\lambda_e^2) \tag{6.53}$$

This vanishes for $k \to 0$ and increases rapidly with increasing wave number, such that for $k\lambda_e > 0.3$ the damping $\omega_i \sim \omega_e$ is so large that such waves are never observed.

In Section 6.7 we shall consider the physical mechanism of Landau damping. Heuristically, this can be considered as follows. Consider a wave with a phase speed $V_\varphi = \omega_r/k$ in a Maxwellian plasma (Fig. 6.2). Those particles with speeds u very close to V_φ interact strongly with the wave. Particles with speeds slightly faster than V_φ are grabbed by the wave and slowed down, giving up energy to the wave, while particles with speeds slightly slower than the wave are sped up, taking energy from the wave. Since in a Maxwellian plasma there are more particles with speeds slightly less than V_φ than with speeds slightly greater than V_φ, the net result is an energy gain by the particles and an energy loss by the wave; this is Landau damping.

On the other hand, consider a wave with phase speed $V_\varphi = \omega_r/k = u$ such that $d_u(g) > 0$ as on the left half of the "bump-on-tail" in Fig. 5.1. Now there are more particles slightly faster than V_φ than slightly slower, the particles lose net energy to the wave, and the wave grows. These ideas will be made quantitative in Section 6.7.

6.6 WAVE ENERGY

In Section 6.5 the real and imaginary parts of the normal mode frequency for longitudinal waves are determined from a knowledge of the dielectric function $\epsilon(k,\omega)$. The importance of the dielectric function for longitudinal waves is due to its equivalence to Poisson's equation

$$ikE = 4\pi\rho \tag{6.54}$$

which is effectively replaced by

$$ik\epsilon(k,\omega)E = 0 \tag{6.55}$$

for purposes of calculating normal modes [see (6.33)]. Thus, all of the physics contained in Poisson's equation is also contained in the dielectric function. It is important to note that the "1" in the dielectric function

$$\epsilon(k,\omega) = 1 - \frac{\omega_e^2}{k^2}\int du\frac{d_u g(u)}{u-(\omega/k)} \tag{6.34}$$

comes from the left side of Poisson's equation (the "vacuum" contribution) whereas the other term comes from the right side of Poisson's equation and represents the contribution of the plasma (plasma = medium = dielectric).

The dielectric function $\epsilon(k,\omega)$ provides a very useful approach to wave energy. In this section, we present a somewhat heuristic demonstration of the relation between the dielectric function and wave energy. A more rigorous development can be found in Landau and Lifshitz [4].

When we deal with energy, we deal with squared quantities, such as electric fields; therefore we must be certain to have only real quantities before squaring. Consider a real oscillatory electric field

$$\tilde{E}(t) = \frac{1}{2} E(t) \exp(-i\omega_r t) + \frac{1}{2} E^*(t) \exp(i\omega_r t) \tag{6.56}$$

at a fixed spatial point. The time-averaged electric field energy density at this spatial point is

$$\begin{aligned} W_E \equiv \frac{\overline{\tilde{E}^2}}{8\pi} &= \frac{1}{32\pi} \overline{[E^2 \exp(-2i\omega_r t) + 2|E|^2 + E^{*2} \exp(2i\omega_r t)]} \\ &= \frac{|E|^2}{16\pi} \end{aligned} \tag{6.57}$$

where the terms at frequencies $\pm 2i\omega_r t$ vanish on averaging over the fast time scale $2\pi/\omega_r$.

With the real electric field $\tilde{E}$ one can associate a real current $\tilde{J}$, such that $\tilde{J}(t) = \frac{1}{2}J(t) \exp(-i\omega_r t) + \frac{1}{2} J^*(t) \exp(i\omega_r t)$. Since we are dealing with linear waves, there is a linear relation between current and electric field,

$$J(t) = \sigma(\omega)E(t) \tag{6.58}$$

where $\sigma(\omega)$ is the *conductivity*. There is a simple relationship between the conductivity $\sigma(\omega)$ and the dielectric function $\epsilon(\omega)$. Ampere's law for longitudinal waves yields

$$\begin{aligned} 0 = \nabla \times \mathbf{B} &= \frac{4\pi}{c} \mathbf{J} + \frac{1}{c} \frac{\partial \mathbf{E}}{\partial t} \\ &= \frac{-i\omega}{c} \left(\frac{4\pi\sigma}{-i\omega} + 1 \right) \mathbf{E} \\ &= \frac{1}{c} \frac{\partial \mathbf{D}}{\partial t} \\ &= \frac{-i\omega}{c} \epsilon(\omega) \mathbf{E} \end{aligned} \tag{6.59}$$

with which we identify

$$\boxed{\epsilon(\omega) = 1 + \frac{4\pi i \sigma(\omega)}{\omega}} \tag{6.60}$$

where in accordance with electromagnetic theory we have introduced the displacement $\mathbf{D} = \epsilon \mathbf{E}$.

Let us now develop an expression for the time rate of change of wave energy. From the first form of Ampere's law in (6.59), we obtain for real fields

$$\frac{\partial \tilde{E}}{\partial t} = - 4\pi \tilde{J} \tag{6.61}$$

Multiplying each side by $\tilde{E}/4\pi$ yields

$$\frac{1}{8\pi} \frac{d}{dt} (\tilde{E})^2 = - \tilde{E}\tilde{J}$$

$$= - \frac{1}{4} [E \exp(-i\omega_r t) + E^* \exp(i\omega_r t)][\sigma E \exp(-i\omega_r t) + \sigma^* E^* \exp(i\omega_r t)] \tag{6.62}$$

where we have used (6.56) and (6.58). Taking the average of both sides over the short time $2\pi/\omega_r$, the second harmonic terms disappear and we have, using (6.57),

$$\frac{1}{16\pi} \frac{d}{dt} |\mathrm{E}|^2 = - \frac{1}{4} (\mathrm{E}\sigma^* E^* + E^* \sigma E) \tag{6.63}$$

At this point we expand $\sigma(\omega)$ exactly as we expanded $\epsilon(\omega)$ in (6.41). From (6.60),

$$\epsilon_r(\omega_r) = 1 - \frac{4\pi\sigma_i(\omega_r)}{\omega_r} \tag{6.64}$$

while

$$\epsilon_i(\omega_r) = \frac{4\pi\sigma_r(\omega_r)}{\omega_r} \tag{6.65}$$

Consistent with the assumptions used in (6.41), we assume $|\sigma_r(\omega_r)| << |\sigma_i(\omega_r)|$ and expand

$$\sigma(\omega) = \sigma(\omega_r) + \left.\frac{d\sigma}{d\omega}\right|_{\omega_r} i\omega_i + \ldots$$

$$\cong \sigma_r(\omega_r) + i\sigma_i(\omega_r) - \omega_i \left.\frac{d\sigma_i}{d\omega}\right|_{\omega_r} \tag{6.66}$$

where we have ignored the term $\omega_i(d\sigma_r/d\omega)|_{\omega_r}$ as being second order in small quantities. By (6.56), the normal mode electric field $E(t)$ has the time dependence $\exp(\omega_i t)$; therefore $dE/dt = \omega_i E$, and we can write

$$\omega_i \left.\frac{d\sigma_i}{d\omega}\right|_{\omega_r} E = \left.\frac{d\sigma_i}{d\omega}\right|_{\omega_r} \frac{d}{dt} E \tag{6.67}$$

whereupon (6.63) becomes

$$\frac{1}{16\pi} \frac{d}{dt} |E|^2 = - \frac{1}{4}\left[2\sigma_r(\omega_r)|E|^2 - E \left.\frac{d\sigma_i}{d\omega}\right|_{\omega_r} \frac{d}{dt} E^* - E^* \left.\frac{d\sigma_i}{d\omega}\right|_{\omega_r} \frac{d}{dt} E\right]$$

$$= - \frac{1}{2} \sigma_r(\omega_r)|E|^2 + \frac{1}{4} \left.\frac{d\sigma_i}{d\omega}\right|_{\omega_r} \frac{d}{dt} |E|^2 \tag{6.68}$$

Moving a term to the left side, we find

$$\boxed{\frac{1}{16\pi}\frac{d}{dt}|E|^2 - \frac{1}{4}\frac{d\sigma_i}{d\omega}\bigg|_{\omega_r}\frac{d}{dt}|E|^2 = -\frac{1}{2}\sigma_r(\omega_r)|E|^2} \tag{6.69}$$

field energy — particle energy — dissipation

where we recognize the first term on the left as the time derivative of the electric field energy. Because the second term on the left involves σ_i, we identify it as the time derivative of the particle energy contained in the current. The right side represents dissipation due to $\sigma_r(\omega_r)$, which is proportional to $\epsilon_i(\omega_r)$ and thus to ω_i by (6.43). Since $\sigma_r(\omega_r)$ represents dissipation, it is known as the *resistive* part of the conductivity, while $\sigma_i(\omega_r)$ represents the particle energy in the wave and is thus called the *reactive* part of the conductivity.

The two terms on the left of (6.69) represent the time derivative of the total wave energy W_{tot}. Thus, as a function of the electric field energy density (6.57),

$$W_{\text{tot}} = \left(1 - 4\pi\frac{d\sigma_i}{d\omega}\bigg|_{\omega_r}\right)W_E \tag{6.70}$$

From (6.64), the constant can be written

$$1 - 4\pi\frac{d\sigma_i}{d\omega}\bigg|_{\omega_r} = \frac{d}{d\omega}[\omega\epsilon_r(\omega)]_{\omega_r} \tag{6.71}$$

so the total *wave energy* is

$$\boxed{W_{\text{tot}} = \frac{d}{d\omega}[\omega\epsilon_r(\omega)]_{\omega_r}W_E} \tag{6.72}$$

For example, for Langmuir waves with $k\lambda_e \to 0$, we have [see (6.48)]

$$\epsilon_r(\omega) = 1 - \frac{\omega_e^2}{\omega^2} \tag{6.73}$$

Therefore

$$\frac{d}{d\omega}(\omega\epsilon_r)|_{\omega_r} = 1 + \frac{\omega_e^2}{\omega^2}\bigg|_{\omega_e} = 2 \tag{6.74}$$

so there are equal amounts of energy in particles and electric field in a Langmuir wave.

EXERCISE Compare this result to your result in Problem 6.4.

Since the wave energy is $\sim |E|^2$, and $E \sim \exp(\omega_i t)$, we must have

$$\frac{dW_{\text{tot}}}{dt} = 2\omega_i W_{\text{tot}} \tag{6.75}$$

Let us verify that (6.75) is indeed satisfied by (6.69), which says

$$\frac{dW_{\text{tot}}}{dt} = -\frac{1}{2}\sigma_r(\omega_r)|E|^2 = -\frac{1}{2}\sigma_r(\omega_r)16\pi W_E$$

$$= -8\pi\sigma_r(\omega_r)\frac{1}{(d/d\omega)(\omega\epsilon_r)|_{\omega_r}}W_{\text{tot}} \tag{6.76}$$

But $(d/d\omega)(\omega\epsilon_r)|_{\omega_r} = \omega_r(d\epsilon_r/d\omega)|_{\omega_r}$ since $\epsilon_r(\omega_r) = 0$ by (6.42), and $\sigma_r(\omega_r) = \omega_r\epsilon_i(\omega_r)/4\pi$ from (6.65); thus (6.76) is

$$\frac{dW_{\text{tot}}}{dt} = -\frac{2\epsilon_i(\omega_r)}{d\epsilon_r/d\omega|_{\omega_r}} W_{\text{tot}} = 2\omega_i W_{\text{tot}} \tag{6.77}$$

where (6.43) has been used to insert ω_i. Thus, (6.75) is indeed satisfied.

The convenient formulas (6.43) for growth rate and (6.72) for wave energy in terms of the dielectric function $\epsilon(k,\omega)$ are applicable to all electrostatic waves in a plasma, and are very useful in practice. Because of the form (6.34), one can plausibly state that a full knowledge of $\epsilon(k,\omega)$ for all values of k and ω implies a full knowledge of the distribution function $g(u)$. In the next section, we consider in detail the effect of an electrostatic wave on the distribution function; this leads to a microscopic understanding of Landau damping.

6.7 PHYSICS OF LANDAU DAMPING

In Section 6.5 the phenomenon of Landau damping (6.52) is introduced as a mathematical consequence of the solution of the dispersion relation. In the present section we discuss the detailed physics of Landau damping [5]. This is done by considering the effect of the small wave, associated with the perturbed distribution function $f_1(x,v,t = 0)$, on the background plasma represented by $f_0(v)$. For convenience in this section, we denote the x-component of velocity by v, and suppress the "electron" subscript on f_1 and f_0.

Consider a linear Langmuir wave of the form $E_1(x',t) = E_0 \sin(kx' - \omega_r t)$ where x' denotes the laboratory frame of reference, E_0 is a small constant, and for the moment we ignore the imaginary part of the frequency. In the frame of reference x moving with the phase speed $\omega_r/k > 0$ with respect to the laboratory frame, the wave field is independent of time and is given by $E_1(x) = E_0 \sin(kx)$, as shown in Fig. 6.9. All of the particles in the background distribution function $f_0(v)$ are affected by this electric field, and some are speeded up while others are slowed down. We focus our attention on only those particles in $f_0(v)$ that have speed v_0 in the lab frame at $t = 0$ and, thus, have speed $\tilde{v} = v_0 - v_\varphi$ in the frame moving with the wave phase speed $v_\varphi \equiv \omega_r/k$ (Fig. 6.10). This "beam" of particles with speed v_0 in the lab frame and speed $\tilde{v}_0$ in the wave frame will see the energy of some

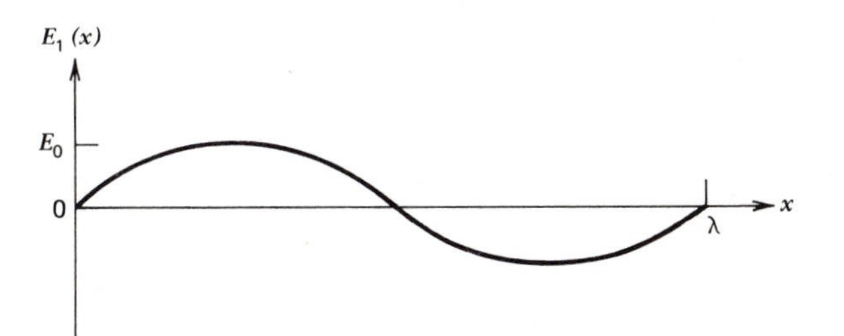

Fig. 6.9 Stationary electric field as seen in the frame $x = x' - (\omega_r/k)t$ moving to the right at speed ω_r/k with respect to the laboratory frame x'.

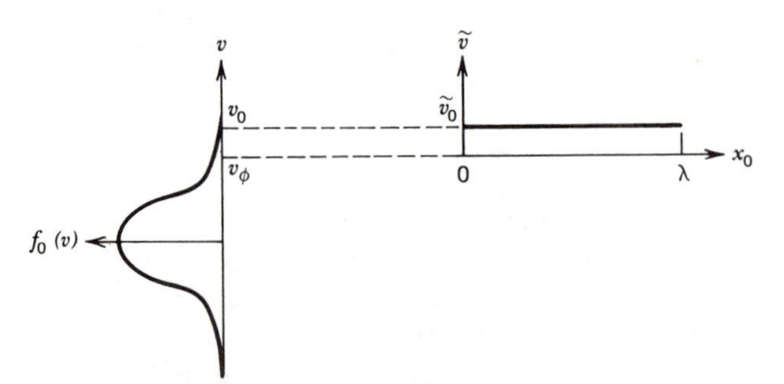

Fig. 6.10 Beam of electrons, all with speed v_0 with respect to the laboratory frame and speed $\tilde{v}_0 = v_0 - v_\varphi$ with respect to the moving frame.

of its members increase with time and the energy of some of its members decrease with time, depending on the initial particle position x_0. After a time t, a particle will have experienced a change in speed Δv (independent of frame) so that the particle's energy, as measured in the lab frame, suffers a change

$$\Delta E = \frac{1}{2} m_e(v_0 + \Delta v)^2 - \frac{1}{2} m_e v_0^2$$
$$= m_e v_0 \Delta v + \mathcal{O}[(\Delta v)^2] \tag{6.78}$$

where we shall ignore the term in $(\Delta v)^2$ in what follows; it can be shown that in the present derivation this term gives a negligible contribution to quantities of interest. We are interested in the average change in energy over a wavelength,

$$\langle \Delta E \rangle_{x_0} = m_e v_0 \langle \Delta v \rangle_{x_0} \tag{6.79}$$

Since Δv can be calculated in any frame, we work in the wave frame. Then

$$\Delta v(t) = \int_0^t \dot{\tilde{v}}(t')\, dt' = \frac{-eE_0}{m_e} \int_0^t dt' \sin k\tilde{x}(t') \tag{6.80}$$

where $\tilde{x}(t)$ is the orbit of the particle. If we insert in (6.80) the unperturbed particle orbit $\tilde{x}(t) = x_0 + \tilde{v}_0 t$, without the effects of the wave field, we would find $\langle \Delta v \rangle_{x_0} = 0$. Thus, we must include the lowest order correction to the particle orbit due to the effect of the wave. This is done as follows:

$$\Delta v(t) = \frac{-eE_0}{m_e} \int_0^t dt' \sin k\tilde{x}(t')$$
$$\tilde{x}(t') \rightarrow x_0 + \int_0^{t'} dt'' \, \tilde{v}(t'')$$
$$\tilde{v}(t'') \rightarrow \tilde{v}_0 - \frac{eE_0}{m_e} \int_0^{t''} dt''' \sin k\tilde{x}(t''')$$
$$\tilde{x}(t''') \rightarrow x_0 + \tilde{v}_0 t''' \tag{6.81}$$

where we have gone far enough to pick up the lowest order correction in E_0. Performing the last integration we find

$$\tilde{v}(t'') = \tilde{v}_0 + \frac{eE_0}{m_e k\tilde{v}_0}[\cos(kx_0 + k\tilde{v}_0 t'') - \cos kx_0] \tag{6.82}$$

The next to last integration yields

$$\tilde{x}(t') = x_0 + \tilde{v}_0 t' - \frac{t' eE_0}{m_e k\tilde{v}_0}\cos(kx_0)$$
$$+ \frac{eE_0}{m_e k^2\tilde{v}_0^2}[\sin(kx_0 + k\tilde{v}_0 t') - \sin(kx_0)] \tag{6.83}$$

The first integrand is of the form $\sin(kx_0 + k\tilde{v}_0 t' + \tilde{\Delta})$ where $\tilde{\Delta}$ is proportional to E_0,

$$\tilde{\Delta} \equiv -\frac{t' eE_0}{m_e\tilde{v}_0}\cos(kx_0)$$
$$+ \frac{eE_0}{m_e k\tilde{v}_0^2}[\sin(kx_0 + k\tilde{v}_0 t') - \sin(kx_0)] \tag{6.84}$$

Since we are looking for the lowest order correction in E_0, we can Taylor expand

$$\sin(a + \tilde{\Delta}) = \sin a + \tilde{\Delta}\cos a \tag{6.85}$$

to lowest order in $\tilde{\Delta}$. Then

$$\Delta v(t) = -\frac{eE_0}{m_e}\int_0^t dt'[\tilde{\Delta}\cos(kx_0 + k\tilde{v}_0 t') + \sin(kx_0 + k\tilde{v}_0 t')] \tag{6.86}$$

We next wish to average Δv over one wavelength, upon which the sin term disappears. The other terms are evaluated using the identities

$$\langle \sin(u - a)\cos(u - b)\rangle_u = -\frac{1}{2}\sin(a - b) \tag{6.87}$$

$$\langle \sin(u - a)\sin(u - b)\rangle_u = \langle \cos(u - a)\cos(u - b)\rangle_u$$
$$= \frac{1}{2}\cos(a - b) \tag{6.88}$$

where $\langle\ \rangle_u$ means an average over one period of the variable u. We find

$$\langle \Delta v(t)\rangle_{x_0} = \left(\frac{eE_0}{m_e}\right)^2 \frac{k}{2}\int_0^t dt'\left\{-\frac{1}{k^2\tilde{v}_0^2}\sin(k\tilde{v}_0 t')\right.$$
$$\left. + \frac{t'}{k\tilde{v}_0}\cos(k\tilde{v}_0 t')\right\} \tag{6.89}$$

The integration can be performed, and yields

$$\boxed{\langle \Delta v(t)\rangle_{x_0} = \left(\frac{eE_0}{m_e}\right)^2 \frac{1}{2k^2\tilde{v}_0^3}\{2[\cos(k\tilde{v}_0 t) - 1] + k\tilde{v}_0 t\sin(k\tilde{v}_0 t)\}} \tag{6.90}$$

Our aim is to form the change of energy, $\langle \Delta E\rangle_{x_0} = m_e v_0\langle \Delta v\rangle_{x_0}$ and integrate over all velocities v_0 in the lab frame to find the total change in energy of the particles. Before doing so, let us evaluate (6.90) at early time, such that $k\tilde{v}t << 1$. Note that early time for one "beam" of velocity $\tilde{v}_0$ may not be early time for another "beam" of velocity $>> \tilde{v}_0$. Using the formulas

$$\sin x \approx x - \frac{x^3}{6} + \ldots \tag{6.91}$$

and

$$\cos x \approx 1 - \frac{x^2}{2} + \frac{x^4}{24} + \ldots \tag{6.92}$$

we find

$$2(\cos x - 1) + x \sin x \approx \frac{-x^4}{12} \tag{6.93}$$

Therefore (6.90) becomes

$$\boxed{\langle \Delta v(t) \rangle_{x_0} = -\left(\frac{eE_0}{m_e}\right)^2 \frac{k^2 \tilde{v}_0}{24} t^4 \qquad k\tilde{v}_0 t << 1} \tag{6.94}$$

Thus, we see that "beams" with $\tilde{v} > 0$, that is, those moving faster than the wave, are indeed slowed down at early times, in the sense of an average over x_0. "Beams" with $\tilde{v} < 0$, those moving slower than the wave, are sped up in an average sense.

We are now ready to integrate the spatially averaged energy change over all velocities, in the lab frame. We cannot use (6.94) for $\langle \Delta v \rangle_{x_0}$, because $k\tilde{v}_0 t << 1$ is not true for all "beams" at a given time t. Rather, we use (6.90) to find the total energy change $W(t)$,

$$W(t) \equiv \int_{-\infty}^{\infty} dv_0\, f_0(v_0) \langle \Delta E \rangle_{x_0} = \int_{-\infty}^{\infty} dv_0\, f_0(v_0) m_e v_0 \langle \Delta v \rangle_{x_0} \tag{6.95}$$

where we have used (6.78). Since (6.90) involves velocities $\tilde{v}_0$ in the wave frame, it is convenient to make the change of variable $\tilde{v}_0 = v_0 - v_\varphi$ in (6.95), which becomes

$$W(t) = m_e \int_{-\infty}^{\infty} d\tilde{v}_0\, \tilde{f}(\tilde{v}_0)(\tilde{v}_0 + v_\varphi) \langle \Delta v \rangle_{x_0} \tag{6.96}$$

where $\tilde{f}(\tilde{v}_0) \equiv f_0(\tilde{v}_0 + v_\varphi)$. We expect the major contribution to (6.96) to come from particles with velocities close to $\tilde{v}_0 \approx 0$; this can be seen from the fact that the expression (6.90) for $\langle \Delta v \rangle_{x_0}$ varies as $(\tilde{v}_0)^{-3}$. We therefore expand $\tilde{f}(\tilde{v}_0) \approx \tilde{f}(0) + \tilde{v}_0 \tilde{f}'(0)$. The product

$$\begin{aligned} \tilde{f}(\tilde{v}_0)(\tilde{v}_0 + v_\varphi) &= \tilde{f}(0)\tilde{v}_0 + \tilde{f}(0) v_\varphi \\ &+ \tilde{v}_0^2 \tilde{f}'(0) + \tilde{v}_0 v_\varphi \tilde{f}'(0) \end{aligned} \tag{6.97}$$

has four terms. Since $\langle \Delta v \rangle_{x_0}$ in (6.90) is odd in $\tilde{v}_0$, the two terms in (6.97) that are even in $\tilde{v}_0$ will give zero when the integration in (6.96) is performed. Only the two odd terms in (6.97), $\tilde{f}(0)\tilde{v}_0$, and $\tilde{v}_0 v_\varphi \tilde{f}'(0)$ contribute. Usually, the latter term makes a much larger contribution.

EXERCISE Verify this statement for a Maxwellian, with $v_\varphi >> v_e$.

If we keep only the latter term, (6.96) becomes

$$W(t) = \frac{m_e v_\varphi \tilde{f}'(0)}{2k^2} \left(\frac{eE_0}{m_e}\right)^2 \int_{-\infty}^{\infty} \frac{d\tilde{v}_0}{\tilde{v}_0^{\,2}} \{2[\cos(k\tilde{v}_0 t) - 1] + k\tilde{v}_0 t \sin(k\tilde{v}_0 t)\} \tag{6.98}$$

With the change of variable $x = k\tilde{v}_0 t$, the integral I in (6.98) becomes

$$I = kt \int_{-\infty}^{\infty} \frac{dx}{x^2} [2(\cos x - 1) + x \sin x] \tag{6.99}$$

The second term is found from an integral table to yield

$$\int_{-\infty}^{\infty} dx \frac{\sin x}{x} = \pi \tag{6.100}$$

while the other term yields -2π.

EXERCISE

(a) Verify the last statement by changing the limits of integration to $(0,\infty)$, carefully integrating by parts, and using (6.100).

(b) Evaluate (6.100) using contour integration. First, move the contour off of the (nonexistent) pole at $z = 0$ and then expand the sine in terms of exponentials.

Thus, we find

$$W(t) = -\frac{\pi}{2} \frac{m_e v_\varphi}{k} \tilde{f}'(0) \left(\frac{eE_0}{m_e}\right)^2 t \tag{6.101}$$

or

$$W(t) = -\frac{\pi}{2} \frac{m_e \omega_r}{k^2} f_0'(v_\varphi) \left(\frac{eE_0}{m_e}\right)^2 t \tag{6.102}$$

or identifying f_0 with $n_0 g$, and taking $\omega_r \approx \omega_e$,

$$\boxed{W(t) = -\frac{\omega_e^{\,3}}{8k^2} g'(v_\varphi) E_0^{\,2} t} \tag{6.103}$$

Equation 6.103 shows that the total particle energy is changing as the first power of time t, and is positive when $g'(v_\varphi) < 0$ and negative when $g'(v_\varphi) > 0$. The energy gained or lost by the particles must come from the wave. The rate of change of wave energy W_{wave} must be equal and opposite to the rate of change of particle energy. From (6.103),

$$\frac{d}{dt} W_{\text{wave}} = -\frac{d}{dt} W(t) = \frac{\omega_e^{\,3}}{8k^2} g'(v_\varphi) E_0^{\,2} \tag{6.104}$$

The total wave energy, averaged over a wavelength, is $W_{\text{wave}} = 2E_0^{\,2} \langle \sin^2 (\cdot kx) \rangle_x / 8\pi = E_0^{\,2}/8\pi$, where the factor of 2 is introduced because a Langmuir wave has equal amounts of energy in electric field energy and in particle kinetic energy. Thus, (6.104) is

$$\frac{d}{dt} W_{\text{wave}} = \frac{\pi \omega_e^{\,3}}{k^2} g'(v_\varphi) W_{\text{wave}} \tag{6.105}$$

If the electric field amplitude is varying with time as $E_0(t) \sim \exp(\gamma t)$, then the wave energy $W_{\text{wave}} \sim \exp(2\gamma t)$; thus

$$\frac{dW_{\text{wave}}}{dt} = 2\gamma W_{\text{wave}} \tag{6.106}$$

Comparing (6.106) and (6.105), we find

$$\boxed{\gamma = \frac{\pi}{2} \frac{\omega_e^3}{k^2} g'(v_\varphi)} \tag{6.107}$$

which is in exact agreement with the formula (6.52) obtained by contour integration of the linearized Vlasov equation.

We see therefore that Landau damping is related to the initial behavior of the background particles, with particles moving slightly faster than the wave being slowed at early times and particles moving slightly more slowly than the wave being sped up at early times; this is true only in a spatially averaged sense. The net Landau damping (or Landau growth) comes from contributions from all particles, averaged over space; however, because of the $\tilde{v}_0^{-3}$ dependence in (6.90), particles close to the wave phase speed give the biggest contribution, which is why $g'(v_\varphi)$ is so important.

The theory just developed is a linear one and thus is exact only for waves of infinitesimal amplitude. In the next section we discuss heuristically the modification of these ideas for waves of finite amplitude.

6.8 NONLINEAR STAGE OF LANDAU DAMPING

In previous sections we have treated linear Landau damping, first by integrating the linearized Vlasov equation, and then by considering the detailed orbits of the background distribution function. Let us next look at the consequences of finite wave amplitude.

We again think in terms of the initial value problem. We consider the background distribution function, in the presence of the wave $\tilde{E}(\tilde{x},t) = E_0(t) \sin(k\tilde{x} - \omega_r t)$. In the wave frame, moving at velocity $v_\varphi = \omega_r/k$ with respect to the lab frame, the electric field is $E(x) = E_0 \sin kx$, with corresponding electrostatic potential $\varphi(x) = (E_0/k) \cos kx$ (Fig. 6.11). In this wave frame, ignoring the slow time dependence of $E_0(t)$, the electrons see a time-independent electric field, so their total energy $H = -e\varphi(x) + \frac{1}{2}mv^2$ is a constant, when v is measured in the wave frame. For each particle, the constant H is determined by the initial position and initial velocity; therefore,

$$\begin{aligned} H &= -e\varphi[x(t)] + \tfrac{1}{2}mv^2(t) \\ &= -e\varphi[x(t=0)] + \tfrac{1}{2}mv^2(t=0) \end{aligned} \tag{6.108}$$

The corresponding equation of motion is

$$m\ddot{x} = -eE_0 \sin kx = -\frac{eE_0}{k} \frac{d}{dx}(-\cos kx) \tag{6.109}$$

Thus, the particles are moving in a potential well $-e\varphi(x) = -(eE_0/k) \cos kx$.

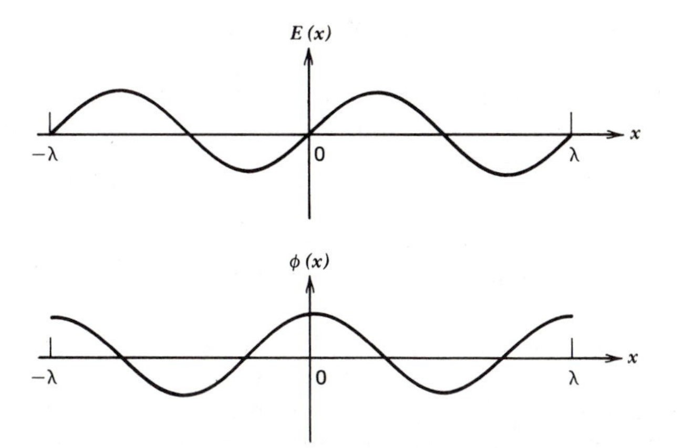

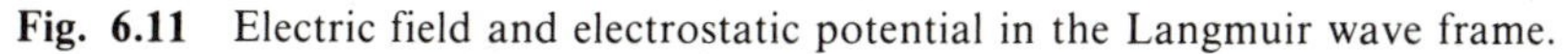

Fig. 6.11 Electric field and electrostatic potential in the Langmuir wave frame.

Consider all particles with $v = 0$ at time $t = 0$. These particles have different energies depending on their position; from (6.108),

$$H = -\, e\varphi[x(t = 0)] \tag{6.110}$$

At $t = 0$, these particles find themselves in a potential field as shown in Fig. 6.12. Particles at A do not move. Particles at B begin to move in the well, with constant energy. Those at C do the same. Those at D are marginally stable; a slight perturbation will allow them to begin moving in the well. Each of these particles oscillates in the well with a certain frequency of oscillation, which decreases as we move up the well, until the frequency at D is zero and the period infinite. Near the bottom of the well, at B for instance, we can find the frequency of oscillation by expanding the force about $x = 0$; thus from (6.109),

$$m\ddot{x} = -\, eE_0 \sin kx \approx -\, eE_0 kx \tag{6.111}$$

from which we identify the characteristic frequency of oscillation,

$$\omega_b{}^2 = \frac{eE_0 k}{m} \tag{6.112}$$

or

$$\boxed{\omega_b = \left(\frac{eE_0 k}{m}\right)^{1/2}} \tag{6.113}$$

where ω_b is known as the *bounce frequency*.

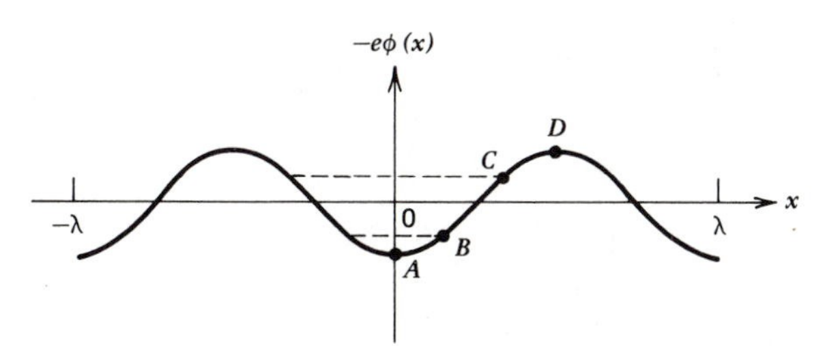

Fig. 6.12 Potential well for electrons in a Langmuir wave.

Linear Landau damping was derived on the basis of only small perturbations in particle orbits. However, after one-half of a bounce period, the particle at B has moved a substantial fraction of a wavelength, and linear theory is invalidated. Thus, we expect linear Landau damping to hold only for short times, such that

$$t \ll \omega_b^{-1} = \left(\frac{m}{eE_0k}\right)^{1/2} \tag{6.114}$$

Another way to look at this phenomenon is to draw the particle orbits in the v-x plane, as shown in Fig. 6.13. The labels A, B, C, and D in this figure correspond to the previous figure, all for particles with initial velocity $v = 0$. The solid lines indicate the orbits of these particles, and also indicate curves of constant energy. Consider the particle at E at the initial time $t = 0$. It has the same position as the particle at B and, thus, the same potential energy. However, because it also has a finite kinetic energy at $t = 0$, its total energy $-\ e\varphi_B + \frac{1}{2}mv^2$ is larger, and it finds itself on the same orbit as particle C. For an even larger initial velocity, we find a particle at position F. This particle is called *untrapped*, since its orbit carries it out of the original wavelength and into the neighboring wavelength to the right. By contrast, the particles at A, B, C, and E are called *trapped*, because their orbits remain forever in the original wavelength. The particle at D is neither trapped nor untrapped; its orbit is called the *separatrix* because it separates the trapped orbits from the untrapped orbits.

At the initial time $t = 0$, the particles take off along their orbits, moving to the right if $v > 0$ and to the left if $v < 0$, as shown in Fig. 6.14. During the time each particle represented by a dot moves from its dot to the tip of its arrow, we have the period of linear Landau damping, or growth. When the slope of the distribution function is negative, this early time behavior results in Landau damping, with the particles on the average gaining energy from the wave; both trapped and untrapped particles contribute. However, after a substantial fraction of a bounce period, the trapped particles are smeared out around their phase space orbits and the stage of linear Landau damping is over. The smearing process is facilitated by

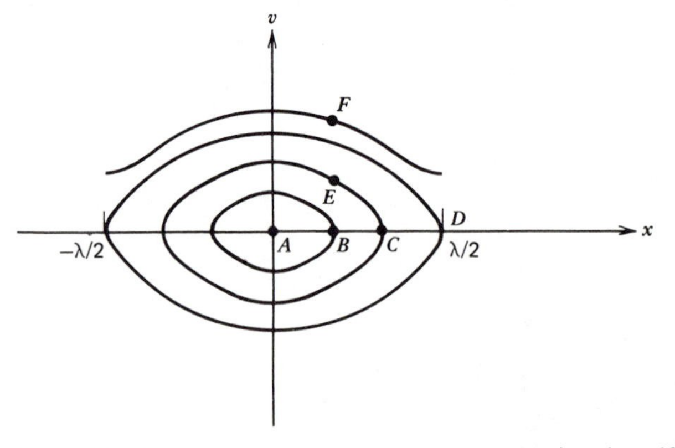

Fig. 6.13 Particle orbits in the v-x plane, neglecting the self-consistent change in the wave electric field due to the motion of the particles.

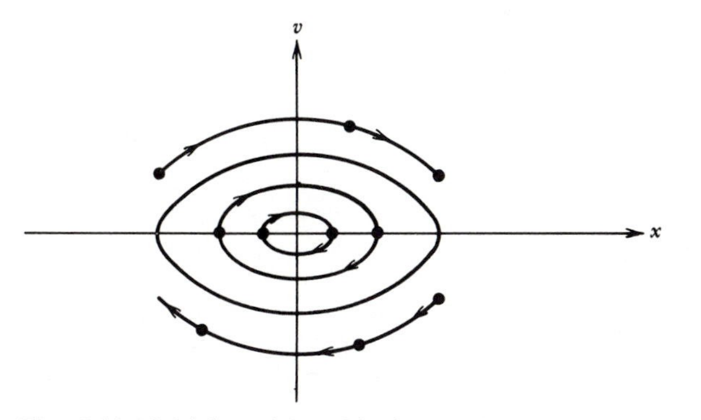

Fig. 6.14 Initial particle orbits in x-v phase space.

the fact that neighboring trapped particle orbits have slightly different bounce periods. At $x = 0$ in the wave frame, the initial distribution might be a Maxwellian, as in Fig. 6.15. At a much later time after many bounce periods, the particles with velocity $v \approx v_\varphi$ are smeared out, and the distribution looks as in Fig. 6.16. The flat region is $v_\varphi - v_t < v < v_\varphi + v_t$, where the *trapping speed* v_t is defined by

$$\tfrac{1}{2}mv_t^2 = 2|e\varphi|_{\max} \tag{6.115}$$

or

$$v_t = 2\left(\frac{e}{m}\right)^{1/2} |\varphi|_{\max}^{1/2} \tag{6.116}$$

or

$$\boxed{v_t = 2\left(\frac{eE_0}{mk}\right)^{1/2}} \tag{6.117}$$

Note the relation between the bounce frequency ω_b from (6.113), and the trapping speed v_t from (6.117),

$$\omega_b = \tfrac{1}{2}kv_t \tag{6.118}$$

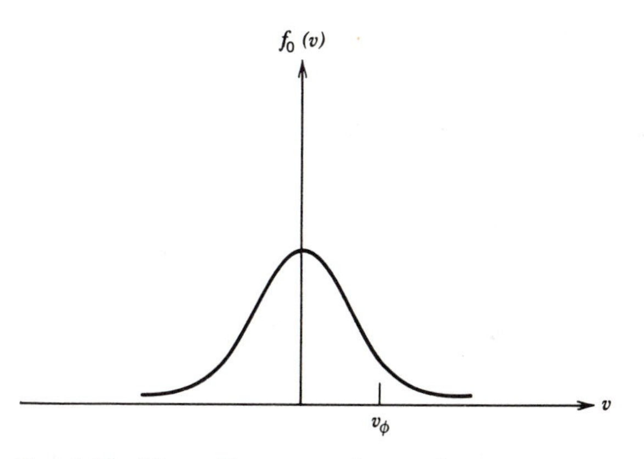

Fig. 6.15 Maxwellian at $t = 0$, $x = 0$.

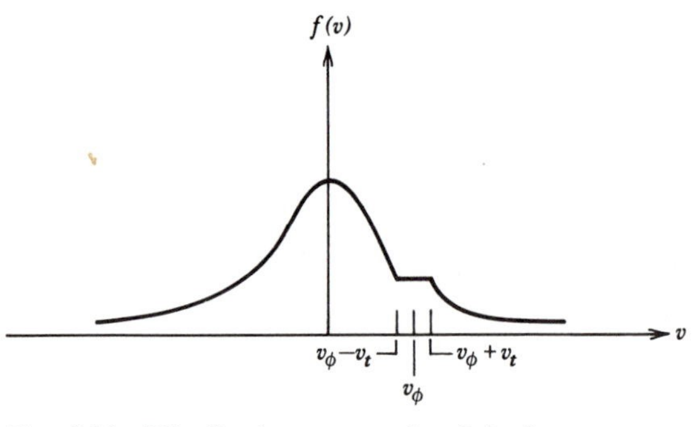

Fig. 6.16 Distribution at $x = 0$, $t >> 0$.

We can now attempt to construct an overall scenario for the initial value problem of a finite amplitude Langmuir wave. At early time $t << \omega_b^{-1}$, we have Landau damping at the appropriate damping rate. At $t \approx \pi/\omega_b$, the trapped particles have gone through a half a bounce and as they start to come through the second half of a bounce, they can put back into the wave some of the energy that they initially took out of it. This reversal of energy is by no means complete, however, because by this time the trapped particles are out of phase with each other. We can then construct a picture of the behavior of wave amplitude versus time as shown in Fig. 6.17. Here, γ_L is the Landau damping rate, and we have assumed $\gamma_L >> \omega_b$. For very large time, far off the right side of the figure, the curve will approach a straight line, all the phases will be completely mixed, and the wave will become a BGK mode [6] (see Section 6.13). Much of the present discussion has been heuristic. A completely self-consistent and nonlinear treatment of Langmuir waves is a very interesting current topic of research; see, for example, References [7] to [9].

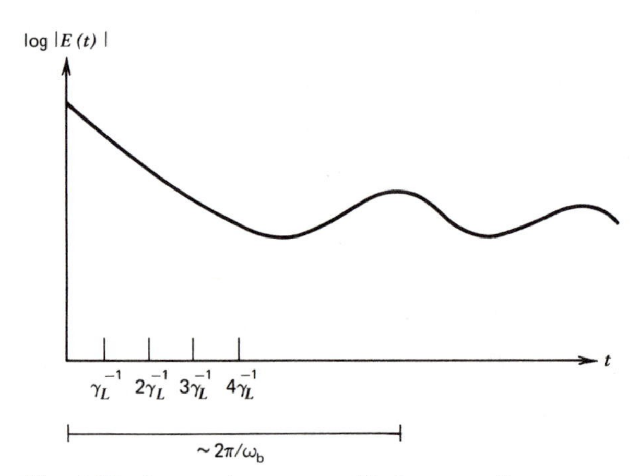

Fig. 6.17 Langmuir wave amplitude versus time.

6.9 STABILITY: NYQUIST METHOD, PENROSE CRITERION

In Section 6.2 we discussed methods of constructing Vlasov equilibria. Once we have found these equilibria, we must ask the question: Are they stable or unstable? For example, we know that when the ions and electrons are both Maxwellian with no relative drift, we expect the system to be stable. On the other hand, when the electrons form a cold beam moving through cold ions, elementary fluid theory (see Chapter 7) predicts instability. The question of whether or not a spatially uniform equilibrium is stable can be answered by the Nyquist method. [Note that the equilibrium must be uniform; this means that the Nyquist method unfortunately cannot determine the stability of BGK modes (see Section 6.13)].

We know that all information concerning the linear stability of an equilibrium to electrostatic perturbations is contained in the dielectric function $\epsilon(k,\omega)$, which is obtained by linearizing the basic physical equations about an equilibrium. Knowledge of $\epsilon(k,\omega)$ everywhere in the complex ω-plane determines all of the electrostatic stability properties of a system.

Consider a general function $\epsilon(k,\omega)$. Regarding k as a fixed real positive constant, we can then consider ϵ to be a function of ω only. Form a new function $(1/\epsilon)\,\partial\epsilon/\partial\omega$. Then it turns out that

$$\frac{1}{2\pi i}\int_{c_\omega}\frac{1}{\epsilon}\frac{\partial\epsilon}{\partial\omega}\,d\omega = N = \#\text{ of zeros of }\epsilon(k,\omega)\text{ inside the contour} \qquad (6.119)$$

where c_ω is any closed contour in the complex ω-plane, the integration is in the counterclockwise direction, and we assume $\partial\epsilon/\partial\omega$ has no poles in the enclosed region, and ϵ has only simple zeros. The derivation of (6.119) is as follows: Near any simple zero ω_0 of ϵ, we have

$$\epsilon(k,\omega) = 0 + \left.\frac{\partial\epsilon}{\partial\omega}\right|_{\omega_0}(\omega-\omega_0) + \ldots \qquad (6.120)$$

while

$$\frac{\partial\epsilon}{\partial\omega} = \left.\frac{\partial\epsilon}{\partial\omega}\right|_{\omega_0} + \left.\frac{\partial^2\epsilon}{\partial\omega^2}\right|_{\omega_0}(\omega-\omega_0) + \ldots \qquad (6.121)$$

Thus, near ω_0, we have

$$\frac{1}{\epsilon}\frac{\partial\epsilon}{\partial\omega} = \frac{1}{\omega-\omega_0} \qquad (6.122)$$

A trivial application of the residue theorem to the left side of (6.119) then yields the right side of (6.119). Thus, (6.119) tells us the number of roots of $\epsilon(\omega,k) = 0$ in a certain region of ω-space. In order to locate all the unstable roots, we simply need to evaluate (6.119) along a contour that includes all of the upper half ω-plane, since having $\epsilon(k,\omega) = 0$ when $\omega_i > 0$ means instability. In Fig. 6.18, (6.119) would yield $N = 2$, while in Fig. 6.19, (6.119) would yield $N = 0$ (left), $N = 1$ (middle), and $N = 3$ (right).

As one integrates around a contour in the ω-plane, it is possible to draw a corresponding contour in the complex ϵ-plane. In that plane we have from (6.119)

$$\frac{1}{2\pi i}\int_{c_\omega}\frac{1}{\epsilon}\frac{\partial\epsilon}{\partial\omega}\,d\omega = \frac{1}{2\pi i}\int_{c_\epsilon}\frac{1}{\epsilon}\,d\epsilon = N \qquad (6.123)$$

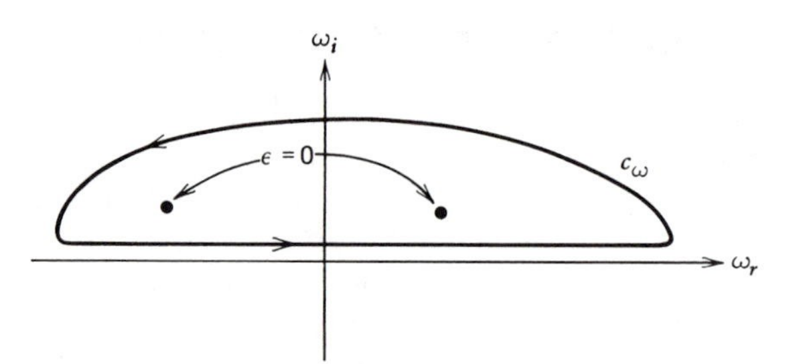

Fig. 6.18 Contour c_ω encircling the upper half of the ω-plane.

where c_ϵ is the contour in the ϵ-plane obtained by evaluating $\epsilon(k,\omega)$ at each point on the contour c_ω in the ω-plane. Thus, the middle term in (6.123) says that the contour c_ϵ must pick up N zeros of ϵ, so that in the ϵ-plane it must circle the origin N times. Three examples are shown in Fig. 6.19.

We thus have a powerful technique, the *Nyquist method*, for determining whether a physical system described by a dielectric function $\epsilon(k,\omega)$ is stable or not. We simply draw the curve c_ϵ in the ϵ-plane, found by mapping the curve c_ω, which encircles the upper half ω-plane. If c_ϵ does not encircle the origin $\epsilon = 0$, then the system is stable. If c_ϵ does encircle the origin one or more times, the system is unstable. The Nyquist method by itself does not tell us the growth rate of the instability.

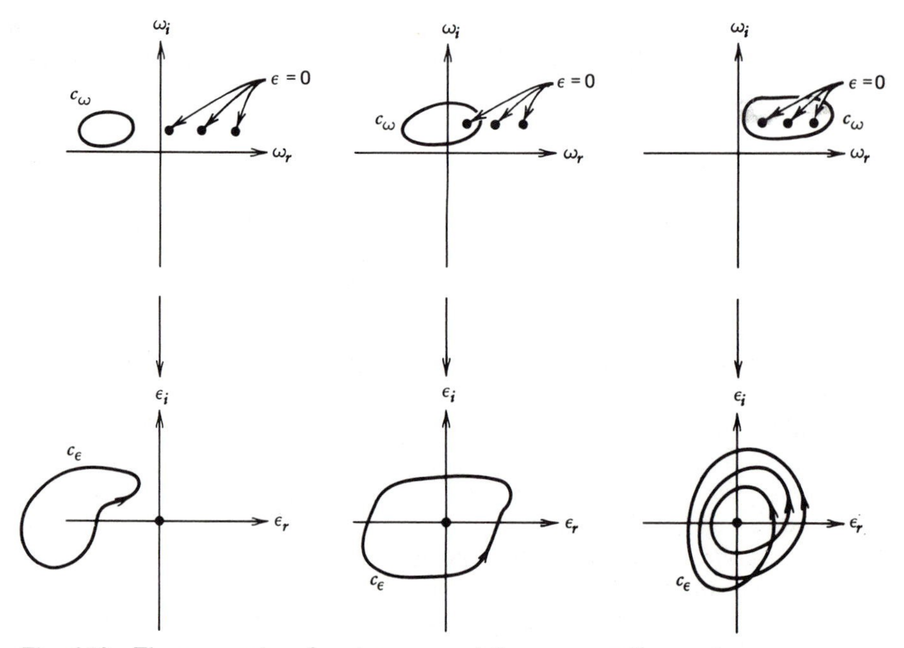

Fig. 6.19 Three examples of contours c_ω and the corresponding contours c_ϵ.

Let us test these ideas using the Vlasov–Poisson system, as represented by $\epsilon(k,\omega)$ in (6.34) for any ω, and in (6.45) for $\omega_i = 0$. We are trying to map the c_ω contour onto the ϵ-plane. First, consider the semicircle at ∞. Then (6.34) yields

$$\epsilon(k,\omega) = 1 \tag{6.124}$$

everywhere on the semicircle, since for $|\omega| \rightarrow \infty$ the second term in (6.34) vanishes. The remainder of the c_ω contour is the path from $\omega = -\infty$ to $\omega = +\infty$ along the real ω-axis. But this is precisely the situation when we can use the form (6.45).

By looking at the sign of the imaginary term in (6.45), we can see that as $\omega \rightarrow +\infty$, $\epsilon_i > 0$, while as $\omega \rightarrow -\infty$, $\epsilon_i < 0$. Also, for large $|\omega|$, $\epsilon_r = 1 - \omega_e^2/\omega^2$. We thus have the beginning of our path c_ϵ in the ϵ-plane as shown in Fig. 6.20. If the remaining part of the path c_ϵ looks as it does in Fig. 6.21, we shall have no instability because the origin is not encircled. However, if the remainder of c_ϵ looks as in Fig. 6.22 we have one unstable mode, because the origin is encircled once in the counterclockwise direction. Note that it is impossible to obtain the contour shown in Fig. 6.23 because this encircles the origin in the clockwise sense, predicting $N = -1$ by (6.123), which is nonsense.

Because of the handy formula (6.45), which describes the entire path c_ϵ except for the point $\epsilon = 1$, we know immediately all the places where c_ϵ crosses the real ϵ-axis. These are just the places where $\epsilon_i = 0$, or by (6.45), where $d_u g(u)|_{\omega/k} = 0$. Thus, a single humped $g(u)$ has only one position u_0 where $d_u g(u) = 0$. In this case, c_ϵ can only cross the real ϵ-axis in one place, and we immediately know that this is a stable system. This is because it is not possible, with what we already know about the contour c_ϵ, to encircle the origin in a counterclockwise sense and only cross the real ϵ-axis once.

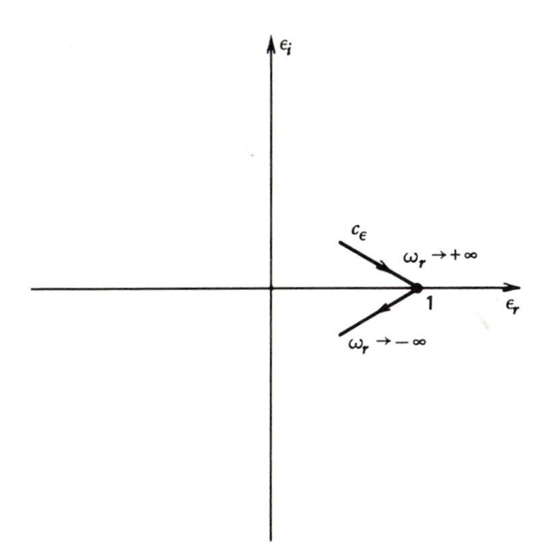

Fig. 6.20 Portion of the contour c_ϵ that comes from all portions of the c_ω contour in Fig. 6.18 with $|\omega| \rightarrow \infty$.

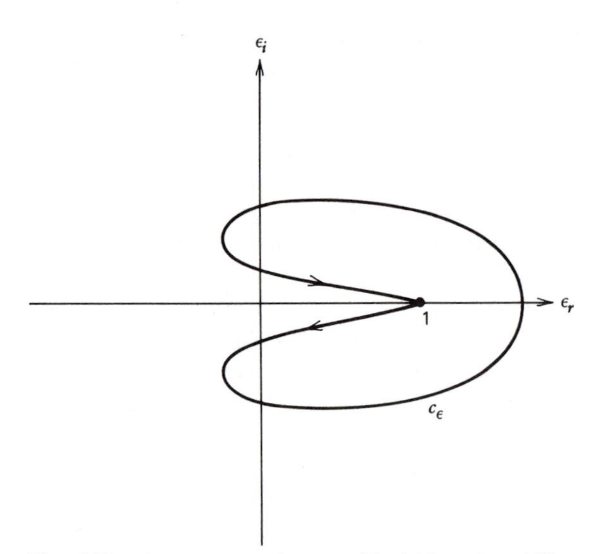

Fig. 6.21 A contour c_ϵ that would yield no instability.

We can verify this conclusion by evaluating ϵ_r at the position $\omega = ku_0$ where ϵ_i vanishes. From (6.45),

$$\epsilon_r = 1 - \frac{\omega_e^2}{k^2} P \int_{-\infty}^{\infty} du \frac{d_u g(u)}{u - \omega/k}$$

$$= 1 - \frac{\omega_e^2}{k^2} P \int_{-\infty}^{\infty} du \frac{d_u g(u)}{u - u_0} \tag{6.125}$$

If in the numerator of the integrand we subtract zero, in the form of $0 = d_u g(u_0)$, we can write

$$\epsilon_r = 1 - \frac{\omega_e^2}{k^2} P \int_{-\infty}^{\infty} du \frac{d_u[g(u) - g(u_0)]}{u - u_0} \tag{6.126}$$

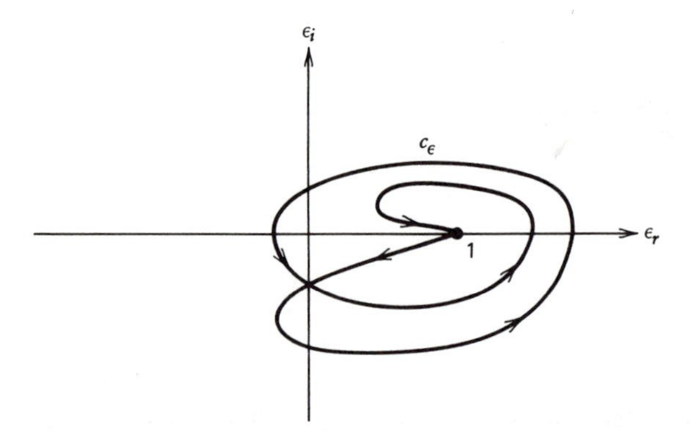

Fig. 6.22 A contour c_ϵ that indicates one unstable mode.

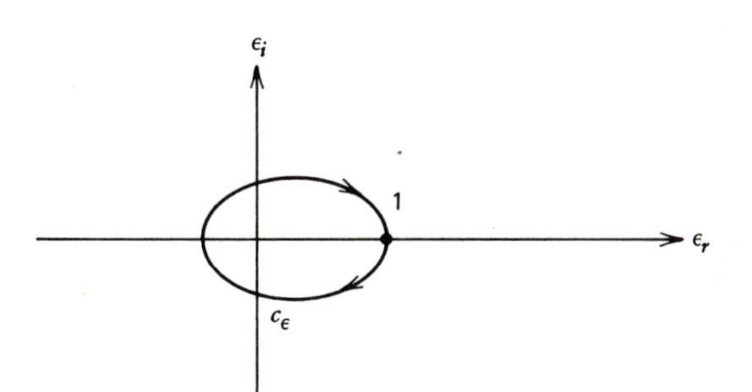

Fig. 6.23 A contour c_ϵ that can never occur because it indicates $N = -1$.

We can integrate (6.126) by parts, because all quantities are well defined at the singularity, to obtain

$$\epsilon_r(\omega = ku_0) = 1 + \frac{\omega_e^2}{k^2} \int_{-\infty}^{\infty} du \, \frac{[g(u_0) - g(u)]}{(u - u_0)^2} \tag{6.127}$$

where the principal value symbol is no longer needed. The integrand in (6.127) is positive definite, since $g(u_0)$ is the maximum value of g. Thus, (6.127) yields

$$\epsilon_r(\omega = ku_0) > 1 \tag{6.128}$$

so that the picture of c_ϵ is as shown in Fig. 6.24, confirming our prediction that the origin is not encircled and that there is no instability for a single-humped distribution. This result is known as Gardner's theorem [3, 10].

In interpreting this result we must remember that $g(u)$ includes an ion contribution as well as an electron contribution. Thus, one situation in which the above result holds is when a single humped electron distribution is moving through a background of infinitely massive ions; when $m_i \to \infty$, (6.22) shows that $g(u)$ depends only on the electron distribution.

Next, consider the case where the distribution has a double hump, with a relative minimum between them (Fig. 6.25). Then (6.45) predicts three values of ω;

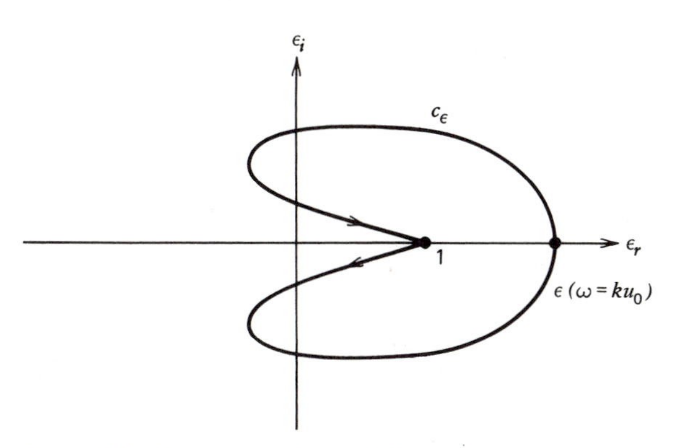

Fig. 6.24 Contour c_ϵ obtained from Vlasov–Poisson theory for a single-humped distribution $g(u)$.

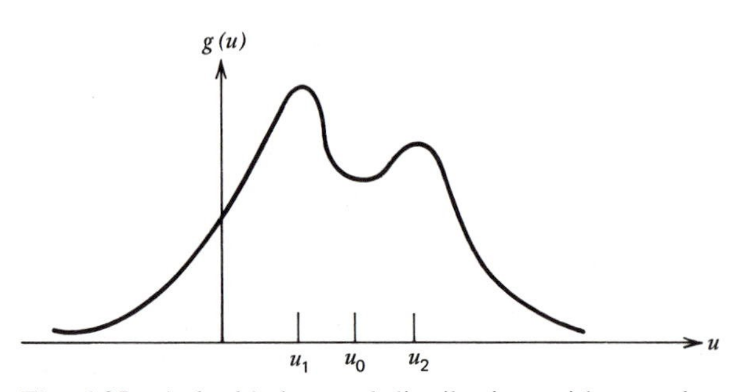

Fig. 6.25 A double-humped distribution, with zero slope at u_1, u_0, and u_2.

$\omega = ku_1$, $\omega = ku_0$, and $\omega = ku_2$; where c_ϵ crosses the real axis. If the absolute maximum of $g(u)$ occurs at $u = u_1$, then it is straightforward to show that $\epsilon_r(\omega = ku_1) > 1$.

EXERCISE Verify this, using the same argument as in (6.127) and (6.128).

It furthermore must be the case that as we move along the ω contour from $\omega = -\infty$ to $\omega = +\infty$, we encounter the crossings of the real ϵ-axis in the order $\epsilon(\omega = ku_1)$, $\epsilon(\omega = ku_0)$, and then $\epsilon(\omega = ku_2)$. Thus, the first part of c_ϵ looks as shown in Fig. 6.26. We are now allowed two more crossings of the real ϵ-axis. One possibility is as shown in Fig. 6.27, which does not give instability. Another possibility is shown in Fig. 6.28, which indicates one unstable root. We see that a necessary condition for instability is

$$\epsilon(\omega = ku_0) < 0 \tag{6.129}$$

which from (6.45) is

$$\epsilon(\omega = ku_0) = 1 - \frac{\omega_e^2}{k^2} P \int_{-\infty}^{\infty} du \frac{d_u g(u)}{u - u_0} \tag{6.130}$$

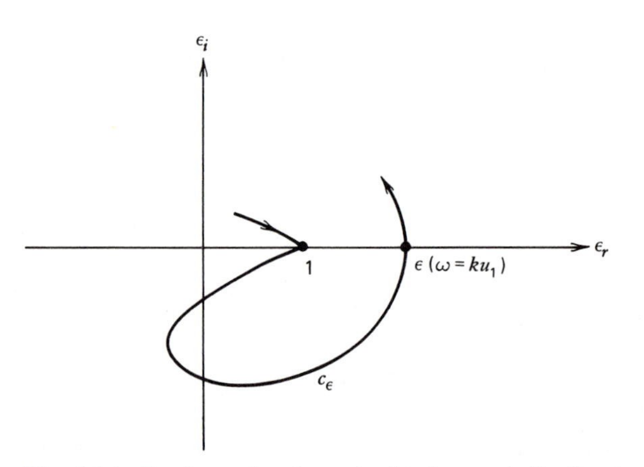

Fig. 6.26 Portions of c_ϵ for a double-humped distribution.

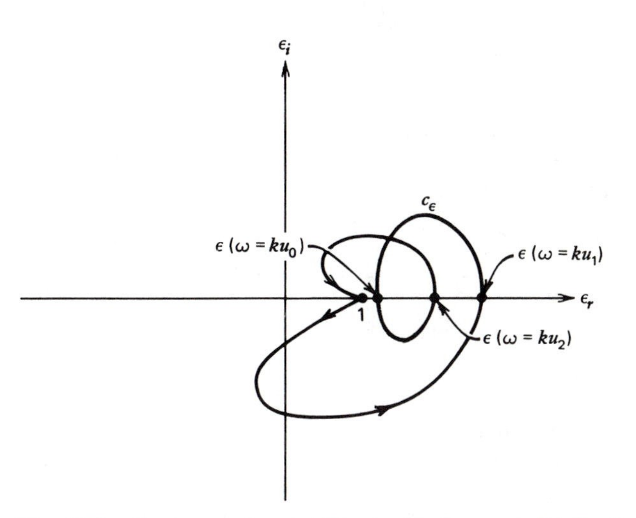

Fig. 6.27 Contour c_ϵ for a stable double-humped distribution.

By once again subtracting $0 = d_u g(u_0)$ in the numerator, we can integrate by parts to obtain

$$\epsilon(\omega = ku_0) = 1 + \frac{\omega_e^2}{k^2} \int_{-\infty}^{\infty} du \frac{[g(u_0) - g(u)]}{(u - u_0)^2} \tag{6.131}$$

where the justification is the same as in (6.127). Now (6.129) says we need $\epsilon(\omega = ku_0) < 0$ for instability. But this will be assured if

$$\boxed{\int_{-\infty}^{\infty} du \frac{[g(u_0) - g(u)]}{(u - u_0)^2} < 0} \tag{6.132}$$

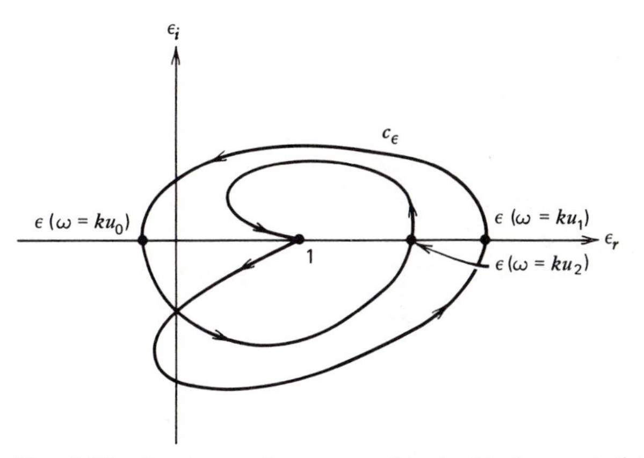

Fig. 6.28 Contour c_ϵ for an unstable double-humped distribution.

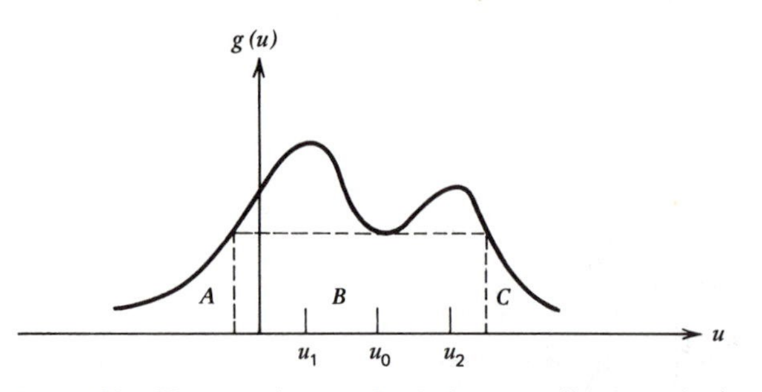

Fig. 6.29 Three regions A, B, C that contribute to the integral in (6.132).

for when (6.132) is true, (6.131) ranges from $\epsilon(\omega = ku_0) = 1$ for $k \to \infty$ to $\epsilon(\omega = ku_0) = -\infty$ for $k \to 0$. Thus, for some value of k, we must have $\epsilon(\omega = ku_0) < 0$ while $\epsilon(\omega = ku_2) > 0$, which is the necessary condition for instability. Equation (6.132) is called the Penrose criterion [11, 12]; it is a necessary and sufficient condition for the linear instability of a Vlasov–Poisson equilibrium.

Consider the integration in (6.132) as applied to the three regions shown in Fig. 6.29. In all of region B the integrand is negative, while in regions A and C the integrand is positive. Thus, the negative contribution in B must exceed the positive contributions from A and C in order that the Penrose criterion be satisfied. Notice that the negative contribution in B is enhanced by a deep hole.

EXERCISE Show that if $g(u_0) = 0$, the Penrose criterion is always satisfied.

The Nyquist method also tells us the range of unstable wave numbers. By requiring $\epsilon(\omega = ku_0) < 0$ and $\epsilon(\omega = ku_2) > 0$, we find that Eq. (6.131) for

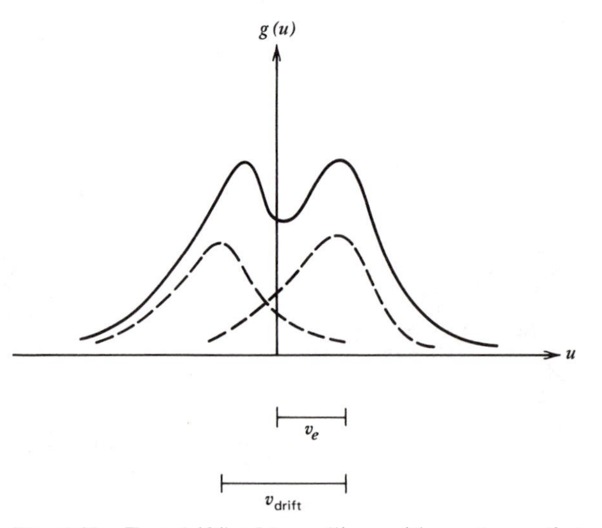

Fig. 6.30 Two drifting Maxwellians with $v_e < v_{\text{drift}}$ that are unstable.

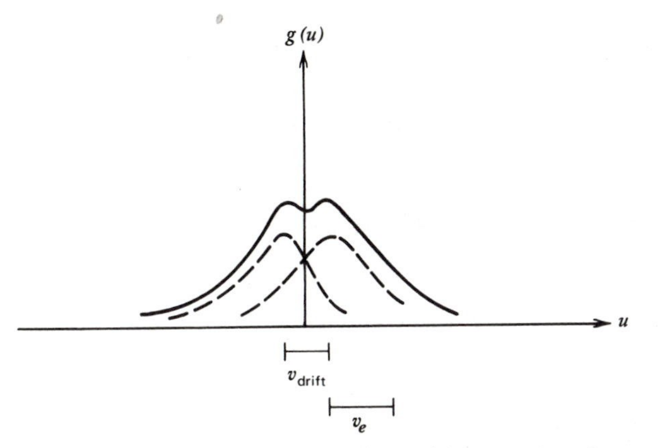

Fig. 6.31 Two drifting Maxwellians with $v_{\text{drift}} < v_e$ that are stable.

$\epsilon(\omega = ku_0)$ and the equivalent expression for $\epsilon(\omega = ku_2)$ yield

$$\omega_e^2 \int_{-\infty}^{\infty} du \frac{[g(u) - g(u_2)]}{(u - u_2)^2} < k^2 < \omega_e^2 \int_{-\infty}^{\infty} du \frac{[g(u) - g(u_0)]}{(u - u_0)^2} \tag{6.133}$$

for the range of unstable wave number.

We have seen that the Penrose criterion needs a deep enough hole to predict instability. Thus, two weakly drifting Maxwellian groups of electrons will be unstable only when, crudely, the drift velocity equals the thermal speed, as indicated in Figs. 6.30 and 6.31.

Other cases of stability or lack thereof will be explored in the problems.

6.10 GENERAL THEORY OF LINEAR VLASOV WAVES

In preceding sections we have discussed linear electrostatic waves in the context of the Vlasov and Poisson equations; there was no background magnetic field. We found Langmuir waves and ion-acoustic waves, and we found a new physical effect, Landau damping. Let us now include a background magnetic field, and set up an approach that will yield all linear waves, including electromagnetic waves. Several new effects will appear. One of these is *cyclotron damping*, which is the magnetized analogue of Landau damping.

The basic equation is the Vlasov equation,

$$\partial_t f_s + \mathbf{v} \cdot \nabla_{\mathbf{x}} f_s + \frac{q_s}{m_s}\left(\mathbf{E} + \frac{\mathbf{v}}{c} \times \mathbf{B}\right) \cdot \nabla_{\mathbf{v}} f = 0 \tag{6.134}$$

where $\mathbf{E}(\mathbf{x},t)$ and $\mathbf{B}(\mathbf{x},t)$ can have internal and external contributions. Linearizing (6.134) about a time and space independent, zero order distribution function, $f_s(\mathbf{x},\mathbf{v},t) = f_{s0}(\mathbf{v}) + f_{s1}(\mathbf{x},\mathbf{v},t)$, we have

$$\frac{q_s}{m_s}\left(\mathbf{E}_0 + \frac{\mathbf{v}}{c} \times \mathbf{B}_0\right) \cdot \nabla_{\mathbf{v}} f_{s0}(\mathbf{v}) = 0 \tag{6.135}$$

for each species, and

$$\partial_t f_{s1} + \mathbf{v} \cdot \nabla_{\mathbf{x}} f_{s1} + \frac{q_s}{m_s} \left(\mathbf{E}_0 + \frac{\mathbf{v}}{c} \times \mathbf{B}_0 \right) \cdot \nabla_{\mathbf{v}} f_{s1}$$

$$= - \frac{q_s}{m_s} \left(\mathbf{E}_1 + \frac{\mathbf{v}}{c} \times \mathbf{B}_1 \right) \cdot \nabla_{\mathbf{v}} f_{s0} \tag{6.136}$$

The total charge density is

$$\rho(\mathbf{x},t) = \sum_s q_s \int d^3v f_{s1}(\mathbf{x},\mathbf{v},t) \tag{6.137}$$

while the total current is

$$\mathbf{J}(\mathbf{x},t) = \sum_s q_s \int d^3v \mathbf{v} f_{s1}(\mathbf{x},\mathbf{v},t) \tag{6.138}$$

where we have taken the zero order charge density and current to vanish. Combining (6.135) to (6.138) with Maxwell's equations

$$\nabla \cdot \mathbf{E}_1 = 4\pi\rho \tag{6.139}$$

$$\nabla \times \mathbf{E}_1 = - \frac{1}{c} \frac{\partial \mathbf{B}_1}{\partial t} \tag{6.140}$$

$$\nabla \cdot \mathbf{B}_1 = 0 \tag{6.141}$$

and

$$\nabla \times \mathbf{B}_1 = \frac{4\pi}{c} \mathbf{J} + \frac{1}{c} \frac{\partial \mathbf{E}_1}{\partial t} \tag{6.142}$$

we have a complete set of linear equations with which to find the dispersion relation for an arbitrary linear wave.

Let us now solve the first order Vlasov equation (6.136). Recall that in Section 6.2 we introduced the concept of unperturbed orbits of hypothetical particles, so that the zero order Vlasov equation (6.135) could be written

$$\frac{D}{Dt} f_{s0}(\mathbf{v}) = 0 \tag{6.143}$$

We then proceeded to find equilibrium distribution functions that were functions of the constants of motion of the hypothetical particles. Consider a hypothetical particle moving in a given force field, consisting of zero order electric and magnetic fields. For the present, we allow the zero order electric and magnetic fields to be functions of both space and time; later we will make them constant. From Newton's laws of motion we can follow the orbit of a particle to find $\mathbf{X}(t)$, its orbit in real space, and $\mathbf{V}(t)$, its orbit in velocity space. The equations for these variables are

$$\dot{\mathbf{X}}(t) = \mathbf{V}(t) \tag{6.144}$$

and

$$\dot{\mathbf{V}}(t) = \frac{q_s}{m_s} \left\{ \mathbf{E}_0[\mathbf{X}(t),t] + \frac{\mathbf{V}(t)}{c} \times \mathbf{B}_0[\mathbf{X}(t),t] \right\} \tag{6.145}$$

In particular, we associate one of these orbits with every point $(\mathbf{x},\mathbf{v},t)$ in seven-dimensional phase space, by choosing the appropriate constant of integration. That is, we choose

$$\mathbf{X}(t') = \mathbf{x} - \int_{t'}^{t} \dot{\mathbf{X}}(t'')\, dt'' \tag{6.146}$$

or

$$\mathbf{X}(t') = \mathbf{x} - \int_{t'}^{t} \mathbf{V}(t'')\, dt'' \tag{6.147}$$

which satisfies (6.144) and also has the property that

$$\mathbf{X}(t) = \mathbf{x} \tag{6.148}$$

Similarly, for velocity, we choose the constant of integration such that

$$\mathbf{V}(t') = \mathbf{v} - \int_{t'}^{t} \dot{\mathbf{V}}(t'')\, dt'' \tag{6.149}$$

Thus, the orbit $[\mathbf{X}(t'),\mathbf{V}(t')]$ is the orbit of that particle which reaches the position $(\mathbf{x},\mathbf{v})$ at time t.

Consider any function $h(\mathbf{x},\mathbf{v},t)\Big|_{\substack{\mathbf{x}=\mathbf{X}(t)\\ \mathbf{v}=\mathbf{V}(t)}}$. Then

$$\begin{aligned}\frac{d}{dt'}\left[h(\mathbf{x}',\mathbf{v}',t')\Big|_{\substack{\mathbf{x}'=\mathbf{X}(t')\\ \mathbf{v}'=\mathbf{V}(t')}}\right] &= \frac{\partial h}{\partial t'}(\mathbf{x}',\mathbf{v}',t')\Big|_{\substack{\mathbf{x}'=\mathbf{X}(t')\\ \mathbf{v}'=\mathbf{V}(t')}} \\ &+ \dot{\mathbf{X}}(t')\cdot\nabla_{\mathbf{x}'} h(\mathbf{x}',\mathbf{v}',t')\Big|_{\substack{\mathbf{x}'=\mathbf{X}(t')\\ \mathbf{v}'=\mathbf{V}(t')}} \\ &+ \dot{\mathbf{V}}(t')\cdot\nabla_{\mathbf{v}'} h(\mathbf{x}',\mathbf{v}',t')\Big|_{\substack{\mathbf{x}'=\mathbf{X}(t')\\ \mathbf{v}'=\mathbf{V}(t')}}\end{aligned} \tag{6.150}$$

Along the unperturbed orbit, we have $\dot{\mathbf{X}}(t') = \mathbf{V}(t') = \mathbf{v}'$, and

$$\dot{\mathbf{V}}(t') = \frac{q_s}{m_s}\left\{\mathbf{E}_0[\mathbf{X}(t'),t'] + \frac{\mathbf{V}(t')}{c}\times\mathbf{B}_0[\mathbf{X}(t'),t']\right\} \tag{6.151}$$

Thus, the right side of (6.150) is just the left side of (6.136) when $h = f_{s1}$, and we can write (6.136) as

$$\begin{aligned}\frac{d}{dt'} f_{s1}(\mathbf{x}',\mathbf{v}',t')\Big|_{\substack{\mathbf{x}'=\mathbf{X}(t')\\ \mathbf{v}'=\mathbf{V}(t')}} &= -\frac{q_s}{m_s}\Big\{\mathbf{E}_1[\mathbf{X}(t'),t'] \\ &+ \frac{\mathbf{V}(t')}{c}\times\mathbf{B}_1[\mathbf{X}(t'),t']\Big\}\cdot\nabla_{\mathbf{v}'} f_{s0}(\mathbf{x}',\mathbf{v}',t')\Big|_{\substack{\mathbf{x}'=\mathbf{X}(t')\\ \mathbf{v}'=\mathbf{V}(t')}}\end{aligned} \tag{6.152}$$

Both sides of (6.152) can be integrated from $t' = -\infty$ to $t' = t$ along the unperturbed orbit that ends up at $\mathbf{X}(t) = \mathbf{x}$ and $\mathbf{V}(t) = \mathbf{v}$. The result is

$$\begin{aligned}f_{s1}(\mathbf{x},\mathbf{v},t) &= f_{s1}[\mathbf{X}(t'),\mathbf{V}(t'),t' = -\infty] \\ &- \frac{q_s}{m_s}\int_{-\infty}^{t} dt'\left\{\mathbf{E}_1[\mathbf{X}(t'),t'] + \frac{\mathbf{V}(t')}{c}\times\mathbf{B}_1[\mathbf{X}(t'),t']\right\} \\ &\cdot\nabla_{\mathbf{v}'} f_{s0}(\mathbf{x}',\mathbf{v}',t')\Big|_{\substack{\mathbf{x}'=\mathbf{X}(t')\\ \mathbf{v}'=\mathbf{V}(t')}}\end{aligned} \tag{6.153}$$

Equation (6.153) is a formal solution for f_{s1}, where in the integrand we must evaluate $\mathbf{E}_1$, $\mathbf{B}_1$, and f_{s0} at the correct point of the unperturbed orbit $[\mathbf{X}(t'),\mathbf{V}(t')]$ of the hypothetical particle at time t'. From now on, we consider only uniform, stationary zero order fields $\mathbf{E}_0 = 0$, $\mathbf{B}_0 =$ constant, and $f_{s0} = f_{s0}(\mathbf{v})$ a function only of velocity, although (6.153) can be used in more general cases.

Let us look for plane wave solutions to (6.153),

$$\mathbf{E}_1(\mathbf{x},t) = \tilde{\mathbf{E}} \exp(-i\omega t + i\mathbf{k}\cdot\mathbf{x}) \tag{6.154}$$

$$\mathbf{B}_1(\mathbf{x},t) = \tilde{\mathbf{B}} \exp(-i\omega t + i\mathbf{k}\cdot\mathbf{x}) \tag{6.155}$$

and

$$f_{s1}(\mathbf{x},\mathbf{v},t) = \tilde{f}_s(\mathbf{v}) \exp(-i\omega t + i\mathbf{k}\cdot\mathbf{x}) \tag{6.156}$$

where $\tilde{\mathbf{E}}$ and $\tilde{\mathbf{B}}$ are constant vectors. For the moment, we take $\mathrm{Im}(\omega) > 0$, so the wave is exponentially growing with respect to time. Then a finite amplitude at time t implies $f_{s1}(t' = -\infty) = 0$, and we can ignore the contribution of $f_{s1}(t' = -\infty)$ in (6.153). The philosophy here is to find a dispersion relation valid for $\mathrm{Im}(\omega) > 0$, and then to analytically continue the dispersion relation for arbitrary ω. This is the same technique used earlier for Langmuir waves, which led in that case to the Landau contour for evaluating the electrostatic dispersion relation.

Equation (6.153) now reads

$$\tilde{f}_s(\mathbf{v}) \exp(-i\omega t + i\mathbf{k}\cdot\mathbf{x}) = -\frac{q_s}{m_s}\int_{-\infty}^{t} dt' \left[\tilde{\mathbf{E}} + \frac{\mathbf{V}(t')}{c} \times \tilde{\mathbf{B}}\right]$$
$$\cdot \nabla_{\mathbf{v}'} f_{s0}(\mathbf{v}')|_{\mathbf{v}'=\mathbf{V}(t')} \exp[-i\omega t' + i\mathbf{k}\cdot\mathbf{X}(t')] \tag{6.157}$$

Moving $\exp(-i\omega t + i\mathbf{k}\cdot\mathbf{x})$ to the right we find

$$\tilde{f}_s(\mathbf{v}) = -\frac{q_s}{m_s}\int_{-\infty}^{t} dt' \left[\tilde{\mathbf{E}} + \frac{\mathbf{V}(t')}{c} \times \tilde{\mathbf{B}}\right]$$
$$\cdot \nabla_{\mathbf{v}'} f_{s0}(\mathbf{v}')|_{\mathbf{v}'=\mathbf{V}(t')} \exp\{-i\omega(t'-t) + i\mathbf{k}\cdot[\mathbf{X}(t') - \mathbf{x}]\} \tag{6.158}$$

Making the change of variable $\tau = t' - t$ we have

$$\tilde{f}_s(\mathbf{v}) = -\frac{q_s}{m_s}\int_{-\infty}^{0} d\tau \left[\tilde{\mathbf{E}} + \frac{\mathbf{V}(\tau)}{c} \times \tilde{\mathbf{B}}\right]$$
$$\cdot \nabla_{\mathbf{v}'} f_{s0}(\mathbf{v}')|_{\mathbf{V}(\tau)} \exp\{-i\omega\tau + i\mathbf{k}\cdot[\mathbf{X}(\tau) - \mathbf{x}]\} \tag{6.159}$$

where we realize that in this notation, $[\mathbf{X}(\tau),\mathbf{V}(\tau)]$ is the orbit that yields $[\mathbf{X}(\tau = 0), \mathbf{V}(\tau = 0)] = (\mathbf{x},\mathbf{v})$.

6.11 LINEAR VLASOV WAVES IN UNMAGNETIZED PLASMA

Let us evaluate (6.159) for the case of an equilibrium plasma with no zero order fields. Then the unperturbed orbits are from (6.147) and (6.149):

$$\mathbf{V}(t') = \mathbf{v} \tag{6.160}$$

and

$$\begin{aligned}\mathbf{X}(t') &= \mathbf{x} - \mathbf{v}(t - t') \\ &= \mathbf{x} + \mathbf{v}\tau\end{aligned} \tag{6.161}$$

Since $\mathbf{V}(\tau)$ does not depend on τ, (6.159) becomes

$$\tilde{f}_s(\mathbf{v}) = -\frac{q_s}{m_s}\left(\tilde{\mathbf{E}} + \frac{\mathbf{v}}{c} \times \tilde{\mathbf{B}}\right) \cdot \nabla_{\mathbf{v}} f_{s0}(\mathbf{v}) \times \int_{-\infty}^{0} d\tau \exp(-i\omega\tau + i\mathbf{k}\cdot\mathbf{v}\tau) \tag{6.162}$$

Since $\mathrm{Im}(\omega) > 0$, the integral is well behaved and we find

$$\int_{-\infty}^{0} d\tau \exp(-i\omega\tau + i\mathbf{k}\cdot\mathbf{v}\tau) = \frac{1}{-i(\omega - \mathbf{k}\cdot\mathbf{v})} \tag{6.163}$$

so that

$$\boxed{\tilde{f}_s(\mathbf{v}) = \frac{(q_s/m_s)}{i(\omega - \mathbf{k}\cdot\mathbf{v})}\left(\tilde{\mathbf{E}} + \frac{\mathbf{v}}{c} \times \tilde{\mathbf{B}}\right) \cdot \nabla_{\mathbf{v}} f_{s0}(\mathbf{v})} \tag{6.164}$$

Taking $\tilde{\mathbf{B}} = 0$ and looking for *electrostatic* waves, we could combine (6.164) for each species with Poisson's equation to obtain our old Vlasov–Poisson dispersion relation, leading to Langmuir waves, ion-acoustic waves, and Landau damping.

Taking $\mathbf{B}_1(\mathbf{x},t) = \tilde{\mathbf{B}} \exp(-i\omega t + i\mathbf{k}\cdot\mathbf{x}) \neq 0$, we find from Ampere's law (6.142) and Faraday's law (6.140) that

$$i\mathbf{k} \times \mathbf{E}_1 = \frac{i\omega}{c}\mathbf{B}_1 \tag{6.165}$$

and

$$i\mathbf{k} \times \mathbf{B}_1 = \frac{4\pi}{c}\mathbf{J} - \frac{i\omega}{c}\mathbf{E}_1 \tag{6.166}$$

or

$$c^2[\mathbf{k} \times (\mathbf{k} \times \mathbf{E}_1)] = -i4\pi\omega\mathbf{J} - \omega^2\mathbf{E}_1 \tag{6.167}$$

For isotropic zero order distribution functions $f_{s0}(\mathbf{v}) = f_{s0}(v)$, the term $(\mathbf{v} \times \tilde{\mathbf{B}}) \cdot \nabla_{\mathbf{v}} f_{s0}$ in (6.164) vanishes.

EXERCISE Verify this.

Then

$$\begin{aligned}\mathbf{J}(\mathbf{x},t) &= \sum_s q_s \int d\mathbf{v}\, \mathbf{v} f_{s1}(\mathbf{x},\mathbf{v},t) \\ &= \sum_s q_s \exp(-i\omega t + i\mathbf{k}\cdot\mathbf{x}) \int d\mathbf{v} \frac{q_s}{m_s} \frac{\tilde{\mathbf{E}} \cdot \nabla_{\mathbf{v}} f_{s0}(\mathbf{v})}{i(\omega - \mathbf{k}\cdot\mathbf{v})}\mathbf{v}\end{aligned} \tag{6.168}$$

Let us look for transverse waves, such that $\tilde{\mathbf{E}} \perp \mathbf{k}$. Then $\mathbf{k} \times (\mathbf{k} \times \tilde{\mathbf{E}}) = -\mathbf{k}^2\tilde{\mathbf{E}}$, and (6.167) becomes

$$(\omega^2 - k^2c^2)\tilde{\mathbf{E}} = -4\pi i\omega \sum_s \frac{q_s^2}{m_s} \int d\mathbf{v} \frac{\tilde{\mathbf{E}} \cdot \nabla_{\mathbf{v}} f_{s0}(\mathbf{v})}{i(\omega - \mathbf{k}\cdot\mathbf{v})}\mathbf{v} \tag{6.169}$$

where the harmonic dependence has been factored out. Suppose the vectors are arranged as shown in Fig. 6.32. Then the numerator in (6.169) has a term $\tilde{\mathbf{E}} \cdot \nabla_{\mathbf{v}} =$

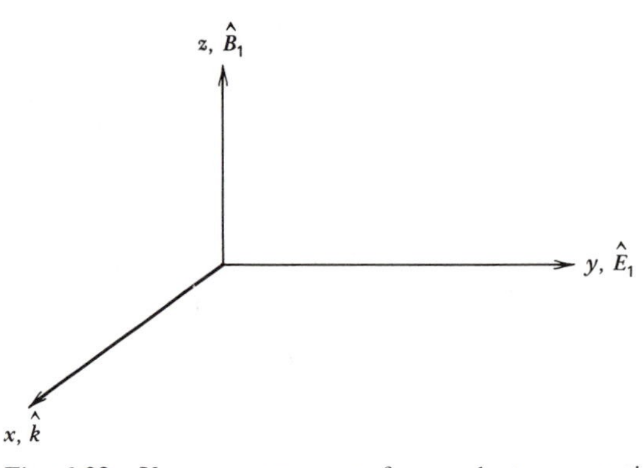

Fig. 6.32 Vector components for an electromagnetic wave in an unmagnetized plasma.

$\tilde{E}\,\partial_{v_y}$, while the denominator is of the form $\omega - kv_x$. Thus, we can integrate by parts in the numerator, picking out only the term $\tilde{E}\,\partial_{v_y}\mathbf{v} = \tilde{E}\hat{y} = \tilde{\mathbf{E}}$. We find

$$(\omega^2 - k^2c^2)\tilde{\mathbf{E}} = 4\pi\omega\tilde{\mathbf{E}}\,\frac{e^2}{m_e}\,n_0 \int dv_x\,\frac{g(v_x)}{\omega - kv_x} \tag{6.170}$$

where $g(v_x)$ is as usual defined by (6.22) and where we have ignored the ion dependence by allowing $m_i \to \infty$. Dividing out the constant vector $\tilde{\mathbf{E}}$, we finally obtain the dispersion relation for electromagnetic waves in unmagnetized plasma,

$$\boxed{\omega^2 = k^2c^2 + \omega_e^2\omega \int_{-\infty}^{\infty} dv_x\,\frac{g(v_x)}{\omega - kv_x}} \tag{6.171}$$

Whenever there might be a problem with the pole in (6.171), the Landau contour must be used, for the same reason as in the discussion of Langmuir waves. The integral in (6.171) can easily be performed with the assumption $\omega/k >> v_x$ for all v_x of interest. Then $\omega - kv_x \approx \omega$, the integral $\int_{-\infty}^{\infty} dv_x\, g(v_x) = 1$, and we find

$$\boxed{\omega^2 = \omega_e^2 + k^2c^2} \tag{6.172}$$

for linear electromagnetic waves in unmagnetized plasma.

EXERCISE Show from (6.172) that even in a more accurate treatment of the integral in (6.171), there is no need to consider a contribution involving $g(v_x)|_{v_x=\omega_r/k}$.

In this section we have seen how electrostatic and electromagnetic waves in a uniform plasma with no external fields are treated via the Vlasov equation. In the next section we shall set up the equivalent procedure for a uniformly magnetized plasma.

6.12 LINEAR VLASOV WAVES IN MAGNETIZED PLASMA

In the previous section we have seen that the linearized Vlasov equation can be solved by integrating along the orbits of hypothetical particles moving in the zero order fields; these orbits are called unperturbed because they do not feel the effect of the wave motion for which one is looking. In an unmagnetized plasma, the unperturbed orbits are simple, and it is straightforward to evaluate the perturbed distribution function. Consider the evaluation of (6.159) in the presence of a uniform background magnetic field. Now the unperturbed orbits are spirals around the magnetic field lines, satisfying Newton's law

$$m_s \dot{\mathbf{V}}(\tau) = \frac{q_s}{c} \mathbf{V}(\tau) \times \mathbf{B}_0 \tag{6.173}$$

Choosing the magnetic field in the $\hat{z}$-direction, the gyro-orbits that satisfy $\mathbf{X}(\tau = 0) = \mathbf{x}$, $\mathbf{V}(\tau = 0) = \mathbf{v}$, are

$$V_z(\tau) = v_z \tag{6.174}$$

$$Z(\tau) = z + v_z \tau \tag{6.175}$$

$$V_x(\tau) = v_\perp \cos(\varphi - \Omega_s \tau) \tag{6.176}$$

$$X(\tau) = x - \frac{v_\perp}{\Omega_s} \sin(\varphi - \Omega_s \tau) + \frac{v_\perp}{\Omega_s} \sin\varphi \tag{6.177}$$

$$V_y(\tau) = v_\perp \sin(\varphi - \Omega_s \tau) \tag{6.178}$$

and

$$Y(\tau) = y + \frac{v_\perp}{\Omega_s} \cos(\varphi - \Omega_s \tau) - \frac{v_\perp}{\Omega_s} \cos\varphi \tag{6.179}$$

where the gyrofrequency is $\Omega_s \equiv q_s B_0 / m_s c$, and φ is a constant $0 \leq \varphi \leq 2\pi$.

EXERCISE Verify that (6.174) to (6.179) satisfy (6.173) with the appropriate boundary conditions.

Inserting the orbit (6.174) to (6.179) into (6.159), we can carry out the integration over τ in (6.159). Then using Maxwell's equations to eliminate **B** in terms of **E**, we could obtain a general dispersion relation for waves in a uniformly magnetized plasma. This dispersion relation would contain all of the waves to be encountered in the next chapter on fluid theory, for example, Alfvén waves, upper-hybrid waves, and extraordinary waves. In addition, entirely new wave modes appear in the Vlasov formulation, which are impossible to obtain from a fluid formulation. Known as *Bernstein modes*, these waves depend on the detailed interaction of the wave motion with the gyro-orbits of the particles.

Because the details of the evaluation of (6.159) are quite tedious (see [13], p. 405 and [14] and [15]), we shall simply sketch the derivation and point to the physically interesting terms. For the zero order distribution, we choose the natural function of the constants of the motion $f_{s0} = f_{s0}(v_\perp, v_z)$. Then with $v_\perp = (v_x^2 + v_y^2)^{1/2}$ we have

$$\begin{aligned} &\nabla_{\mathbf{v}} f_{s0}(v_\perp, v_z)|_{\mathbf{v}=\mathbf{V}(t')} \\ &= (\hat{x}\, \partial_{v_x} f_{s0} + \hat{y}\, \partial_{v_y} f_{s0} + \hat{z}\, \partial_{v_z} f_{s0})_{\mathbf{v}=\mathbf{V}(t')} \\ &= (V_x \hat{x} + V_y \hat{y}) \frac{1}{v_\perp} \partial_{v_\perp} f_{s0} + \hat{z}\, \partial_{v_z} f_{s0} \end{aligned} \tag{6.180}$$

EXERCISE Verify this equation.

Every term in (6.180), except V_x and V_y, is a constant of motion of a particle orbit and can be taken outside the integration in (6.159). In general, the perturbed magnetic field **B** can have three components. However, the combination

$$(\mathbf{V} \times \tilde{\mathbf{B}}) \cdot \nabla_{\mathbf{v}} f_{s0} = (-V_x v_z \tilde{B}_y + v_z V_y \tilde{B}_x) \frac{1}{v_\perp} \partial_{v_\perp} f_{s0}$$
$$+ (V_x \tilde{B}_y - \tilde{B}_x V_y) \partial_{v_z} f_{s0} \tag{6.181}$$

has only single terms, V_x, V_y that depend on τ and must be kept inside the integral. After taking the constants $\tilde{\mathbf{E}}$ and $\tilde{\mathbf{B}}$ (expressed in terms of $\tilde{\mathbf{E}}$ through Maxwell's equations if desired) as well as $v_\perp$ and $f_{s0}(v_\perp, v_z)$ outside the integral, all remaining terms are of the form

$$I = \int_{-\infty}^{0} d\tau \begin{bmatrix} V_x(\tau) \\ V_y(\tau) \\ 1 \end{bmatrix} \exp\{-i\omega\tau + i\mathbf{k} \cdot [\mathbf{X}(\tau) - \mathbf{x}]\} \tag{6.182}$$

The integrals (6.182) can be evaluated in terms of the identities

$$e^{ia \sin\theta} = \sum_{n=-\infty}^{\infty} J_n(a) \exp(in\theta) \tag{6.183}$$

and

$$e^{-ia \sin\theta} = \sum_{n=-\infty}^{\infty} J_n(a) \exp(-in\theta) \tag{6.184}$$

where J_n is the Bessel function of order n. Without loss of generality, we choose the wave number to be $\mathbf{k} = k_x\hat{x} + k_z\hat{z}$; then (6.175) and (6.177) yield

$$\mathbf{k} \cdot [\mathbf{X}(\tau) - \mathbf{x}] = k_z v_z \tau$$
$$- \frac{k_x v_\perp}{\Omega_s} \sin(\varphi - \Omega_s \tau) + \frac{k_x v_\perp}{\Omega_s} \sin\varphi \tag{6.185}$$

Thus, choosing the factor unity in (6.182) as an example, (6.182) becomes

$$I = \int_{-\infty}^{0} d\tau\, e^{-i\omega\tau + ik_z v_z \tau} e^{-ia \sin(\varphi - \Omega_s \tau)} e^{ia \sin\varphi} \tag{6.186}$$

where $a \equiv k_x v_\perp / \Omega_s$. With (6.183) and (6.184) we have

$$I = \int_{-\infty}^{0} d\tau\, e^{-i\omega\tau + ik_z v_z \tau} \sum_{l=-\infty}^{\infty} J_l(a)$$
$$\times e^{-il\varphi + il\Omega_s \tau} \sum_{n=-\infty}^{\infty} J_n(a) e^{in\varphi} \tag{6.187}$$

The integration can now be performed because only three exponential factors depend on τ. We find

$$I = \sum_{l=-\infty}^{\infty} J_l(a) \sum_{n=-\infty}^{\infty} J_n(a) e^{i(n-l)\varphi} \frac{1}{-i(\omega - l\Omega_s - k_z v_z)} \tag{6.188}$$

A glance back at (6.159) shows that we have just calculated the term involving $\tilde{E}_z$ $\partial_{v_z} f_{s0}(v_\perp, v_z)$; this term is therefore

$$\tilde{f}_s(\mathbf{v}) = \frac{q_s}{m_s} \tilde{E}_z \, \partial_{v_z} f_{s0}(v_\perp, v_z) \sum_{l=-\infty}^{\infty} \sum_{n=-\infty}^{\infty} \frac{J_l(a) J_n(a) e^{i(n-l)\varphi}}{i(\omega - l\Omega_s - k_z v_z)} \quad \{+ \text{ other terms involving } \tilde{\mathbf{E}} \text{ and } \tilde{\mathbf{B}}\} \tag{6.189}$$

where $a \equiv k_x v_\perp / \Omega_s$. The other terms in (6.189) are no harder to calculate than the one shown; for example, the integral (6.182) involving $V_x(\tau) = v_\perp \cos(\varphi - \Omega_s \tau)$ is easily calculated using $V_x(\tau) = v_\perp[\exp(i\varphi - i\Omega_s\tau) + \exp(-i\varphi + i\Omega_s\tau)]/2$, which fits naturally into the form (6.187) and has the effect of shifting the indices of the Bessel functions up and down by one. Eliminating the perturbed magnetic field using Maxwell's equations, we find as in the previous section [Eqs. (6.167), (6.169)]

$$\omega^2 \tilde{\mathbf{E}} + c^2[\mathbf{k} \times (\mathbf{k} \times \tilde{\mathbf{E}})] = -4\pi i \omega \sum_s q_s \int_L d^3 v \mathbf{v} \tilde{f}_s(\mathbf{v}) \tag{6.190}$$

where $\tilde{f}_s(\mathbf{v})$ from (6.189) is linearly proportional to the various components of $\tilde{\mathbf{E}}$. Thus, (6.190) is three equations in three unknowns; setting the determinant of the coefficients equal to zero yields the horrendous dispersion relation found in Ref. 13, Eqs. (8.10.10) and (8.10.11). In (6.190), the velocity integral must, as usual, be performed along the Landau contour.

By taking appropriate limits, we could obtain from (6.190) all of the waves of fluid theory. In addition, there occur waves called *Bernstein modes* [16]. These modes propagate across the magnetic field, so that $k_z \to 0$ in the denominator of (6.189), and are predominantly electrostatic. Then the denominator can be taken outside the integral of (6.189), and the dispersion relation will contain sums of terms like $(\omega - l\Omega_s)^{-1}$. This term goes from $-\infty$ to $+\infty$ as ω changes by a small amount Ω_s. The result is a set of modes, one for each harmonic of the cyclotron frequency of each species. One of these modes corresponds to the upper-hybrid mode (see Chapter 7) in the limit of small wave number. The dispersion diagram is qualitatively as shown in Fig. 6.33. This figure is for the case $\omega_{UH} = (\Omega_e^2 + \omega_e^2)^{1/2}$ between $2|\Omega_e|$ and $3|\Omega_e|$. Notice that there are waves for any frequency such that $|\Omega_e| < \omega < \omega_{UH}$, but that for $\omega > \omega_{UH}$ there are "stop" bands where no waves exist.

There are other interesting features of (6.189) and (6.190). Recall that in the unmagnetized case, Landau damping came from a resonant denominator of the form $(\omega - kv)^{-1}$, representing strong interaction with particles with speeds equal to the wave phase speed, $v = \omega/k$. From (6.189), we get the same kind of resonant denominator, but involving only the component of velocity v_z along the magnetic field. The same procedure as in the unmagnetized case will yield damping involving those paticles with parallel speeds

$$v_z \approx \frac{\omega - l\Omega_s}{k_z}, \qquad l = 0, \pm 1, \pm 2, \ldots . \tag{6.191}$$

If $k_z \to 0$, and $\omega \neq l\Omega_s$ for any l, there is no damping.

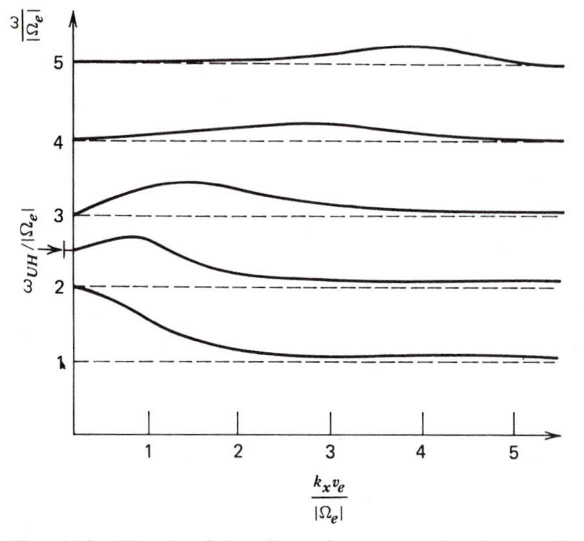

Fig. 6.33 Sketch of the dispersion curves for electron Bernstein modes.

When $l = 0$ in (6.191), we have our old friend Landau damping. When $l \neq 0$, we have *cyclotron damping*. Physically, cyclotron damping occurs when the particle sees a wave whose Doppler shifted frequency is the gyrofrequency or some harmonic thereof:

$$\omega - k_z v_z = l\Omega_s, \qquad l = \pm 1, \pm 2, \ldots . \tag{6.192}$$

Suppose the wave is circularly polarized, or has at least one component that is circularly polarized. For example, consider $l = 1$. Then the particle might see the field shown in Fig. 6.34 as it goes around its gyro-orbit. Evidently, the particle can be continuously accelerated and thus the wave is damped.

The concept of cyclotron damping has an interesting extension to relativistic plasmas. Then $\Omega_s = q_s B_0 / m_s c$ is a function of particle speed, since the relativistic

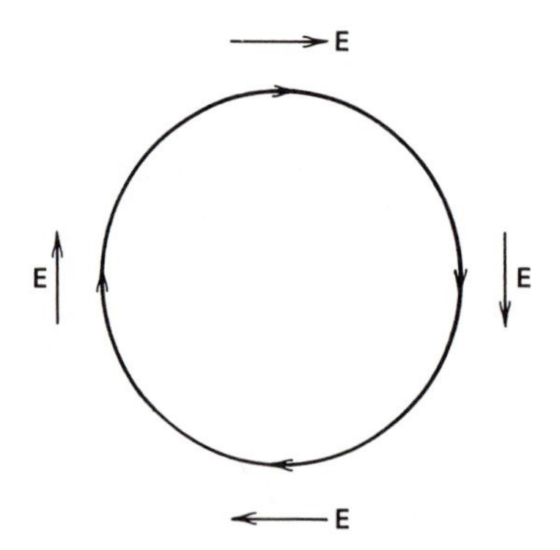

Fig. 6.34 Orientation of the electric field of a wave, and the position of a particle, at four points around the particle's gyro-orbit.

mass is a function of particle speed. The resonance condition (6.192) then depends on $v_\perp$ as well as v_z.

This completes our discussion of linear waves in uniform Vlasov plasma. In the next section, we encounter our first example of a nonlinear wave.

6.13 BGK MODES

In preceding sections we have studied the linear waves that can exist in a Vlasov plasma. As the intensity of such waves is increased, nonlinear effects that are ignored in the linear derivation become important. There are many different nonlinear effects, and much current research in plasma physics is devoted to the theoretical, experimental, and numerical study of nonlinear waves. In this section, we introduce an important class of nonlinear waves, the *BGK modes*, named after *B*ernstein, *G*reene, and *K*ruskal [6].

In Case C of Section 6.2 we encountered an equilibrium distribution function in the presence of a spatially varying electrostatic potential. A BGK mode involves just such a distribution function, where the electrostatic potential is produced self-consistently by the distribution function through Poisson's equation.

For simplicity, let us consider the time independent situation with spatial variation only in the $\hat{x}$-direction. Then for each species the Vlasov equation is

$$\left(v\,\partial_x - \frac{q_s}{m_s}\frac{d\varphi}{dx}\,\partial_v\right) f_s(x,v) = 0 \tag{6.193}$$

where $v \equiv v_x$. The potential $\varphi(x)$ must be determined self-consistently through Poisson's equation

$$\frac{\partial^2\varphi}{\partial x^2} = 4\pi e\left[\int_{-\infty}^{\infty} dv\, f_e(x,v) - \int_{-\infty}^{\infty} dv\, f_i(x,v)\right] \tag{6.194}$$

where it is understood that the v_y and v_z dependencies of $f_s(x,v)$ have been integrated over. We already know the solutions of (6.193); these are just the equilibrium distribution functions (6.12). Thus, we can pick two arbitrary functions $f_s[v_x^2 + 2q_s\varphi(x)/m_s]$, one for each species, insert these in (6.194), and solve the resulting equation, which is

$$\frac{\partial^2\varphi}{\partial x^2} = 4\pi e\int_{-\infty}^{\infty} dv\,\left\{f_e\left[v^2 - 2e\varphi(x)/m_e\right]\right.$$

$$\left. - f_i\left[v^2 + 2e\varphi(x)/m_i\right]\right\} \tag{6.195}$$

This equation must be solved for $\varphi(x)$ subject to appropriate boundary conditions. For example, we may wish to look for periodic wavelike solutions, or for localized soliton solutions. It turns out that there exists a huge number of solutions to the nonlinear integro-differential equation (6.195).

Let us begin to study (6.195) by looking at a very simple case, where each species is a cold beam of particles, each particle of species s having the same speed at a given position. Thus, we choose

$$f_e(x,v) = 2n_0 v_e\,\delta[v^2 - 2e\varphi(x)/m_e - v_e^2] \tag{6.196}$$

where for definiteness we choose only the positive root inside the delta function; that is, we recall the relation

$$\delta[f(y)] = \frac{\delta(y - y_0)}{\left|\frac{df}{dy}\right|_{y=y_0}} \tag{6.197}$$

where y_0 is the solution of $f(y_0) = 0$. Then (6.196) becomes

$$f_e(x,v) = n_0 \frac{v_e}{v} \delta(v - \tilde{v}_e) \tag{6.198}$$

where

$$\tilde{v}_e(x) = [v_e^2 + 2e\varphi(x)/m_e]^{1/2} \tag{6.199}$$

Similarly, for the ions we take

$$f_i(x,v) = n_0 \frac{v_i}{v} \delta(v - \tilde{v}_i) \tag{6.200}$$

with

$$\tilde{v}_i(x) = [v_i^2 - 2e\varphi(x)/m_i]^{1/2} \tag{6.201}$$

Here, v_i and v_e are arbitrary constants that we choose large enough that (6.199) and (6.201) always yield real positive values for $\tilde{v}_e$ and $\tilde{v}_i$. We have chosen the normalization constants n_0 the same for ions and electrons; we must check at the end of the calculation that this gives an overall neutral plasma.

We now look for spatially periodic solutions to (6.195). Integrating f_e, f_i over all velocity space, we find

$$n_e(x) = n_0 \frac{v_e}{\tilde{v}_e} \tag{6.202}$$

and

$$n_i(x) = n_0 \frac{v_i}{\tilde{v}_i} \tag{6.203}$$

so that (6.195) becomes

$$\frac{d^2\varphi}{dx^2} = 4\pi n_0 e \left(\frac{v_e}{\tilde{v}_e} - \frac{v_i}{\tilde{v}_i}\right) \tag{6.204}$$

or

$$\frac{d^2\varphi}{dx^2} = 4\pi n_0 e \left\{\left(1 + \frac{2e\varphi(x)}{m_e v_e^2}\right)^{-1/2} - \left(1 - \frac{2e\varphi(x)}{m_i v_i^2}\right)^{-1/2}\right\} \tag{6.205}$$

We notice the fortunate circumstance that (6.205) is in the form of a pseudopotential equation; that is, it is in the form

$$\frac{d^2\varphi}{dx^2} = -\frac{\partial}{\partial\varphi} V(\varphi) \tag{6.206}$$

with

$$V(\varphi) = -4\pi n_0 \left\{m_e v_e^2 \left(1 + \frac{2e\varphi}{m_e v_e^2}\right)^{1/2} + m_i v_i^2 \left(1 - \frac{2e\varphi}{m_i v_i^2}\right)^{1/2}\right\} \tag{6.207}$$

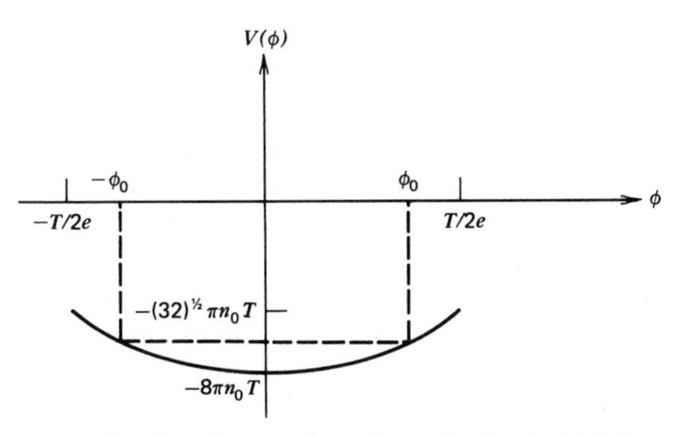

Fig. 6.35 Pseudopotential well used in finding BGK modes for an electron beam traveling through an ion beam.

As a specific example, let us choose $m_e v_e^2 = m_i v_i^2 \equiv T$. Then

$$V(\varphi) = -\ 4\pi n_0 T \left\{ \left(1 + \frac{2e\varphi}{T}\right)^{1/2} + \left(1 - \frac{2e\varphi}{T}\right)^{1/2} \right\} \tag{6.208}$$

The sketch of $V(\varphi)$ is as shown in Fig. 6.35. Equations of the form (6.206) are called pseudopotential equations because of their resemblance to Newton's law of motion $m\ddot{x} = F(x) = -\ dV(x)/dx$. With an initial choice of the "pseudoenergy" somewhere between $-(32)^{1/2}\ \pi n_0 T$ and $-8\pi n_0 T$, the "pseudoparticle" oscillates forever in the pseudopotential well, producing a spatially periodic potential that oscillates between $-\varphi_0$ and φ_0, as shown in Fig. 6.36. The function $\varphi(x)$ is a periodic function but is not a sine function; it becomes a sine function in the limit of very small φ_0.

In the limit of small φ_0, we can make analytic progress by expanding the square roots in (6.205), assuming $e\varphi(x)/T << 1$ for all x. We obtain

$$\frac{d^2\varphi}{dx^2} + \frac{8\pi n_0 e^2}{T}\ \varphi = 0 \tag{6.209}$$

with solution

$$\varphi(x) = \varphi_0 \sin\,(2^{1/2}\ x/\lambda_{\text{eff}}) \tag{6.210}$$

where we have defined an effective Debye length

$$\lambda_{\text{eff}} = v_e/\omega_e \tag{6.211}$$

Recall that v_e here is a constant and not a thermal speed.

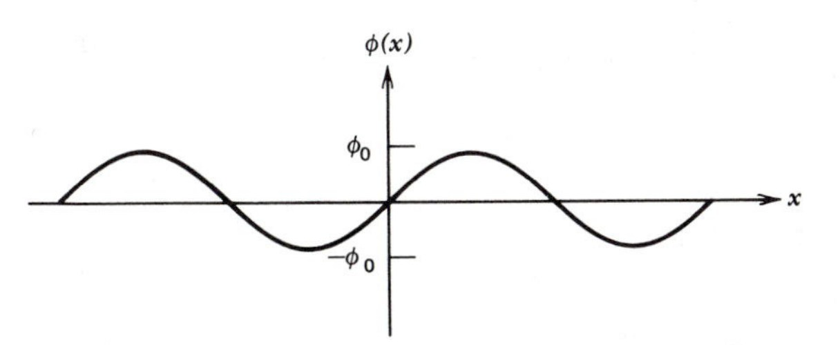

Fig. 6.36 Periodic BGK modes for an electron beam moving through an ion beam.

Our physical picture of this BGK mode, both the nonlinear version (6.206) and the linear limit (6.210), is as follows. A spatially periodic potential exists. The ion beam is accelerated through regions of large negative potential and thus has a lower density there, while the electron beam is decelerated in regions of large negative potential and thus has a higher density there. The net result is a negative net charge in regions of negative potential, of exactly the right amount to *produce* the negative potential. The opposite argument produces the regions of positive potential. The potential and densities thus have the phase relationships shown in Fig. 6.37. The important point is that this physical process works not only in the linear regime of (6.210), but also the nonlinear regime of (6.206).

In the preceding discussion we assumed that the ion and electron velocities were large enough so that none of them were trapped in the electrostatic potential wells. In other words, we had ion energy $> e\varphi(x)$ for all x, and electron energy $> -e\varphi(x)$ for all x. We can also consider the case where some of the electrons or ions are trapped in the potential wells. Amazingly, it turns out that almost any potential $\varphi(x)$ can be constructed by choosing appropriate distributions of trapped electrons, untrapped electrons, trapped ions, and untrapped ions.

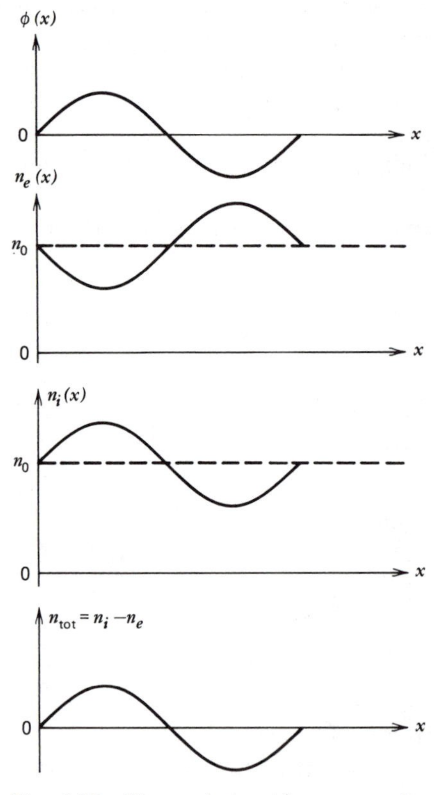

Fig. 6.37 Phase relationships among electrostatic potential, electron density, ion density, and net charge density in the BGK mode of Fig. 6.36.

For this discussion it is convenient to think in terms of distribution functions that depend on energy H rather than distribution functions that depend on velocity v. With the substitution $H = \frac{1}{2}m_s v^2 + q_s\varphi$, we have $dH = m_s v\, dv = m_s[(2/m_s)(H - q_s\varphi)]^{1/2}\, dv = [2m_s(H - q_s\varphi)]^{1/2}\, dv$ and

$$f_s(v)\, dv = f_s[v(H)]\, dv = \frac{f_s(H)\, dH}{[2m_s(H - q_s\varphi)]^{1/2}} \tag{6.212}$$

Equation 6.212 applies to particles with positive speeds. If we assume for convenience (we do not need to do this in general) that there are equal numbers of left-going and right-going particles, then we have

$$n_s(x) = \int_{-\infty}^{\infty} dv\, f_s(x,v) = 2\int_0^{\infty} dv\, f_s(x,v)$$

$$= 2\int_{q_s\varphi}^{\infty} dH\, \frac{f_s(x,H)}{[2m_s(H - q_s\varphi)]^{1/2}} \tag{6.213}$$

where the lower limit on the energy integration must be taken to be $q_s\varphi$; no particle can have energy less than $q_s\varphi(x)$ for then its velocity would be imaginary since $\frac{1}{2}m_s v^2 = H - q_s\varphi$.

Consider a periodic potential, as shown in Fig. 6.38. Then any ion with total energy H less than $e\varphi_{\max}$ will be trapped between the potential hills; the ion with energy $e\varphi_0$ oscillates forever on the solid line shown. An ion with total energy H greater than $e\varphi_{\max}$ will travel forever to the left or to the right. The electrons, however, see the potential upside down, because of their negative charge. Thus, any electron with energy H less than $-e\varphi_{\min}$ will be trapped between the potential minima, while electrons with energies greater than $-e\varphi_{\min}$ travel forever to the left or to the right.

Suppose we are given a completely arbitrary periodic potential $\varphi(x)$, a given distribution $f_i(H)$ of ions (both trapped and untrapped), and a given distribution $f_e(H)$, $H > -e\varphi_{\min}$. Then it turns out that the distribution of trapped electrons $f_e(H)$, $-e\varphi < H < -e\varphi_{\min}$, can always be chosen so that Poisson's equation is satisfied and the given potential $\varphi(x)$ is indeed produced. Details of this calculation can be found on p. 436 of Ref. [13]. Note that f_s can be different for $v < 0$ and $v > 0$.

There are many practical applications of BGK modes, including the nonlinear stage of a Landau damped Langmuir wave (Section 6.8), and the theories of shock

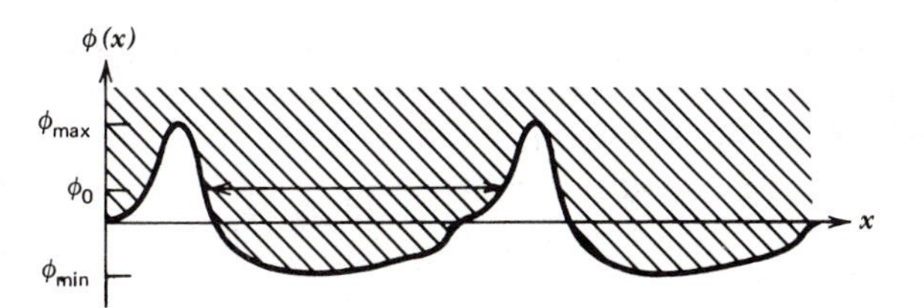

Fig. 6.38 Periodic potential of a BGK mode that contains trapped and untrapped particles.

waves and double layers. By contrast, the theory of Case–Van Kampen modes, to be presented in the next section, has very little practical application; nevertheless, this theory teaches us much about the analytic structure of the Vlasov equation.

6.14 CASE–VAN KAMPEN MODES

In Sections 6.1 to 6.5 we studied Langmuir waves and Landau damping by linearizing the Vlasov and Poisson equations, eliminating the perturbed distribution function f_1, and solving the initial value problem for the electric field E. This led us to look for normal modes of E, which were found by setting the dielectric function $\epsilon(\omega,k)$ equal to zero. In this way we found one normal mode for every value of k, with $\omega(k) \approx \omega_e(1 + \frac{3}{2} k^2\lambda_e^2) + i\gamma$ where γ is the Landau damping rate.

There is another way to approach this problem, and that is to eliminate E and look for normal modes of f_1. In this way we find the *Case–Van Kampen modes* [17, 18].

The Vlasov–Poisson system is

$$\partial_t f + v\,\partial_x f - \frac{eE}{m}\,\partial_v f = 0 \tag{6.214}$$

and

$$\partial_x E = 4\pi e\left[n_0 - \int_{-\infty}^{\infty} dv\, f(v)\right] \tag{6.215}$$

where one-dimensional variations are considered, $v \equiv v_x$, and $f(v)$ is the one-dimensional electron distribution. Equation (6.214) refers to electrons only, the ions being fixed ($m_i \to \infty$). Linearizing (6.214) and (6.215) exactly as in Section 6.3 we find

$$\partial_t f_1 + v\,\partial_x f_1 = \frac{n_0 eE}{m}\frac{\partial g}{\partial v} \tag{6.216}$$

and

$$\partial_x E(x,t) = -\,4\pi e\int_{-\infty}^{\infty} dv\, f_1(x,v,t) \tag{6.217}$$

where $f_0(v) = n_0 g(v)$, $\int_{-\infty}^{\infty} dv\, g(v) = 1$. We can now look for normal modes of f_1 that have the spatial and time dependence $\exp(-i\omega t + ikx)$. Note that this is *not* the same procedure as used previously in Section 6.4 for the electric field E. There, we assumed only the spatial dependence $\exp(ikx)$, and used Laplace transform techniques to consider the complete time evolution. At late times, we found that only the normal modes given by the zeros of $\epsilon(\omega,k)$ were important. Here we are looking immediately for normal modes in space and time. We are not considering an initial value problem; the connection between the normal modes found here and an initial value problem must be established separately.

Looking for solutions $f_1(x,v,t) = \tilde{f}_1(v)\exp(-i\omega t + ikx)$, $E = E_0\exp(-i\omega t + ikx)$, (6.216) and (6.217) become

$$(-i\omega + ikv)\tilde{f}_1 = \frac{n_0 eE_0}{m}\frac{\partial g}{\partial v} \tag{6.218}$$

and

$$E_0 = \frac{-4\pi e}{ik} \int_{-\infty}^{\infty} dv\, \tilde{f}_1(v) \tag{6.219}$$

from which we can eliminate E_0 to obtain

$$\left(v - \frac{\omega}{k}\right)\tilde{f}_1 = \frac{\omega_e^2}{k^2} \frac{\partial g}{\partial v} \int_{-\infty}^{\infty} dv'\, \tilde{f}_1(v') \tag{6.220}$$

Defining $\eta(v) \equiv (\omega_e^2/k^2)\, \partial g/\partial v$, this is

$$\left(v - \frac{\omega}{k}\right) \tilde{f}_1(v) = \eta(v) \int_{-\infty}^{\infty} dv'\, \tilde{f}_1(v') \tag{6.221}$$

Equation (6.221) is a linear integral equation for $\tilde{f}_1(v)$, with nonconstant coefficients (the v on the left and the $\eta(v)$ on the right). One good approach to solving such equations is to guess the solution. We guess

$$\boxed{\tilde{f}_1(v) = P\left[\frac{\eta(v)}{v - \omega/k}\right] + \delta(v - \omega/k)\left[1 - P\int_{-\infty}^{\infty} \frac{\eta(v')\, dv'}{v' - \omega/k}\right]} \tag{6.222}$$

with which $f_1(x,v,t) = \tilde{f}_1(v) \exp(-i\omega t + ikx)$ is the *Case–Van Kampen mode* [17, 18]. In (6.222), P stands for principal value, and is defined by

$$P\left(\frac{1}{x - a}\right) = \left\{\begin{array}{ll} \dfrac{1}{x - a} & x \neq a \\ \lim\limits_{x \to a^{\pm}} \dfrac{1}{x - a} & x = a \end{array}\right\} \tag{6.223}$$

so that this expression is either $+\infty$ or $-\infty$ at $x = a$, depending on from which side the limit is taken. This definition has the important consequence that

$$(x - a)\, P\left(\frac{1}{x - a}\right) = 1 \tag{6.224}$$

since the left side of (6.224) is unity at $x = a$ no matter from which side the limit is taken.

Note that (6.222) says that for any wave number k, there are an infinite number of normal modes, one for each value of real ω. This is in contrast to the initial value problem for $E(x,t)$, where only one normal mode was found for each value of k (if negative frequencies are counted, then two normal modes were found for each value of k). Note further that the normal modes are not damped, but exist for all time with real frequency ω.

Let us verify that (6.222) is indeed a solution of (6.221). The left side is simplified because one of its terms is of the form $u\delta(u)$, which is zero. Thus, in (6.221),

$$\text{left side} = \eta(v) \tag{6.225}$$

On the right side of (6.221), the two terms involving principal values cancel each other when integrated over v, and we are left with only the term

$$\text{right side} = \eta(v) \tag{6.226}$$

so that the left and right sides of (6.221) are indeed equal and (6.222) is indeed a normal mode solution. Since (6.221) is a linear equation, $\tilde{f}(v)$ in (6.222) can be multiplied by any constant.

EXERCISE Verify (6.225) and (6.226).

What is the electric field associated with this normal mode? Using (6.219), we find

$$E_0 = \frac{-4\pi e}{ik} \int_{-\infty}^{\infty} dv\, \tilde{f}_1(v) \tag{6.227}$$

or

$$E_0 = \frac{-4\pi e}{ik} \tag{6.228}$$

so that

$$\boxed{E(x,t) = \frac{-4\pi e}{ik}\, e^{-i\omega t + ikx}} \tag{6.229}$$

is the electric field associated with the normal mode (6.222).

These normal modes are peculiar, both mathematically and physically. Mathematically, we have $f_1(v = \omega/k) \to \infty$ because of the δ-function in (6.222). But it is not consistent to linearize the Vlasov equation with $f = f_0 + f_1$ and then find f_1 infinite! Physically, (6.222) says that we must have a finite number of particles per unit spatial volume with velocity exactly equal to ω/k, which is impossible to do. We conclude that the individual Case–Van Kampen modes as given by (6.222) are not physically relevant. What then is the importance of the modes in (6.222)?

The importance of the modes in (6.222) lies in the possibility of creating a physically and mathematically acceptable disturbance by adding up many such modes. Consider a fixed wave number k. Since the basic linearized Vlasov–Poisson system is indeed linear, we may construct a solution by taking any linear combination of the solutions in (6.222). The general solution is

$$f_1(x,v,t) = e^{ikx} \int_{-\infty}^{\infty} d\omega\, e^{-i\omega t} \tilde{f}_1(v,\omega) c(\omega) \tag{6.230}$$

where we label the normal modes of (6.222) by their frequency ω, and $c(\omega)$ is an arbitrary weighting function. For sufficiently well behaved $c(\omega)$, the singularities in $\tilde{f}_1(v,\omega)$ will be smoothed out, and $f_1(x,v,t)$ will be a mathematically and physically nice function. For a given initial condition $f_1(x,v,t = 0)$, the function $c(\omega)$ must be chosen such that (6.230) yields the correct solution at $t = 0$.

Inserting (6.222) for $\tilde{f}_1(v)$ into (6.230), we find

$$\begin{aligned} f_1(x,v,t) = {} & e^{ikx}\, \eta(v)\, P \int_{-\infty}^{\infty} d\omega\, \frac{e^{-i\omega t} c(\omega)}{v - (\omega/k)} \\ & + e^{ikx} k\, e^{-ikvt}\, c(\omega = kv) \\ & - k\, e^{ikx - ikvt}\, c(\omega = kv)\, P \int_{-\infty}^{\infty} \frac{\eta(v')dv'}{v' - v} \end{aligned} \tag{6.231}$$

where we have used $\delta(v - \omega/k) = k\delta(\omega - kv)$. Consider the middle term in (6.231). This term does not damp away at late times, but rather oscillates in velocity faster and faster with increasing time, as shown in Fig. 6.39. This behavior is due to the free streaming of particles, and would occur for $f_1(x,v,t)$ even if the charge of the electrons were zero. For example, consider the linearized Vlasov equation (6.216) when the charge is zero;

$$\partial_t \tilde{f}_1 + ikv\tilde{f}_1 = 0 \tag{6.232}$$

where spatial dependence e^{ikx} is assumed. The solution of (6.232) is

$$\tilde{f}_1(v,t) = \tilde{f}_1(v,t = 0)e^{-ikvt} \tag{6.233}$$

which becomes more and more pathological with increasing time. Physically, a small number of collisions would wipe out this behavior at late time.

Returning to the general case with a nonzero charge, we can ask: Does the electric field behave more reasonably? Yes it does. From Poisson's equation,

$$E(x,t) = \frac{-4\pi e}{ik}\int_{-\infty}^{\infty} dv\, f_1(v,x,t) \tag{6.234}$$

or

$$\boxed{E(x,t) = 4\pi ei\, e^{ikx}\int_{-\infty}^{\infty} dv\, e^{-ikvt}c(\omega = kv)} \tag{6.235}$$

where the first and last terms on the right of (6.231) cancel upon integration.

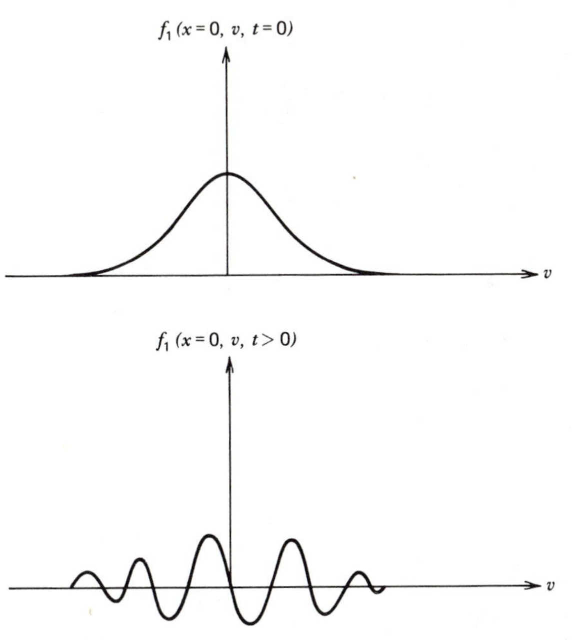

Fig. 6.39 A portion of the distribution function of a Case–Van Kampen mode at two different times.

EXERCISE Verify this.

The right side of (6.235) is in the form of a Fourier transform. It turns out that when $c(\omega)$ is correctly chosen to represent the initial $f_1(x,v,t)$ through (6.230), then (6.235) is exactly equivalent to the expression (6.35) together with (6.34). Thus, the right side of (6.235) would produce transients, as before, together with the correct Landau damped normal modes. Thus, there is complete agreement between the normal mode picture for $E(x,t)$ of Sections 6.4 and 6.5, and the normal mode picture for $f_1(x,v,t)$ of the present section. The latter is somewhat more complete, since $f_1(x,v,t)$ determines $E(x,t)$ and therefore determines its own future at all times, while $E(x,t)$ could be produced by many different functions $f_1(x,v,t)$ and therefore does not determine its own future at all times. In practice, either approach may be used because $E(x,t)$ does pick out the slowly Landau damped normal modes of the system at late times, after the transients have died away.

This brings us to the end of our discussion of the Vlasov equation. We recall that the Vlasov equation is obtained as an approximate theory from the Klimontovich and Liouville approaches of Chapters 3 to 5 by neglecting the physics of collisions. In the next two chapters, we take moments of the Vlasov equation to obtain even simpler and more approximate theories of a plasma; these are the *fluid theory* and *magnetohydrodynamics*.

REFERENCES

[1] A. A. Vlasov, *J. Phys. (U.S.S.R.)*, *9*, 25 (1945).

[2] L. D. Landau, *J. Phys. (U.S.S.R.)*, *10*, 25 (1946).

[3] J. D. Jackson, *J. Nucl. Energy, Part C*, *1*, 171 (1960).

[4] L. D. Landau and E. M. Lifshitz, *Electrodynamics of Continuous Media*, Pergamon, Oxford, 1960.

[5] J. Dawson, *Phys. Fluids*, *4*, 869 (1961).

[6] I. B. Bernstein, J. M. Greene, and M. D. Kruskal, *Phys. Rev.*, *108*, 546 (1957).

[7] T. M. O'Neil, J. H. Winfrey, and J. H. Malmberg, *Phys. Fluids*, *14*, 1204 (1971).

[8] G. J. Morales and J. H. Malmberg, *Phys. Fluids*, *17*, 609 (1974).

[9] T. O'Neil, *Phys. Fluids*, *8*, 2255 (1965).

[10] C. S. Gardner, *Phys. Fluids*, *6*, 839 (1963).

[11] R. Penrose, *Phys. Fluids*, *3*, 258 (1960).

[12] P. Noerdlinger, *Phys. Rev.*, *118*, 879 (1960).

[13] N. A. Krall and A. W. Trivelpiece, *Principles of Plasma Physics*, McGraw-Hill, New York, 1973.

[14] T. H. Stix, *The Theory of Plasma Waves*, McGraw-Hill, New York, 1962, Chs. 8 and 9.

[15] G. Bekefi, *Radiation Processes in Plasmas*, Wiley, New York, 1966.

[16] I. B. Bernstein, *Phys. Rev.*, *109*, 10 (1958).
[17] K. M. Case, *Ann. Phys. (N.Y.)*, *7*, 349 (1959).
[18] N. G. Van Kampen, *Physica*, *21*, 949 (1955).

PROBLEMS

6.1 Resistive vs. Reactive Instabilities

The type of instability found in (6.52) when $d_u g(u)|_{u=\omega_r/k} > 0$ is often called a "resistive" instability; these are instabilities to which (6.42) and (6.43) apply. Show why (6.42) and (6.43) do not apply to the dielectric function $\epsilon(\omega) = 1 + \omega_0^2/\omega^2$, which yields a "reactive" instability.

6.2 Ion-Acoustic Waves

The dispersion relation (6.21) has a branch describing ion-acoustic waves as well as one describing Langmuir waves. We consider $g(u)$ given by two Maxwellians as in (6.24), with $T_i << T_e$, and look for a wave with phase speed such that

$$v_i << \omega/k << v_e$$

For the ion contribution, expand the denominator in (6.21) as in (6.26). For the electron contribution, approximate the integral by ignoring $\omega/k << u$ in the denominator. Then solve the dispersion relation $\epsilon(k,\omega) = 0$ to find $\omega = \omega(k)$ for ion-acoustic waves.

6.3 Electrostatic Waves

The dispersion relation (6.21) for electrostatic waves must be solved using the Landau contour of Section 6.4; alternatively, the integral can be evaluated for $\omega_i > 0$ and the result analytically continued to $\omega_i < 0$. Evaluate (6.21) and find the normal modes $\omega(k)$ for the following distribution functions $g(u)$.

(a) Cold plasma, $g(u) = \delta(u)$.
(b) Cold beam, $g(u) = \delta(u - u_0)$.
(c) Square distribution, $g(u) = (2c)^{-1}$ for $|u| < c$, $g(u) = 0$ for $|u| > c$; c is a real positive constant.
(d) Cauchy distribution, $g(u) = (c/\pi)(u^2 + c^2)^{-1}$. (Why can there never be a true Cauchy distribution?)

6.4 Ion-Acoustic Wave Energy

Apply the wave energy formula (6.72) to the ion-acoustic dielectric function derived in Problem 6.2. Is most of the energy in electric field energy or in particle energy? Explain physically and in detail why this is so (see discussion in Chapter 7).

6.5 Two Drifting Cauchy Distributions

Consider a Vlasov equilibrium consisting of infinitely massive ions and two counterstreaming Cauchy distributions (see Problem 6.3) such that

$$g(u) = \frac{\Delta}{2\pi}\left[\frac{1}{(u-a)^2+\Delta^2} + \frac{1}{(u+a)^2+\Delta^2}\right]$$

(a) Sketch $g(u)$.

(b) Apply Gardner's theorem to show that stability is guaranteed for $a < \Delta/\sqrt{3}$.

(c) Apply the Penrose criterion to show that this equilibrium is stable for $a < \Delta$, and unstable for $a > \Delta$.

6.6 Isotropic Stability

(a) Consider a plasma with infinitely massive ions and an electron distribution function that is the surface of a sphere in three-dimensional velocity space,

$$f_e(\mathbf{v}) = C\,\delta(v - v_0) = C\,\delta[(v_x{}^2 + v_y{}^2 + v_z{}^2)^{1/2} - v_0]$$

Calculate $g(u)$, sketch it, and use Gardner's theorem to show that this distribution is stable to electrostatic perturbations.

(b) Consider any isotropic distribution function

$$f_e(\mathbf{v}) = f_e(v)$$

Use Gardner's theorem to show that such a distribution is stable to electrostatic perturbations.

CHAPTER 7

Fluid Equations

7.1 INTRODUCTION

There are many phenomena in plasma physics that can be studied by thinking of the plasma as two interpenetrating fluids, an electron fluid and an ion fluid. In this approach, it is not necessary to consider the fact that each species consists of particles with different velocities. The advantage of this approach is its simplicity; it leads to equations in three spatial dimensions and time rather than in the seven-dimensional phase space of Vlasov theory (Chapter 6). The disadvantage of this approach is that it misses velocity-dependent effects such as Landau damping.

In this section, we introduce the fluid equations heuristically, for the benefit of those readers who have not yet studied Chapter 6 on Vlasov theory. In the next section, we present a more rigorous derivation of the fluid equations from the Vlasov equation.

The first equation of fluid theory is the *continuity equation*, which expresses the fact that the fluid is not being created or destroyed, so that the only way that the fluid density $n_s(\mathbf{x},t)$ of fluid species s can change at a point $\mathbf{x} = (x,y,z)$ is by having a net amount of fluid enter or leave a small spatial volume including that point. The density n_s is the number of particles of species s per unit volume. To every element of fluid there corresponds a velocity vector $\mathbf{V}_s(\mathbf{x},t)$ that gives the velocity of the fluid element at the point $\mathbf{x}$ at time t. Mathematically, the continuity equation for fluid species s is

$$\boxed{\partial_t n_s(\mathbf{x},t) + \nabla \cdot (n_s \mathbf{V}_s) = 0} \tag{7.1}$$

where $\nabla = (\partial_x, \partial_y, \partial_z)$ is the usual gradient operator in three-dimensional configuration space. A derivation of the fluid continuity equation can be found in most undergraduate mechanics books (see, for example, Ref. [1]).

The second equation of fluid theory is the *force equation*, which is simply Newton's second law of motion for a fluid. This can be written for fluid species s as

$$n_s m_s \dot{\mathbf{V}}_s(\mathbf{x},t) = \mathbf{F}_s(\mathbf{x},t) \tag{7.2}$$

where $\mathbf{F}_s(\mathbf{x},t)$ is the force per unit volume acting on the fluid element at position $\mathbf{x}$ at time t. The time derivative on the left of Newton's law refers to the fluid element as an entity and therefore must be taken along the orbit of the fluid element. Thus,

$$\begin{aligned}\dot{\mathbf{V}}_s(\mathbf{x},t) &= \partial_t \mathbf{V}_s + \left(\frac{d\mathbf{x}}{dt}\right)\bigg|_{\text{orbit}} \cdot \nabla \mathbf{V}_s \\ &= \partial_t \mathbf{V}_s + (\mathbf{V}_s \cdot \nabla)\mathbf{V}_s \end{aligned} \tag{7.3}$$

On the right side of (7.2) are all of the forces that act on a fluid element. One such force is the pressure gradient force. A fluid of charged particles has a pressure $P_s(\mathbf{x},t) = n_s(\mathbf{x},t)T_s(\mathbf{x},t)$ and an associated force per unit volume $-\nabla P_s$. Another force is the Lorentz force per unit volume, $q_s n_s(\mathbf{x},t)\mathbf{E}(\mathbf{x},t) + (q_s/c)n_s(\mathbf{x},t)\mathbf{V}_s(\mathbf{x},t) \times \mathbf{B}(\mathbf{x},t)$. With these forces, Eq. (7.2) becomes

$$n_s m_s\, \partial_t \mathbf{V}_s + n_s m_s \mathbf{V}_s \cdot \nabla \mathbf{V}_s = -\nabla P_s + q_s n_s \mathbf{E} + \frac{q_s}{c} n_s \mathbf{V}_s \times \mathbf{B} \tag{7.4}$$

or

$$\boxed{\partial_t \mathbf{V}_s + \mathbf{V}_s \cdot \nabla \mathbf{V}_s = \frac{-1}{n_s m_s} \nabla P_s + \frac{q_s}{m_s} \mathbf{E} + \frac{q_s}{m_s c} \mathbf{V}_s \times \mathbf{B}} \tag{7.5}$$

which can be thought of as the force equation per particle. The fields $\mathbf{E}(\mathbf{x},t)$ and $\mathbf{B}(\mathbf{x},t)$ are the macroscopic fields (those which would be measured by a probe), as discussed in Chapter 3.

With the given fluid quantities, the total charge density ρ is defined by

$$\rho(\mathbf{x},t) = \sum_s q_s n_s(\mathbf{x},t) \tag{7.6}$$

while the total current density $\mathbf{J}$ is defined by

$$\mathbf{J}(\mathbf{x},t) = \sum_s q_s n_s(\mathbf{x},t)\mathbf{V}_s(\mathbf{x},t) \tag{7.7}$$

When combined with Maxwell's equations

$$\nabla \cdot \mathbf{E}(\mathbf{x},t) = 4\pi\rho \tag{7.8}$$

$$\nabla \cdot \mathbf{B}(\mathbf{x},t) = 0 \tag{7.9}$$

$$\nabla \times \mathbf{E}(\mathbf{x},t) = -\frac{1}{c}\,\partial_t \mathbf{B} \tag{7.10}$$

$$\nabla \times \mathbf{B}(\mathbf{x},t) = \frac{4\pi}{c}\mathbf{J} + \frac{1}{c}\,\partial_t \mathbf{E} \tag{7.11}$$

the fluid equations provide a complete, but approximate, description of plasma physics. A more careful development of the fluid equations from the Vlasov equation is provided in the next section. The reader who has not yet studied the Vlasov equation (Chapter 6) can proceed directly to Section 7.3.

7.2 DERIVATION OF THE FLUID EQUATIONS FROM THE VLASOV EQUATION

Except for the neglect of collisions, the Vlasov equation (Chapter 6) is an exact description of a plasma. By taking velocity moments of the Vlasov equation in seven-dimensional $(\mathbf{x},\mathbf{v},t)$ space, an infinite hierarchy of equations in four-dimensional $(\mathbf{x},t)$ space can be derived. When an appropriate truncation of this infinite hierarchy is carried out, the standard *two-fluid theory* of plasma physics is obtained. This procedure is reminiscent of the truncation of the BBGKY hierarchy in Chapter 4 that led to the plasma kinetic equation and thence to the Vlasov equation.

The Vlasov equation (6.5) is

$$\underbrace{\partial_t f_s(\mathbf{x},\mathbf{v},t)}_{①} + \underbrace{\mathbf{v}\cdot\nabla_{\mathbf{x}} f_s}_{②} + \underbrace{\frac{q_s}{m_s}\mathbf{E}\cdot\nabla_{\mathbf{v}} f_s}_{③} + \underbrace{\frac{q_s}{m_s}\frac{\mathbf{v}}{c}\times\mathbf{B}\cdot\nabla_{\mathbf{v}} f_s}_{④} = 0 \tag{6.5}$$

i.e., $\partial_t f_s(\mathbf{x},\mathbf{v},t) + \mathbf{v}\cdot\nabla_{\mathbf{x}} f_s + \frac{q_s}{m_s}\left(\mathbf{E} + \frac{\mathbf{v}}{c}\times\mathbf{B}\right)\cdot\nabla_{\mathbf{v}} f_s = 0$

We use the normalization

$$n_s(\mathbf{x},t) = \int d\mathbf{v}\, f_s(\mathbf{x},\mathbf{v},t) \tag{7.12}$$

and note that the fluid velocity $\mathbf{V}_s$ is

$$\mathbf{V}_s(\mathbf{x},t) = \frac{1}{n_s}\int d\mathbf{v}\, \mathbf{v} f_s(\mathbf{x},\mathbf{v},t) \tag{7.13}$$

The first fluid equation (the continuity equation) is obtained by integrating (6.5) over all velocity space (i.e., we first multiply by "unity"). The first term yields $\partial_t n_s(\mathbf{x},t)$. The second term is

$$② = \int d\mathbf{v}\, \mathbf{v}\cdot\nabla_{\mathbf{x}} f_s = \nabla_{\mathbf{x}}\cdot\int d\mathbf{v}\, \mathbf{v} f_s = \nabla_{\mathbf{x}}\cdot(n_s\mathbf{V}_s) \tag{7.14}$$

The third and fourth terms vanish upon performing the velocity integration.

EXERCISE Show the above.

The result is the exact *continuity equation*

$$\partial_t n_s(\mathbf{x},t) + \nabla_{\mathbf{x}}\cdot(n_s\mathbf{V}_s) = 0 \tag{7.15}$$

which agrees with (7.1). (Except in this section, $\nabla_{\mathbf{x}}$ is denoted by ∇ in this chapter.)

The force equation is obtained by multiplying (6.5) by $\mathbf{v}$ and integrating over all velocity space. This yields

$$\underbrace{\frac{\partial}{\partial t}\int d\mathbf{v}\, \mathbf{v} f_s}_{①} + \underbrace{\int d\mathbf{v}\, \mathbf{v}\mathbf{v}\cdot\nabla_{\mathbf{x}} f_s}_{②} + \underbrace{\frac{q_s}{m_s}\int d\mathbf{v}\, \mathbf{v}\left[\mathbf{E}\cdot\nabla_{\mathbf{v}} f_s\right]}_{③} + \underbrace{\frac{q_s}{m_s}\int d\mathbf{v}\, \mathbf{v}\left[\frac{1}{c}\mathbf{v}\times\mathbf{B}\cdot\nabla_{\mathbf{v}} f_s\right]}_{④} = 0 \tag{7.16}$$

i.e., $\frac{\partial}{\partial t}\int d\mathbf{v}\, \mathbf{v} f_s + \int d\mathbf{v}\, \mathbf{v}\mathbf{v}\cdot\nabla_{\mathbf{x}} f_s + \frac{q_s}{m_s}\int d\mathbf{v}\, \mathbf{v}\left[\left(\mathbf{E} + \frac{1}{c}\mathbf{v}\times\mathbf{B}\right)\cdot\nabla_{\mathbf{v}} f_s\right] = 0$

In term ①, we have $\partial_t(n_s \mathbf{V}_s)$ by (7.13). In ②, we perform the manipulation $\mathbf{vv} \cdot \nabla_{\mathbf{x}} f_s = \mathbf{v} \cdot \nabla_{\mathbf{x}}(\mathbf{v} f_s) = \nabla_{\mathbf{x}} \cdot (\mathbf{vv}f_s)$. Since $f_s(\mathbf{x},\mathbf{v},t)$ is a probability distribution, the ensemble average of any quantity is

$$\langle g \rangle = \frac{\int d\mathbf{v}\, g f_s}{\int d\mathbf{v}\, f_s} = \frac{1}{n_s} \int d\mathbf{v}\, g f_s \tag{7.17}$$

Thus, term ② is

$$\nabla_{\mathbf{x}} \cdot \int d\mathbf{v}\; \mathbf{vv} f_s = \nabla_{\mathbf{x}} \cdot (n_s \langle \mathbf{v}\, \mathbf{v} \rangle) \tag{7.18}$$

Term ③ is easily evaluated by an integration by parts, yielding $-(q_s/m_s)\mathbf{E} n_s$.

EXERCISE Verify this result for at least one component of the combination $\mathbf{vE} \cdot \nabla_{\mathbf{v}}$.

In term ④, it is useful to move $\nabla_{\mathbf{v}}$ to the left, obtaining $\nabla_{\mathbf{v}} \cdot [(\mathbf{v} \times \mathbf{B}) f_s]$; an integration by parts then yields

$$④ = -\frac{q_s}{m_s c} \int d\mathbf{v}\, (\mathbf{v} \times \mathbf{B}) f_s = \frac{-q_s}{m_s c}\, n_s (\mathbf{V}_s \times \mathbf{B}) \tag{7.19}$$

where we have evaluated each component in (7.19) using (7.13). Combining all terms, (7.16) becomes

$$\boxed{\partial_t(n_s \mathbf{V}_s) + \nabla \cdot (n_s \langle \mathbf{v}\, \mathbf{v} \rangle) = \frac{q_s}{m_s}\, n_s \left(\mathbf{E} + \frac{1}{c}\, \mathbf{V}_s \times \mathbf{B} \right)} \tag{7.20}$$

which is the *fluid force equation* for species s. Multiplying through by the mass m_s, we see that each term has units of (force/volume). Equation (7.20) is also called the *momentum equation*, since it determines the time rate of change of momentum per unit volume.

Note that the continuity equation (7.15) for n_s involves the function $\mathbf{V}_s$, and the force equation (7.20) for $\mathbf{V}_s$ involves the function $\langle \mathbf{v}\, \mathbf{v} \rangle$. It is clear that every equation for n factors of $\mathbf{v}$ will involve a term with $n + 1$ factors of $\mathbf{v}$. Thus, to obtain a complete description of a plasma, we need an infinite number of moment equations as derived from the Vlasov equation. This is equivalent to replacing the seven-dimensional Vlasov equation by an infinite number of four-dimensional fluid equations. In practice, we seek to truncate this series of equations by using a physical argument to evaluate the term with $n + 1$ factors of $\mathbf{v}$, rather than using the fluid equation for that term. For example, we shall use physical arguments to evaluate the $\langle \mathbf{v}\, \mathbf{v} \rangle$ term in (7.20), so that the force equation (7.20), the continuity equation (7.15), and Maxwell's equations become a complete description of the plasma. In detailed descriptions of plasmas found, for example, in magnetic confinement devices and in the solar wind, the fluid equation for $\langle \mathbf{v}\, \mathbf{v} \rangle$ is used and physical arguments are used to evaluate terms with three components of velocity [2]. The equation for the time derivative of $\langle \mathbf{v}\, \mathbf{v} \rangle$ is known as the *energy equation*.

There are various circumstances where it is easy to evaluate $\langle \mathbf{v}\, \mathbf{v} \rangle$. For example, suppose the species is cold, so that all particles have the same macroscopic velocity $\mathbf{V}_s$. Then $f_s(\mathbf{x},\mathbf{v},t) = n_s(\mathbf{x},t)\delta(\mathbf{v} - \mathbf{V}_s)$; thus

$$\langle \mathbf{v}\,\mathbf{v} \rangle = \frac{1}{n_s} \int d\mathbf{v}\, n_s \mathbf{v}\mathbf{v}\, \delta(\mathbf{v} - \mathbf{V}_s) = \mathbf{V}_s \mathbf{V}_s \tag{7.21}$$

EXERCISE Verify (7.21), recalling that $\delta(\mathbf{v} - \mathbf{V}_s) = \delta(v_x - V_{sx})\,\delta(v_y - V_{sy})\,\delta(v_z - V_{sz})$.

Another case is where the distribution function $f_s(\mathbf{x},\mathbf{v},t)$ is isotropic at each point in space. Then with

$$\langle \mathbf{v}\,\mathbf{v} \rangle = \left\langle \begin{pmatrix} v_x v_x & v_x v_y & v_x v_z \\ v_y v_x & v_y v_y & v_y v_z \\ v_z v_x & v_z v_y & v_z v_z \end{pmatrix} \right\rangle \tag{7.22}$$

we have $\mathbf{V}_s = 0$, and upon taking the average, all of the off-diagonal terms in (7.22) vanish.

EXERCISE Prove this case.

The diagonal terms are $\langle v_x^2 \rangle = \langle v_y^2 \rangle = \langle v_z^2 \rangle = v_s^2$ where $v_s(\mathbf{x})$ is the thermal speed. Equation (7.22) becomes

$$\langle \mathbf{v}\,\mathbf{v} \rangle = v_s^2(\mathbf{x})\, \overleftrightarrow{\mathbf{I}} \tag{7.23}$$

where $\overleftrightarrow{\mathbf{I}}$ is the unit tensor, and we take into account the possibility that the temperature ($T_s \equiv m_s v_s^2$) is spatially dependent.

The second term in (7.20) than takes the form

$$\nabla \cdot (n_s \langle \mathbf{v}\,\mathbf{v} \rangle) = \nabla \cdot (n_s v_s^2 \overleftrightarrow{\mathbf{I}}) = \nabla (n_s v_s^2) = \nabla P_s / m_s \tag{7.24}$$

where the pressure $P_s \equiv n_s m_s v_s^2 = n_s T_s$.

More generally, we might have a distribution that has a net velocity $\mathbf{V}_s$ in a certain direction, and has an isotropic velocity distribution in the frame moving with velocity $\mathbf{V}_s$; therefore $\langle (v_x - V_{sx})^2 \rangle = \langle (v_y - V_{sy})^2 \rangle = \langle (v_z - V_{sz})^2 \rangle = v_s^2 = P_s / m_s n_s$. Then

$$\langle \mathbf{v}\,\mathbf{v} \rangle = \mathbf{V}_s \mathbf{V}_s + \frac{P_s}{m_s n_s} \overleftrightarrow{\mathbf{I}} \tag{7.25}$$

and

$$\begin{aligned} \nabla \cdot (n_s \langle \mathbf{v}\,\mathbf{v} \rangle) &= \nabla \cdot (n_s \mathbf{V}_s \mathbf{V}_s) + \frac{1}{m_s} \nabla P_s \\ &= (\nabla \cdot \mathbf{V}_s)(n_s \mathbf{V}_s) + (\mathbf{V}_s \cdot \nabla)(n_s \mathbf{V}_s) + \frac{1}{m_s} \nabla P_s \end{aligned} \tag{7.26}$$

EXERCISE By writing out components, or by any other method, justify the manipulations in (7.26).

The force equation (7.20) becomes, multiplying by m_s,

$$\begin{aligned} \partial_t (m_s n_s \mathbf{V}_s) &+ (\nabla \cdot \mathbf{V}_s)(m_s n_s \mathbf{V}_s) + (\mathbf{V}_s \cdot \nabla)(m_s n_s \mathbf{V}_s) \\ &= -\nabla P_s + q_s n_s \left(\mathbf{E} + \frac{1}{c} \mathbf{V}_s \times \mathbf{B} \right) \end{aligned} \tag{7.27}$$

Equation (7.27) can be simplified by subtracting the continuity equation (7.15), multiplied by $\mathbf{V}_s$, from the left side. We find

$$m_s n_s \, \partial_t \mathbf{V}_s + m_s n_s (\mathbf{V}_s \cdot \nabla)\mathbf{V}_s = -\nabla P_s + q_s n_s \left(\mathbf{E} + \frac{1}{c} \mathbf{V}_s \times \mathbf{B} \right) \quad (7.28)$$

in agreement with (7.5).

When combined with Maxwell's equations, and when some means is found for describing the pressure P_s, Eqs. (7.15) and (7.28) provide a complete description of fluid plasma behavior. We shall find several different means for describing the pressure. For variations in one direction only, the pressure is $P_s = n_s T_s = n_s m_s \langle v^2 \rangle$ where we evaluate $\langle v^2 \rangle$ along the direction of variation. If we are dealing with a motion, a wave for example, which is slowly varying compared to the equilibration time of species s, we might have isothermal behavior so

$$\nabla P_s = \nabla(n_s T_s) = T_s \nabla n_s \quad (7.29)$$

On the other hand, a rapidly varying compression may involve adiabatic motion, so

$$\nabla P_s = \nabla(n_s T_s) = n_s \nabla T_s + T_s \nabla n_s = \gamma_s T_s \nabla n_s \quad (7.30)$$

where $\gamma_s = (2 + D)/D$ is the so-called "ratio of specific heats." Here, D is the number of dimensions that share in the increased temperature, and it is assumed that the motion in (7.30) involves only small departures of the density and temperature from their unperturbed values n_s, T_s. In succeeding sections on wave motions in fluid plasma, we shall apply these ideas.

EXERCISE Verify (7.30) using the basic ideas of adiabatic compressions.

7.3 LANGMUIR WAVES

Now that we have developed a complete description of a plasma, in the form of the two-fluid equations, what can we do with it? The first thing we can do is to study the various kinds of waves that can propagate through a plasma. Waves are very important. They propagate energy from one part of a plasma to another. They send information out of the plasma that enables an external observer to know what is occurring inside. They can become unstable, growing as they propagate, to such large amplitudes that they disrupt the confinement of a plasma.

Our first example of a wave is a very simple one, the electron plasma wave, or Langmuir wave (also called a space-charge wave). Suppose the ions are infinitely massive, so that they do not contribute to the motion, but have a fixed particle density n_0 and a fixed charge density en_0. Then we need only three equations to describe the electron motion: the electron continuity equation, the electron force equation, and Poisson's equation. These are (in one dimension, with $\mathbf{B}_0 = 0$)

$$\partial_t n_e + \partial_x (n_e V_e) = 0 \quad (7.31)$$

$$m_e n_e (\partial_t V_e + V_e \, \partial_x V_e) = -\partial_x P_e - e n_e E \quad (7.32)$$

and

$$\partial_x E = 4\pi e(n_0 - n_e) \tag{7.33}$$

We seek solutions to (7.31) to (7.33) in the form of small amplitude waves, where the electric field has a sinusoidal spatial variation. In order to look for such small amplitude waves, we first linearize (7.31) to (7.33). With

$$n_e = n_0 + n_1 \tag{7.34}$$

$$E = E_1 \tag{7.35}$$

and

$$V_e = v_1 \tag{7.36}$$

we first neglect the pressure P_e, assuming that the electrons are cold. Then the only zeroth order contribution from (7.31) to (7.33) is

$$\partial_t n_0 = 0 \tag{7.37}$$

which is trivially satisfied. The first order terms are

$$\partial_t n_1 + n_0\, \partial_x v_1 = 0 \tag{7.38}$$

$$m_e n_0\, \partial_t v_1 = -en_0 E_1 \tag{7.39}$$

and

$$\partial_x E_1 = -4\pi e n_1 \tag{7.40}$$

These equations are now linear, and we may look for wave solutions in which each variable has the form $\cos(kx - \omega t + \theta)$,

$$E_1(x,t) = \tilde{E}_1 \cos(kx - \omega t) \tag{7.41}$$

$$n_1(x,t) = \tilde{n}_1 \cos(kx - \omega t + \theta_n) \tag{7.42}$$

and

$$v_1(x,t) = \tilde{v}_1 \cos(kx - \omega t + \theta_v) \tag{7.43}$$

where $\tilde{E}_1$, $\tilde{n}_1$, and $\tilde{v}_1$ are real constants and θ_n, θ_v are possible phase shifts. It turns out that it is very awkward to use sin's and cos's, and it is very convenient to use solutions that vary as $\exp(-i\omega t + ikx)$. We may do this by noting that if (7.41) to (7.43) is a solution of the linearized equations (7.38) to (7.40), then the expression obtained by giving each of (7.41) to (7.43) a phase shift of $\pi/2$ will also be a solution, where sin replaces cos in (7.41) to (7.43). Any linear combination of these two sets of solutions will also be a solution; in particular, [cos solution] + i[sin solution] is a solution, of the form

$$E_1(x,t) = \tilde{E}_1 \exp(-i\omega t + ikx) \tag{7.44}$$

$$n_1(x,t) = \tilde{n}_1 \exp(i\theta_n) \exp(-i\omega t + ikx) \tag{7.45}$$

and

$$v_1(x,t) = \tilde{v}_1 \exp(i\theta_v) \exp(-i\omega t + ikx) \tag{7.46}$$

If we next absorb the phase factors $\exp(i\theta_n)$ and $\exp(i\theta_v)$ into new complex constants $\tilde{\tilde{n}}_1 = \tilde{n}_1 \exp(i\theta_n)$, $\tilde{\tilde{v}}_1 = \tilde{v}_1 \exp(i\theta_v)$, we have

$$E_1(x,t) = \tilde{E}_1 \exp(-i\omega t + ikx) \tag{7.47}$$

$$n_1(x,t) = \tilde{\tilde{n}}_1 \exp(-i\omega t + ikx) \tag{7.48}$$

and

$$v_1(x,t) = \tilde{\tilde{v}}_1 \exp(-i\omega t + ikx) \tag{7.49}$$

After we have obtained the solutions (7.47) to (7.49), we can add them to their complex conjugate to obtain the physically relevant real solutions, if desired.

EXERCISE Why is the complex conjugate of (7.47) to (7.49) also a solution?

Inserting the assumed wave solution (7.47) to (7.49) into the linearized equations (7.38) to (7.40), one obtains

$$-i\omega\tilde{\tilde{n}}_1 + ikn_0\tilde{\tilde{v}}_1 = 0 \tag{7.50}$$

$$-i\omega m_e n_0 \tilde{\tilde{v}}_1 = -en_0\tilde{E}_1 \tag{7.51}$$

and

$$ik\tilde{E}_1 = -4\pi e\tilde{\tilde{n}}_1 \tag{7.52}$$

where we have divided each side by $\exp(-i\omega t + ikx)$. Solving (7.50) for $\tilde{\tilde{v}}_1$ in terms of $\tilde{\tilde{n}}_1$, and solving (7.52) for $\tilde{\tilde{n}}_1$ in terms of $\tilde{E}_1$, and inserting in (7.51), we find

$$\frac{\omega^2 m_e \tilde{E}_1}{4\pi n_0 e^2} = \tilde{E}_1 \tag{7.53}$$

Upon dividing by the constant $\tilde{E}_1$, we see that

$$\omega^2 = \frac{4\pi n_0 e^2}{m_e} = \omega_e^2 \tag{7.54}$$

or

$$\boxed{\omega = \pm\,\omega_e} \tag{7.55}$$

for our cold plasma oscillations; thus the wave frequency is just our old friend the electron plasma frequency.

EXERCISE If we had kept the ion component with mass m_i, can you guess what the wave frequency would be?

Thus, we have shown that any expression of the form (7.44) to (7.46) is a solution of (7.38) to (7.40), provided the frequency is given by (7.55). Note that this is true for arbitrary wave number k.

EXERCISE Using (7.50) to (7.52), determine the phase shifts between E_1, n_1, and v_1. Choosing a value of k, add (7.47) to (7.49) to its complex conjugate to obtain real solutions, and sketch these solutions with their appropriate phase shifts.

The expression (7.55) for ω is called a dispersion relation, because it is supposed to represent the relation between frequency ω and wave number k. In this case the dispersion relation is trivial, because it does not depend on k. The group velocity

$$V_g \equiv \frac{\partial \omega}{\partial k} = 0 \tag{7.56}$$

does not depend on wave number k, so these waves are dispersionless. An initial wave packet will not propagate, but merely oscillates forever at frequency $\omega = \omega_e$. Thus, the physics of these waves is as simple as the physics of the oscillating slabs used to derive the plasma frequency in Chapter 1.

It is always useful to sketch dispersion relations. In this case, the sketch of (7.55) is simple, consisting of two straight lines at $\omega = \pm\, \omega_e$, as shown in Fig. 7.1.

The most serious assumption in our derivation of the cold plasma waves is the neglect of the pressure term in (7.32). Let us repeat the derivation, including the pressure P_e. We have $\nabla P_e = \nabla(n_e T_e) = \nabla[(n_0 + n_1)(T_0 + T_1)] = n_0 \nabla T_1 + T_0 \nabla n_1$, keeping only first order terms. Now in order to relate the first order temperature change T_1 in the wave to the first order density change n_1, we must go outside the fluid theory. We consider long wavelength waves, such that a typical electron travels only a fraction of a wavelength λ in one wave period; then the compression of the wave will be an adiabatic one. Thus, the assumption $v_e \omega^{-1} << \lambda$, or

$$\omega / k >> v_e \tag{7.57}$$

leads us to consider adiabatic compressions. If we further assume that the collision frequency is small,

$$\nu_c << \omega \tag{7.58}$$

then the change in temperature during the compression along the direction of wave propagation will not be transmitted to the other two directions. We conclude that our compression is a *one-dimensional adiabatic* compression, which means that

$$\nabla P_e = n_0 \nabla T_1 + T_0 \nabla n_1 = 3 T_0 \nabla n_1 \tag{7.59}$$

by the expression below (7.30).

With the expression (7.59) for ∇P_e, our previous derivation goes through with the addition of one term; thus (7.50) to (7.52) are replaced by

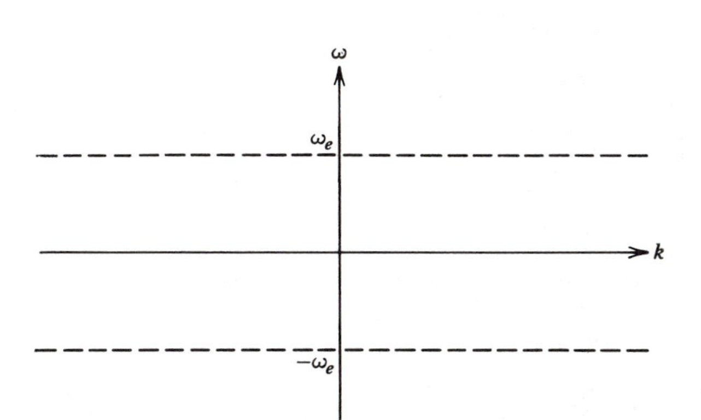

Fig. 7.1 Dispersion relation for electron plasma waves when the pressure is ignored.

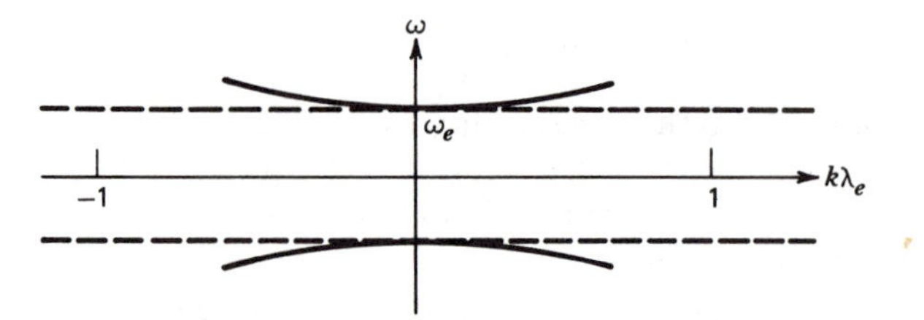

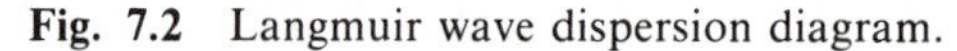

Fig. 7.2 Langmuir wave dispersion diagram.

$$-i\omega\tilde{\tilde{n}}_1 + ikn_0\tilde{\tilde{v}}_1 = 0 \tag{7.60}$$

$$-i\omega m_e n_0 \tilde{\tilde{v}}_1 = -en_0\tilde{E}_1 - ik3T_0\tilde{\tilde{n}}_1 \tag{7.61}$$

$$ik\tilde{E}_1 = -4\pi e\tilde{\tilde{n}}_1 \tag{7.62}$$

Solving (7.60) for $\tilde{\tilde{v}}_1$, solving (7.62) for $\tilde{\tilde{n}}_1$, and inserting in (7.61), one finds

$$(\omega^2 - \omega_e^2 - 3k^2T_0/m_e)\tilde{E}_1 = 0 \tag{7.63}$$

or

$$\boxed{\omega^2 = \omega_e^2 + 3k^2T_0/m_e = \omega_e^2 + 3k^2v_e^2} \tag{7.64}$$

which is the famous *Langmuir wave* dispersion relation (see Eq. 6.28).

Since we have used the assumption (7.57), the dispersion relation (7.64) has a limited range of validity, restricted to

$$v_e << \frac{\omega}{k} \approx \frac{\omega_e}{k} \tag{7.65}$$

or

$$k\lambda_e << 1 \tag{7.66}$$

Thus, another useful form of (7.64) is

$$\omega = \pm(\omega_e^2 + 3k^2v_e^2)^{1/2} \cong \pm\, \omega_e \left(1 + \frac{3}{2}\, k^2\lambda_e^2\right) \tag{7.67}$$

A graph of the dispersion relation (7.67) is shown in Fig. 7.2. The Langmuir wave is seen to have dispersion, since the group velocity is

$$V_g \equiv \frac{d\omega}{dk} = 3(\mathrm{k}\lambda_e)v_e \tag{7.68}$$

and this depends on wave number.

EXERCISE Returning to (7.60) to (7.62), find the density perturbation and velocity perturbation associated with the wave. Show that for $k \to 0$ we regain the cold plasma wave, while for $k > 0$ the pressure acts as an additional restoring force which gives the wave a higher frequency.

7.4 DIELECTRIC FUNCTION

It is often useful to draw an analogy between a plasma and a dielectric medium. Recall that in ordinary dielectric theory, we are able to replace Poisson's equation for charges in a vacuum,

$$\nabla \cdot \mathbf{E} = 4\pi\rho \tag{7.69}$$

by

$$\nabla \cdot \mathbf{D} = 0 \tag{7.70}$$

where

$$\mathbf{D} = \overleftrightarrow{\epsilon} \cdot \mathbf{E} \tag{7.71}$$

Here $\overleftrightarrow{\epsilon}$ is the dielectric tensor, the properties of the medium have been incorporated into the displacement **D**, and we assume there are no additional free charges. This same operation can of course be performed for a plasma as well as for a dielectric medium. In one dimension, and assuming plane wave fields, Eq. (7.69) is

$$ikE = 4\pi\rho \tag{7.72}$$

so that if we can write ρ in terms of E it will be easy to identify the dielectric function:

$$ik\left(E - \frac{4\pi\rho}{ik}\right) \equiv ik\epsilon E = 0 \tag{7.73}$$

For cold plasma waves this has been done in (7.50) to (7.52), which are easily written in the form

$$ik\left(1 - \frac{\omega_e^2}{\omega^2}\right) E = 0 \tag{7.74}$$

so that the dielectric function is

$$\boxed{\epsilon(\omega) = 1 - \frac{\omega_e^2}{\omega^2}} \tag{7.75}$$

and we notice that the dispersion relation (7.55) is precisely equivalent to equating $\epsilon(\omega)$ to zero,

$$\epsilon(\omega) = 0 \Rightarrow \omega = \pm\, \omega_e \tag{7.76}$$

Thus, we see that *the normal modes of a plasma are obtained from the zeros of the dielectric function* (see Chapter 6).

EXERCISE Verify (7.74).

In a similar fashion, because (7.60) to (7.62) for Langmuir waves can be written in the form

$$ik\left[1 - \frac{\omega_e^2}{\omega^2 - 3k^2 v_e^2}\right] E = 0 \tag{7.77}$$

we identify the dielectric function

$$\epsilon(\omega,k) = 1 - \frac{\omega_e^2}{\omega^2 - 3k^2 v_e^2} \tag{7.78}$$

the zeros of which yield the normal modes (7.67):

$$\epsilon(\omega,k) = 0 \Rightarrow \omega(k) \tag{7.79}$$

$$\omega(k) = \omega_e(1 + 3k^2\lambda_e^2)^{1/2} \tag{7.80}$$

The dielectric function has been studied in more detail in Chapter 6.

7.5 ION PLASMA WAVES

In Section 7.3 we studied high frequency electron plasma oscillations, with frequency ω near the electron plasma frequency ω_e. For these waves, the ion motion is negligible and irrelevant. Here, we consider low frequency waves, $\omega \lesssim \omega_i$, where the ion motion dominates the wave physics. (In an unmagnetized plasma, the words "low frequency" refer to $\omega \lesssim \omega_i$ while "high frequency" refers to $\omega \gtrsim \omega_e$. In a magnetized plasma, we often have the ordering $\Omega_i << \omega_i << \omega_e < |\Omega_e|$. (Why?) Then "low frequency" means $\omega \lesssim \Omega_i$, and other frequencies are called low or high depending on what they are being compared to.)

To include both electron and ion physics, we need five fluid equations: electron and ion continuity, electron and ion force, and Poisson's equation. These are, from (7.1) and (7.5),

$$\partial_t n_e + \partial_x (n_e V_e) = 0 \tag{7.81}$$

$$m_e n_e \, \partial_t V_e + m_e n_e V_e \, \partial_x V_e = -\partial_x P_e - e n_e E \tag{7.82}$$

$$\partial_t n_i + \partial_x (n_i V_i) = 0 \tag{7.83}$$

$$m_i n_i \, \partial_t V_i + m_i n_i V_i \, \partial_x V_i = -\partial_x P_i + e n_i E \tag{7.84}$$

and

$$\partial_x E = 4\pi e (n_i - n_e) \tag{7.85}$$

This set of equations is not as bad as it looks. First, because we intend to linearize, all $V \, \partial_x V$ terms disappear. Second, all pressure terms can be written $\partial_x P_{e,i} = \gamma_{e,i} T_{e,i} \, \partial_x n_{e,i}$, where temporarily we do not specify the coefficients $\gamma_{e,i}$, and $T_{e,i}$ are the unperturbed electron and ion temperatures. Then linearizing (7.81) to (7.85) with $n_s = n_0 + n_{s1}$, and all other quantities first order, we obtain

$$\partial_t n_{e1} + n_0 \, \partial_x V_e = 0 \tag{7.86}$$

$$m_e n_0 \, \partial_t V_e = - \gamma_e T_e \, \partial_x n_{e1} - e n_0 E \tag{7.87}$$

$$\partial_t n_{i1} + n_0 \, \partial_x V_i = 0 \tag{7.88}$$

$$m_i n_0 \, \partial_t V_i = - \gamma_i T_i \, \partial_x n_{i1} + e n_0 E \tag{7.89}$$

and

$$\partial_x E = 4\pi e (n_{i1} - n_{e1}) \tag{7.90}$$

Before solving (7.86) to (7.90), let us guess the properties of the wave we are looking for. The ions will have a sinusoidal density perturbation as shown in Fig. 7.3. Since the frequency is very low, the electrons see an almost static ion density perturbation, and they will try to obtain the same density in order to prevent huge electric fields. However, since the electrons are flying about very fast, the attempt

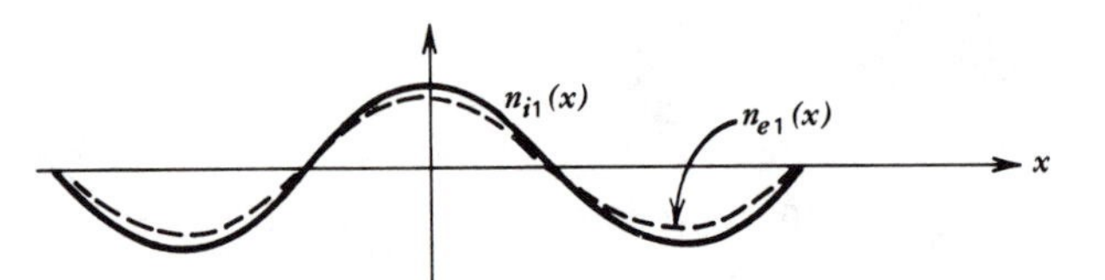

Fig. 7.3 Electron and ion density perturbations in an ion plasma wave.

to exactly cancel the ion charge distribution will not totally succeed; rather, the electrons try to smear themselves out more smoothly. Thus, the electron density perturbation is slightly smaller than the ion density perturbation, and the resulting density difference produces the electric field of the waves.

EXERCISE Draw the electric field produced by the densities in Fig. 7.3.

We have seen the tendency for this cancellation of electron and ion charge before, in Problem 1.4, "Plasma in a gravitational field." This important property is called *quasineutrality*, and it is found in almost all low frequency plasma behavior.

This discussion encourages us to look for a solution of (7.86) to (7.90) with $n_{i1} \approx n_{e1}$. We therefore ignore Poisson's equation; and we find from (7.86) and (7.88) that we must have $V_e \approx V_i$. We eliminate the electric field by adding (7.87) and (7.89) to obtain

$$(m_e + m_i)n_0 \frac{\partial V_e}{\partial t} = -(\gamma_e T_e + \gamma_i T_i) \frac{\partial n_{e1}}{\partial x} \tag{7.91}$$

Next, we eliminate the velocity V_e by taking the time derivative of (7.86) and inserting the result in the spatial derivative of (7.91), to obtain

$$(m_e + m_i) \frac{\partial^2 n_{e1}}{\partial t^2} = (\gamma_e T_e + \gamma_i T_i) \frac{\partial^2 n_{e1}}{\partial x^2} \tag{7.92}$$

or, neglecting m_e compared to m_i,

$$\frac{\partial^2 n_{e1}}{\partial t^2} = c_s^2 \frac{\partial^2 n_{e1}}{\partial x^2} \tag{7.93}$$

where we have defined the sound speed

$$c_s \equiv \left(\frac{\gamma_e T_e + \gamma_i T_i}{m_i} \right)^{1/2} \tag{7.94}$$

Assuming a plane wave solution of the form $\exp(-i\omega t + ikx)$, (7.93) yields the *ion-acoustic dispersion relation*

$$\boxed{\omega^2 = k^2 c_s^2} \tag{7.95}$$

The name ion-acoustic arises from the similarity between the dispersion relation (7.95) and the equivalent relation for sound waves traveling through a gas.

It is difficult to determine the regime of validity of (7.95) from the foregoing discussion. We do know, however, that we have neglected the difference $(n_{i1} - n_{e1})$, which by Poisson's equation is proportional to $\partial_x E \sim kE$. We thus expect that (7.95) is limited to small k; we shall see in a moment that this is so.

What shall we take for γ_e, γ_i? In practice, this depends on the region of density, temperature, and wave number in which we are working. It may be the case that the ion motions are adiabatic in one dimension, so $\gamma_i = 3$. It may also be that collisions are important enough to redistribute the wave compressional energy in three dimensions, so that $\gamma_i = (D + 2)/D = 5/3$ for adiabatic compressions in three dimensions. As for the electrons, it is the case that a typical electron travels many wavelengths in one wave period; that is, the distance traveled in one period $v_e/\omega \sim v_e/kc_s >> k^{-1}$, since

$$\frac{v_e}{c_s} \sim \left(\frac{T_e}{m_e} \frac{m_i}{\gamma_e T_e + \gamma_i T_i}\right)^{1/2} >> 1$$

Thus, the electrons are communicating over many wavelengths during one wave period so that they remain isothermal; we therefore choose the iosthermal $\gamma_e = 1$. When $T_e >> T_i$, we then have the very simple and easy to remember formula

$$\boxed{c_s = (T_e/m_i)^{1/2}} \tag{7.96}$$

In other words, the sound speed is the thermal speed that the ions would have if they had the electron temperature.

Let us now return to a more exact solution of (7.86) to (7.90) that does not assume quasineutrality. Because the electron mass is very small, we ignore the electron "inertia" term on the left side of (7.87), upon which (7.87) yields

$$\frac{\partial n_{e1}}{\partial x} = \frac{-en_0}{\gamma_e T_e} E \tag{7.97}$$

For the ions, we eliminate V_i from the spatial derivative of (7.89) by using the time derivative of (7.88), to obtain

$$-m_i \frac{\partial^2 n_{i1}}{\partial t^2} = -\gamma_i T_i \frac{\partial^2 n_{i1}}{\partial x^2} + en_0 \frac{\partial E}{\partial x} \tag{7.98}$$

Looking for plane wave solutions to (7.90), (7.97), and (7.98) (or Fourier transforming, if you like), we find

$$ikE = 4\pi e(n_{i1} - n_{e1}) \tag{7.99}$$

$$ikn_{e1} = \frac{-en_0}{\gamma_e T_e} E \tag{7.100}$$

and

$$\left(\omega^2 - k^2 \frac{\gamma_i T_i}{m_i}\right) n_{i1} = \frac{en_0}{m_i} ikE \tag{7.101}$$

Inserting (7.100) and (7.101) into (7.99) we find

$$ik \left[1 - \frac{\omega_i^2}{\omega^2 - k^2\gamma_i T_i/m_i} + \frac{\omega_e^2}{k^2\gamma_e T_e/m_e}\right] E = 0 \tag{7.102}$$

from which we identify the dielectric function

$$\epsilon(\omega,k) = 1 - \frac{\omega_i^2}{\omega^2 - k^2\gamma_i T_i/m_i} + \frac{\omega_e^2}{k^2\gamma_e T_e/m_e} = 0 \tag{7.103}$$

the zeros of which yield the dispersion relation $\omega(k)$. Solving (7.103), we find

$$\boxed{\omega^2 = k^2 \frac{\gamma_i T_i}{m_i} + \frac{k^2\gamma_e T_e/m_i}{1 + \gamma_e k^2 \lambda_e^2}} \tag{7.104}$$

which is the general dispersion relation for ion plasma waves.

EXERCISE Solve (7.103) to obtain (7.104).

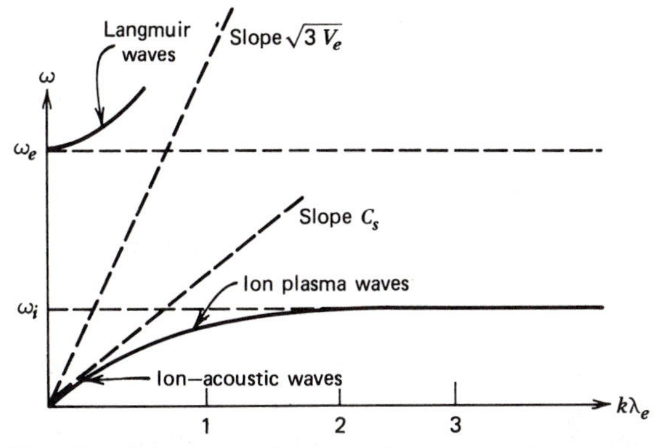

Fig. 7.4 Dispersion diagrams for electron plasma waves and ion plasma waves ($T_i = 0$).

In the small $k\lambda_e$ limit, we regain the ion-acoustic dispersion relation (7.95). We further expect that (7.104) is only valid when the second term on the right is larger than the first. If this were not so, we would have $\omega/k \sim (T_i/m_i)^{1/2} \sim v_i$, which would mean that many ions would have speeds of the same order as the wave phase speed. When this is the case, we do not expect fluid theory to be valid; rather, we must use the Vlasov equation to properly treat those particles which can interact resonantly with the wave.

Another interesting limit of (7.104) is reached when $T_i \rightarrow 0$ and $k\lambda_e >> 1$; we then find

$$\omega^2 \approx \omega_i^2 \tag{7.105}$$

which are ion plasma waves oscillating at the ion plasma frequency. Because the wavelength is short compared to the electron Debye length, the electrons are incapable of shielding, and we have ions oscillating in a uniform background of negative charge. This is quite analogous to our cold electron plasma oscillations at $\omega = \omega_e$, which are electrons oscillating in a uniform background of positive charge.

We can now draw the dispersion diagrams of electron plasma waves and ion plasma waves on the same diagram; we do this schematically for the case $T_i \rightarrow 0$ in Fig. 7.4. Note that the dispersive ($k^2\lambda_e^2$) term in the denominator of (7.104) becomes more important at larger k, leading to a transition from the acoustic behavior at small k (7.95) to the oscillations at $\omega = \omega_i$ (7.105) at large k.

7.6 ELECTROMAGNETIC WAVES

The only other waves in unmagnetized homogeneous plasma are electromagnetic waves. We shall find that these waves are high frequency, $\omega \geq \omega_e$, so that we can ignore ion motion. We shall further find that these waves are transverse, $\mathbf{k} \cdot \mathbf{E} = 0$ and $\mathbf{k} \cdot \mathbf{B} = 0$, so that we can ignore Poisson's equation and the $\nabla \cdot \mathbf{B} = 0$ equation. The fluid equations that we then need are

$$\nabla \times \mathbf{E} = -\frac{1}{c}\,\partial_t \mathbf{B} \tag{7.106}$$

$$\nabla \times \mathbf{B} = \frac{4\pi}{c}\,\mathbf{J} + \frac{1}{c}\,\partial_t \mathbf{E} \tag{7.107}$$

$$\mathbf{J} = -\,en_e\mathbf{V}_e \tag{7.108}$$

$$m_e n_e \partial_t \mathbf{V}_e + m_e n_e(\mathbf{V}_e \cdot \nabla)\mathbf{V}_e = -\,\nabla P_e - en_e\mathbf{E} - \frac{en_e}{c}\,\mathbf{V}_e \times \mathbf{B} \tag{7.109}$$

$$\partial_t n_e + \nabla \cdot (n_e \mathbf{V}_e) = 0 \tag{7.110}$$

We have assumed that the wave is transverse, $\mathbf{k} \cdot \mathbf{E} = 0$. If $\mathbf{k}$ is in the $\hat{x}$-direction, we may choose $\mathbf{E}$ in the $\hat{y}$-direction; then by Faraday's law (7.106), we have $\mathbf{B}$ along $\hat{z}$ (Fig. 7.5). We next assume that there is no zero order component of $\mathbf{V}_e$, and that we can neglect the $\mathbf{V}_e \times \mathbf{B}$ force; this assumption must be checked at the end of the calculation. We look for solutions that have $\mathbf{k} \cdot \mathbf{V}_e = 0$; then (7.110) predicts $\partial n_e/\partial t = 0$, so $n_e = n_0$ everywhere and we can ignore ∇P_e in (7.109). With no further assumption, the $(\mathbf{V}_e \cdot \nabla)\mathbf{V}_e$ term in (7.109) also vanishes. We have left the equations

$$\nabla \times \mathbf{E} = -\frac{1}{c}\,\partial_t \mathbf{B} \tag{7.111}$$

$$\nabla \times \mathbf{B} = -\frac{4\pi e}{c}\,n_0\mathbf{V}_e + \frac{1}{c}\,\partial_t \mathbf{E} \tag{7.112}$$

$$m_e n_0\,\partial_t \mathbf{V}_e = -\,en_0\mathbf{E} \tag{7.113}$$

Taking the time derivative of (7.112) and the curl of (7.111), we eliminate $\partial_t \mathbf{V}_e$ and $\mathbf{B}$ in (7.112) to obtain

$$-\,c\nabla \times (\nabla \times \mathbf{E}) = \frac{4\pi n_0 e^2}{m_e c}\,\mathbf{E} + \frac{1}{c}\,\frac{\partial^2 \mathbf{E}}{\partial t^2} \tag{7.114}$$

or taking a plane wave solution, $\nabla \times \nabla \times \rightarrow k^2$, we get

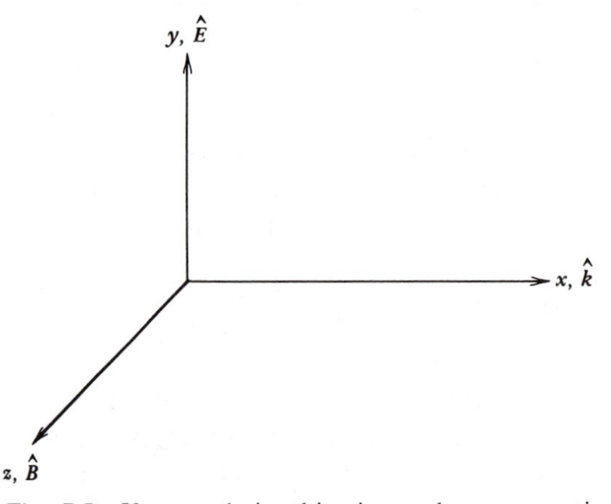

Fig. 7.5 Vector relationships in an electromagnetic wave in unmagnetized plasma.

$$(\omega^2 - k^2c^2 - \omega_e^2)\mathbf{E} = 0 \tag{7.115}$$

which yields the *electromagnetic dispersion relation*

$$\boxed{\omega^2 = \omega_e^2 + k^2c^2} \tag{7.116}$$

Letting the plasma density approach zero we regain the free space light waves with $\omega = kc$.

Note the similarity between the electromagnetic dispersion relation and the Langmuir dispersion relation, where c^2 is replaced by $3v_e^2$. On the same dispersion diagram, the two branches look as shown in Fig. 7.6.

Recall that in the theory of optical media (air, water, etc.) it is useful to define an index of refraction

$$n \equiv \frac{ck}{\omega} \tag{7.117}$$

for light traveling through the medium. From (7.116) we see that in a plasma the index of refraction is

$$\boxed{n \equiv \frac{\sqrt{\omega^2 - \omega_e^2}}{\omega} = \sqrt{1 - \omega_e^2/\omega^2}} \tag{7.118}$$

According to (7.118), the index of refraction becomes imaginary when $\omega < \omega_e$. Thus, for real ω one obtains imaginary k, corresponding to evanescence. The result of this evanescence is that when an electromagnetic wave impinges on an inhomogeneous plasma, it reflects at the point where $\omega = \omega_e$, called the position of *critical density*. This effect is important in laser fusion and in the interaction of radio waves with the ionosphere (Fig. 7.7).

We can now reconsider our neglect of the $(q/c)\mathbf{V}_e \times \mathbf{B}$ force in this derivation. By Faraday's law, $B \sim (ck/\omega)E \sim nE$, but $n < 1$ always by (7.118), so $B < E$. Thus, $(q/c)|\mathbf{V}_e \times \mathbf{B}| < q|V_e/c||\mathbf{E}|$ so that the magnetic force will always be

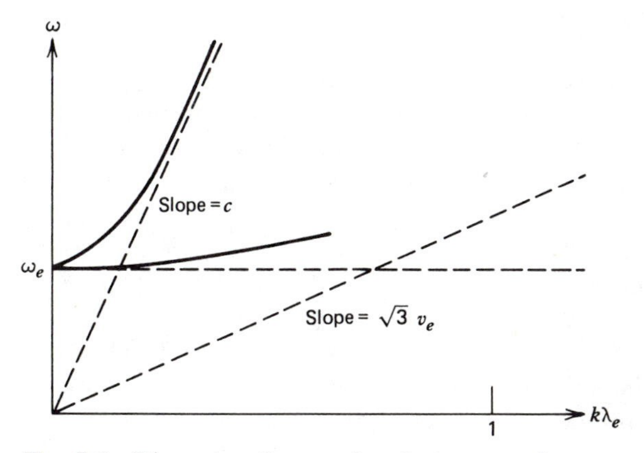

Fig. 7.6 Dispersion diagram for electromagnetic waves and Langmuir waves in unmagnetized plasma.

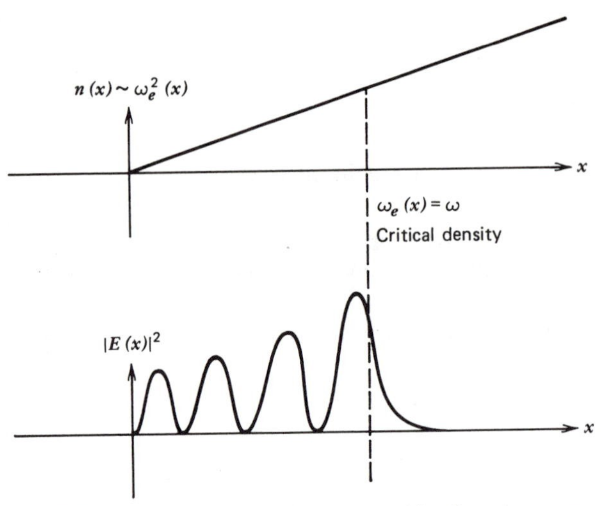

Fig. 7.7 Sketch of the standing wave Airy function pattern for an electromagnetic wave incident from the left on an inhomogeneous plasma.

negligible compared to the electric force as long as the motion produced by the electric field, V_e, is always nonrelativistic.

EXERCISE Sketch the group speed $d\omega/dk$ and the phase speed ω/k, both vs. k, for electromagnetic waves.

This completes our discussion of linear fluid waves for uniform unmagnetized plasma. We have found only three different waves for a given wave number k: the ion plasma wave or ion-acoustic wave, the electron plasma wave or Langmuir wave, and the electromagnetic wave. The addition of inhomogeneity or a magnetic field will greatly multiply the number of linear modes, as we shall see shortly.

7.7 UPPER HYBRID WAVES

Up to this point we have considered waves in unmagnetized plasma. We now wish to consider linear waves in magnetized plasma. In general, there will be two important directions for any wave motion, the direction of the external magnetic field $\hat{B}_0$, and the direction of the wave number $\hat{k}$. Since we are looking for linear waves, there will be two important wave quantities, the first order electric field $\mathbf{E}_1$, and the first order magnetic field $\mathbf{B}_1$. There are six terms that are used to describe relations among the four quantities $\hat{B}_0$, $\hat{k}$, $\mathbf{E}_1$, and $\mathbf{B}_1$. If $\hat{k}$ is along $\hat{B}_0$, $\hat{k} \cdot \hat{B}_0 = 1$, we call the wave *parallel*; if $\hat{k} \cdot \hat{B}_0 = 0$ the wave is *perpendicular*. If $\hat{k} \cdot \hat{E}_1 = 1$ the wave is *longitudinal*, while if $\hat{k} \cdot \hat{E}_1 = 0$, the wave is *transverse*. When $\mathbf{B}_1 = 0$, the wave is *electrostatic*, while if $\mathbf{B}_1 \neq 0$ the wave is *electromagnetic*. Of course, not all waves deserve one or another of these terms; that is, a wave with $\mathbf{k}$ at a 45° angle to $\mathbf{B}_0$ is neither parallel nor perpendicular. We can often relate the last two terms using Faraday's law $\nabla \times \mathbf{E}_1 = -(1/c)\,\partial_t \mathbf{B}_1$, or $\mathbf{k} \times \mathbf{E}_1 = (\omega/c)\mathbf{B}_1$. Since for *longitudinal* waves, $\mathbf{k} \times \mathbf{E}_1 = 0$, thus the wave is *electrostatic*, and vice versa. Similarly, all *transverse* waves are *electromagnetic* (but not vice versa; why not?).

Let us first look for high frequency waves traveling across the magnetic field, with $\hat{E}_1 \cdot \hat{k} = 1$; these perpendicular longitudinal electrostatic waves are known as *upper hybrid waves*. We take $T_e = T_i = 0$ and $m_i \to \infty$; therefore the ions do not move, but form a fixed background of positive charge. The relevant equations are then

$$\partial_t n_e + \nabla \cdot (n_e \mathbf{V}_e) = 0 \tag{7.119}$$

$$m_e n_e \partial_t \mathbf{V}_e + m_e n_e \mathbf{V}_e \cdot \nabla \mathbf{V}_e = -en_e \mathbf{E}_1 - \frac{en_e}{c} \mathbf{V}_e \times \mathbf{B} \tag{7.120}$$

$$\nabla \cdot \mathbf{E}_1 = 4\pi e(n_0 - n_e) \tag{7.121}$$

EXERCISE Why aren't the other Maxwell equations relevant?

With $\mathbf{E}_1 = E_1\hat{x}$, $\mathbf{k} = k\hat{x}$, $\mathbf{B}_0 = B_0\hat{z}$, (7.120) will produce components of $\mathbf{V}_e$ in both the $\hat{x}$ and $\hat{y}$ directions, so $\mathbf{V}_e = (v_x, v_y, 0)$ (Fig. 7.8). Then linearizing (7.119) to (7.121) and looking for plane wave solutions yields

$$-i\omega n_{e1} + ikn_0 v_x = 0 \tag{7.122}$$

$$-i\omega m_e v_x = -eE_1 - \frac{e}{c} v_y B_0 \tag{7.123}$$

$$-i\omega m_e v_y = \frac{e}{c} v_x B_0 \tag{7.124}$$

and

$$ikE_1 = -4\pi e n_{e1} \tag{7.125}$$

Equations 7.122 and 7.125 yield v_x in terms of E_1, and (7.124) yields v_y in terms of v_x; substituting the result into (7.123) yields

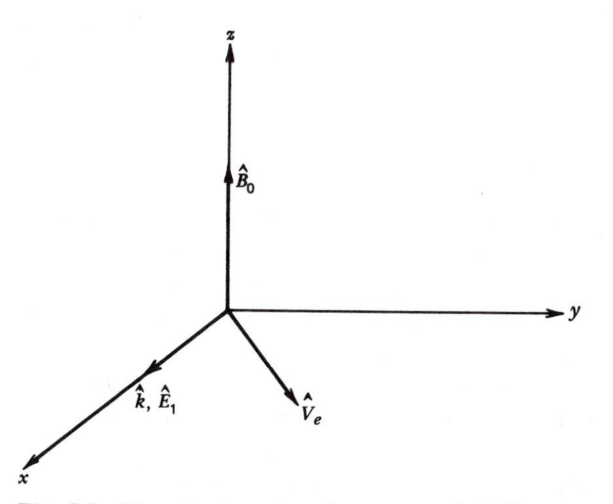

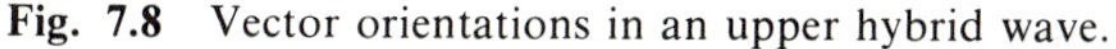

Fig. 7.8 Vector orientations in an upper hybrid wave.

$$\left(\frac{-\omega^2}{\omega_e^2} + 1 + \frac{\Omega_e^2}{\omega_e^2}\right) E_1 = 0 \tag{7.126}$$

from which, upon dividing out E_1, one obtains the dispersion relation

$$\boxed{\omega^2 = \omega_e^2 + \Omega_e^2 \equiv \omega_{UH}^2} \tag{7.127}$$

for *upper hybrid waves*. As in the case of cold plasma waves with no magnetic field [to which (7.127) reduces when $B_0 \to 0$] this frequency is independent of the wave number.

EXERCISE Obtain (7.126) as indicated.

Physically, these waves have a density variation similar to the cold plasma ($B_0 = 0$) waves, but now the electrons are performing a sort of gyromotion in the wave; the electric field tries to make them move in the $\hat{x}$-direction, but the $\mathbf{V}_e \times \mathbf{B}_0$ force produces a velocity in the $\hat{y}$-direction. Suppose we take $v_x = v_{x0} \exp(-i\omega t + ikx)$; then by (7.124) we have

$$v_y = -i\frac{\Omega_e}{\omega} v_x = -i\frac{\Omega_e}{\omega} v_{x0} \exp(-i\omega t + ikx) \tag{7.128}$$

With v_{x0} real, we can obtain a real solution by adding this solution to its complex conjugate and multiplying by 1/2; we find

$$v_x = v_{x0} \cos(kx - \omega t) \tag{7.129}$$

and

$$v_y = \frac{\Omega_e}{\omega} v_{x0} \sin(kx - \omega t) \tag{7.130}$$

By (7.127), $\Omega_e/\omega < 1$ always, so $(v_y)_{\max} < (v_x)_{\max}$ always. Because the Lorentz force acts as an extra restoring force for the wave, the frequency is higher than the cold plasma ($B_0 = 0$) wave. As the magnetic field $B_0 \to 0$ we regain the cold plasma wave, while as the density goes to zero ($\omega_e \to 0$) we have a wave consisting of particles gyrating in the magnetic field, $(v_y)_{\max} = (v_x)_{\max}$.

7.8 ELECTROSTATIC ION WAVES

We next look for low frequency electrostatic waves whose physics is dominated by the ions. Because we are looking for electrostatic waves, the only one of Maxwell's equations needed is Poisson's equation. However, if we assume quasineutrality, $n_{e1} \approx n_{i1}$, then we can avoid using Poisson's equation, and we have only the four fluid equations (electron and ion continuity equations, electron and ion force equation). (Note that here with a low frequency wave, we assume $n_{ei} \approx n_{i1}$ because the electrons have time to adjust to the ions; whereas in the previous section, the high frequency motions are dominated by n_{e1}, and the massive ions do not have time to follow; therefore, $n_{i1} \approx 0$ and $n_{e1} \neq n_{i1}$.) Thus, we need only the four fluid equations, which we linearize immediately, obtaining

$$m_e n_0 \, \partial_t \mathbf{V}_e = -\gamma_e T_e \nabla n_{e1} - e n_0 \mathbf{E} - \frac{e n_0}{c} \mathbf{V}_e \times \mathbf{B}_0 \tag{7.131}$$

$$\partial_t n_{e1} + n_0 \nabla \cdot \mathbf{V}_e = 0 \tag{7.132}$$

$$m_i n_0 \, \partial_t \mathbf{V}_i = -\gamma_i T_i \nabla n_{e1} + e n_0 \mathbf{E} + \frac{e n_0}{c} \mathbf{V}_i \times \mathbf{B}_0 \tag{7.133}$$

$$\partial_t n_{e1} + n_0 \nabla \cdot \mathbf{V}_i = 0 \tag{7.134}$$

where n_{e1}, V_e, V_i, and **E** are all first order quantities. We are interested in waves traveling at any angle to the magnetic field, which we take in the $\hat{z}$-direction. We can take the wave vector **k** to lie in the x-z plane (Fig. 7.9).

Adding (7.131) to (7.133) to eliminate **E**, one obtains

$$-i\omega n_0(m_e \mathbf{V}_e + m_i \mathbf{V}_i) = -i\mathbf{k}(\gamma_e T_e + \gamma_i T_i) n_{e1} + \frac{e n_0}{c} (\mathbf{V}_i - \mathbf{V}_e) \times \mathbf{B}_0 \tag{7.135}$$

Looking for plane wave solutions, taking the dot product of the wave number **k** with (7.135), and inserting $\mathbf{k} \cdot \mathbf{V}_{e,i} = \omega n_{e1}/n_0$ from (7.132) and (7.134), yields

$$-i\omega n_0(m_e + m_i)\omega n_{e1}/n_0 = -ik^2(\gamma_e T_e + \gamma_i T_i) n_{e1} + \frac{e n_0}{c} \mathbf{k} \cdot [(\mathbf{V}_i - \mathbf{V}_e) \times \mathbf{B}_0] \tag{7.136}$$

The last term is

$$\mathbf{k} \cdot [(\mathbf{V}_i - \mathbf{V}_e) \times \mathbf{B}_0] = k_x(V_{iy} - V_{ey}) B_0 \tag{7.137}$$

In order to express V_{ey}, V_{iy} in terms of n_{e1}, we go back to (7.131) and take its cross product with the wave number **k**, obtaining

$$-i\omega m_e n_0(\mathbf{k} \times \mathbf{V}_e) = -\frac{e n_0}{c} \mathbf{k} \times (\mathbf{V}_e \times \mathbf{B}_0) \tag{7.138}$$

or

$$\mathbf{k} \times \mathbf{V}_e = \frac{\Omega_e}{-i\omega} \mathbf{k} \times (\mathbf{V}_e \times \hat{z}) \tag{7.139}$$

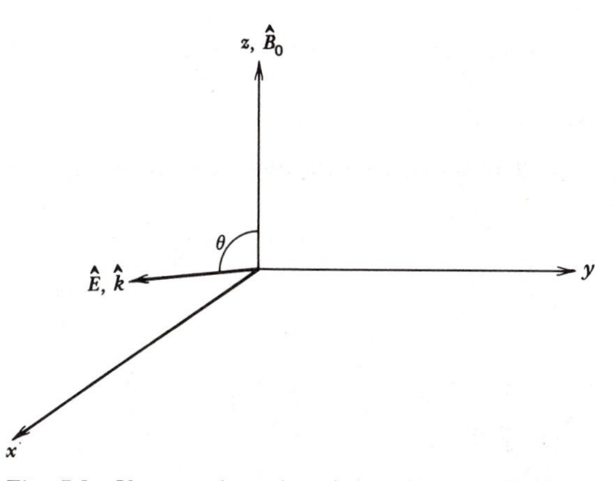

Fig. 7.9 Vector orientations in an electrostatic ion wave.

which gives three equations

$$-V_{ey}k_z = i\,\frac{\Omega_e}{\omega}\,V_{ex}k_z \tag{7.140}$$

$$-k_xV_{ez} + V_{ex}k_z = i\,\frac{\Omega_e}{\omega}\,V_{ey}k_z \tag{7.141}$$

and

$$k_xV_{ey} = -\,i\,\frac{\Omega_e}{\omega}\,V_{ex}k_x \tag{7.142}$$

These can be solved for V_{ex}, V_{ez} in terms of V_{ey},

$$V_{ex} = \frac{i\omega}{\Omega_e}\,V_{ey} \tag{7.143}$$

which is just what we had for the upper hybrid wave, and

$$V_{ez} = -V_{ey}\,\frac{k_z}{k_x}\,i\,\frac{\Omega_e}{\omega}\left(1 - \frac{\omega^2}{\Omega_e^2}\right) \tag{7.144}$$

By the continuity equation (7.132),

$$k_xV_{ex} + k_zV_{ez} = \omega n_{e1}/n_0 \tag{7.145}$$

or

$$\frac{\omega n_{e1}}{n_0} = \left[k_x\left(\frac{i\omega}{\Omega_e}\right) + k_z\left(\frac{-i\Omega_e}{\omega}\right)\frac{k_z}{k_x}\left(1 - \frac{\omega^2}{\Omega_e^2}\right)\right]V_{ey} \tag{7.146}$$

Finally, invert (7.146) for V_{ey}, insert in (7.136), and obtain [ignoring $m_e << m_i$, and using (7.137)]

$$-i\omega^2 m_i n_{e1} = -\,ik^2(\gamma_eT_e + \gamma_iT_i)n_{e1} + \frac{en_0}{c}\,k_xB_0\,\frac{\omega n_{e1}}{n_0}$$

$$\times\left\{\frac{1}{k_x\dfrac{i\omega}{\Omega_i} - \dfrac{i\Omega_i}{\omega}\dfrac{k_z^2}{k_x}\left(1 - \dfrac{\omega^2}{\Omega_i^2}\right)} - \frac{1}{k_x\dfrac{i\omega}{\Omega_e} - \dfrac{i\Omega_e}{\omega}\dfrac{k_z^2}{k_x}\left(1 - \dfrac{\omega^2}{\Omega_e^2}\right)}\right\} \tag{7.147}$$

where we have obtained V_{iy} by replacing Ω_e with Ω_i everywhere in (7.146).

EXERCISE Justify the last step.

We can divide n_{e1} from every term in (7.147) to obtain the dispersion relation [with $c_s^2 = (\gamma_eT_e + \gamma_iT_i)/m_i$ as usual]

$$\boxed{1 - \frac{k^2c_s^2}{\omega^2} + \frac{\Omega_i}{\omega}\left\{\frac{1}{\dfrac{\omega}{\Omega_e} - \dfrac{\Omega_e}{\omega}\dfrac{k_z^2}{k_x^2}\left(1 - \dfrac{\omega^2}{\Omega_e^2}\right)} - \frac{1}{\dfrac{\omega}{\Omega_i} - \dfrac{\Omega_i}{\omega}\dfrac{k_z^2}{k_x^2}\left(1 - \dfrac{\omega^2}{\Omega_i^2}\right)}\right\} = 0} \tag{7.148}$$

which is the dispersion relation for *electrostatic ion waves.*

Let us look at this monster in various limits. First, let $\mathbf{k}$ be along $\hat{B}_0$, $\mathbf{k} = k_z\hat{z}$. Then $k_x \to 0$, and each denominator becomes infinite, provided $\omega \neq \pm\,\Omega_e, \pm\,\Omega_i$. Then

$$1 - \frac{k_z^2 c_s^2}{\omega^2} = 0 \tag{7.149}$$

or

$$\boxed{\omega = \pm\, k_z c_s} \tag{7.150}$$

which is our old friend the ion-acoustic wave; we would have hoped to recover this wave for parallel propagation, since then the magnetic field does not influence the wave properties, and we should recover all unmagnetized waves. (The magnetic field may, however, affect the values of γ_e, γ_i, which are now hidden in c_s.)

We assumed $\omega \neq \pm\,\Omega_e, \pm\,\Omega_i$. Let us return to consider that possibility. First, we let $\omega \to \Omega_i$ in (7.148), and take the limit $k_x \to 0$. The first denominator becomes infinite, while the remainder of (7.148) yields

$$1 - \frac{k^2 c_s^2}{\Omega_i^2} - \frac{1}{1 - \dfrac{k_z^2}{k_x^2}\left(1 - \dfrac{\omega^2}{\Omega_i^2}\right)} = 0 \tag{7.151}$$

Then we can allow $k_x \to 0$ and $\omega \to \Omega_i$ in such a way that this equation is satisfied; that is, $\omega = \Omega_i$ is a solution, and can be called an *ion-cyclotron wave.*

EXERCISE Convince yourself that $\omega = \Omega_e$ is also a solution as $k_x \to 0$; this is an *electron-cyclotron wave.*

Let us now look in the other direction, at perpendicular propagation with $\mathbf{k} = k_x\hat{x}$. The limit $k_z \to 0$ in (7.148) yields

$$1 - \frac{k_x^2 c_s^2}{\omega^2} + \frac{\Omega_i\Omega_e}{\omega^2} - \frac{\Omega_i^2}{\omega^2} = 0 \tag{7.152}$$

Ignoring $\Omega_i^2 << |\Omega_i\Omega_e|$, we find

$$\boxed{\omega^2 = k_x^2 c_s^2 + |\Omega_i\Omega_e|} \tag{7.153}$$

which are *lower hybrid waves* propagating perpendicular to the magnetic field. For small k_x, we have

$$\boxed{\omega = \sqrt{|\Omega_i\Omega_e|} \equiv \omega_{LH}} \tag{7.154}$$

where ω_{LH} is the *lower hybrid frequency.*

The physical interpretation of lower hybrid waves is quite simple. Since $\mathbf{E} \parallel \mathbf{k} \perp \mathbf{B}_0$, we might suppose that the massive ions move along $\mathbf{E}$, while the light electrons perform an $\mathbf{E} \times \mathbf{B}_0$ drift in the $\hat{y}$-direction. It turns out that the $\hat{x}$ displacement of the ions is equal to the $\hat{x}$ displacement of the electrons (because of the polarization drift) only if $\omega = \omega_{LH} = \sqrt{|\Omega_i\Omega_e|}$.

EXERCISE Show that $\omega = 0$ is also a solution of (7.148) as $k_z \to 0$.

Let us now ask what happens if we are propagating almost, but not quite, perpendicular to $\mathbf{B}_0$, so that $k_x >> k_z$. Then, since we look for very low frequency waves, $\omega \sim \Omega_i$, $\omega/|\Omega_e| << 1$, we discard all terms of order $\omega/|\Omega_e|$. Equation (7.148) becomes

$$1 - \frac{k^2 c_s^2}{\omega^2} - \frac{\Omega_i k_x^2}{\Omega_e k_z^2} - \frac{1}{\dfrac{\omega^2}{\Omega_i^2} - \dfrac{k_z^2}{k_x^2}\left(1 - \dfrac{\omega^2}{\Omega_i^2}\right)} = 0 \tag{7.155}$$

Suppose $k_x/k_z << (m_i/m_e)^{1/2}$; then we can discard the third term compared to unity. Likewise, we can discard the term proportional to $(k_z/k_x)^2 << 1$ in the denominator. We find

$$1 - \frac{1}{\omega^2}(k^2 c_s^2 + \Omega_i^2) = 0 \tag{7.156}$$

or

$$\boxed{\omega^2 = k^2 c_s^2 + \Omega_i^2} \tag{7.157}$$

which is the dispersion relation for *electrostatic ion cyclotron waves*, valid for $k_z/k_x >> (m_e/m_i)^{1/2}$; that is, for an electron-proton plasma, at angles greater than 2° away from perpendicular to the magnetic field.

Let us summarize our results. Define an angle θ as the angle between $\mathbf{k}$ and $\mathbf{B}_0$, as in Fig. 7.9. Then we have found:

$$\theta = 0, \quad k_x = 0: \quad \omega^2 = k^2 c_s^2$$

$$\omega^2 = \Omega_i^2$$

$$\omega^2 = \Omega_e^2 \tag{7.158}$$

$$\theta < \pi/2, \quad 1 >> k_z/k_x >> (m_e/m_i)^{1/2}: \quad \omega^2 = k^2 c_s^2 + \Omega_i^2 \tag{7.159}$$

$$\theta = \pi/2, \quad k_z = 0: \quad \omega^2 = k^2 c_s^2 + |\Omega_i \Omega_e|$$

$$\omega^2 = 0 \tag{7.160}$$

We may suppose from (7.158) to (7.160), or by looking at the basic equation (7.148), that there are three branches of solutions.

7.9 ELECTROMAGNETIC WAVES IN MAGNETIZED PLASMAS

We wish to extend the treatment of electromagnetic waves in a uniform unmagnetized plasma (Section 7.6) to the case of a magnetized plasma. Because we again expect high frequency waves, we ignore ion motion. For simplicity, consider a cold plasma, $T_e = 0$. The relevant equations are then Maxwell's equations together with the electron force equation, used to calculate the current. We have, linearizing immediately,

$$m_e n_0 \frac{\partial \mathbf{V}}{\partial t} = -e n_0 \mathbf{E}_1 - \frac{e}{c} n_0 \mathbf{V} \times \mathbf{B}_0 \tag{7.161}$$

$$\nabla \times \mathbf{E}_1 = -\frac{1}{c}\frac{\partial \mathbf{B}_1}{\partial t} \tag{7.162}$$

$$\nabla \times \mathbf{B}_1 = \frac{4\pi}{c}\mathbf{J} + \frac{1}{c}\frac{\partial \mathbf{E}_1}{\partial t} \tag{7.163}$$

$$\mathbf{J} = -en_0\mathbf{V} \tag{7.164}$$

where $\mathbf{V} \equiv \mathbf{V}_e$. Since we already have as many equations as unknowns, we shall avoid using Poisson's equation or the electron continuity equation.

With the coordinate system shown in Fig. 7.10, let us first look for waves traveling perpendicular to $\hat{B}_0$, $\mathbf{k} = k\hat{x}$. Then there are two possibilities: the electric field can be along $\hat{B}_0$ (the *ordinary* wave), and the electric field can be in the x-y plane, perpendicular to $\hat{B}_0$ (the *extraordinary* wave). In the first case, the electric field $\mathbf{E}_1 = E_1\hat{z}$ induces an electron velocity in the $\hat{z}$-direction, and the $\mathbf{V}_e \times \mathbf{B}_0$ force vanishes. Thus, Eqs. (7.161) to (7.164) reduce to the equivalent equations for an unmagnetized plasma, (7.111) to (7.113), and we can immediately write down the dispersion relation (7.116), which is

$$\boxed{\omega^2 = \omega_e^2 + k^2c^2} \tag{7.165}$$

which describes the *ordinary* wave; this wave propagates as if there were no magnetic field.

Next suppose the electric field is in the x-y plane. Then the electric field will create a $\mathbf{V}$ in the x-y plane, and the $\mathbf{V} \times \mathbf{B}_0$ force will produce another component of velocity in the x-y plane; note, however, that no component of velocity is produced in the $\hat{z}$-direction (Fig. 7.11). It is for this reason that we can consider the *ordinary* mode (7.165) and the *extraordinary* mode (currently being derived) separately.

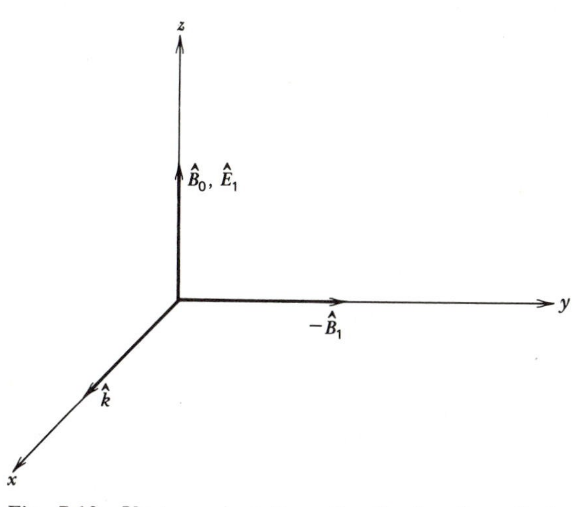

Fig. 7.10 Vector orientations for the "ordinary" electromagnetic wave in magnetized plasma.

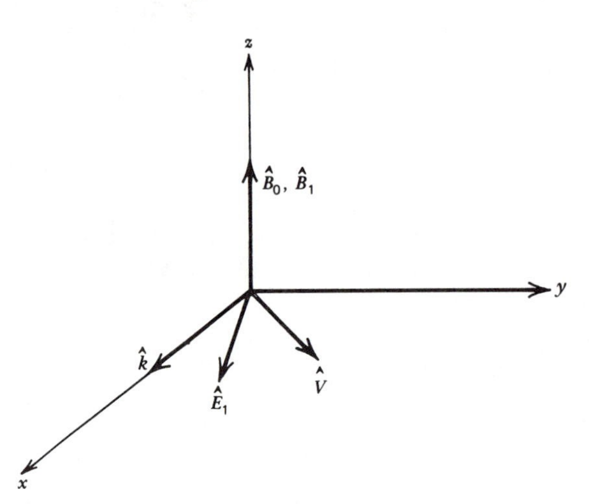

Fig. 7.11 Vector orientations for the "extraordinary" electromagnetic wave in magnetized plasma.

The relevant components of (7.161) to (7.164) become, with $\mathbf{E}_1 = E_x\hat{x} + E_y\hat{y}$,

$$-i\omega m_e V_x = -eE_x - \frac{e}{c} V_y B_0 \tag{7.166}$$

$$-i\omega m_e V_y = -eE_y + \frac{e}{c} V_x B_0 \tag{7.167}$$

$$ikE_y = \frac{i\omega}{c} B_1 \tag{7.168}$$

$$-ikB_1 = \frac{-4\pi e n_0}{c} V_y - \frac{i\omega}{c} E_y \tag{7.169}$$

$$0 = \frac{-4\pi e n_0}{c} V_x - \frac{i\omega}{c} E_x \tag{7.170}$$

EXERCISE Reconvince yourself that $\nabla \cdot \mathbf{A} = i\mathbf{k} \cdot \mathbf{A}$, $\partial_t \mathbf{A} = -i\omega\mathbf{A}$, and $\nabla \times \mathbf{A} = i\mathbf{k} \times \mathbf{A}$ if $\mathbf{A} = \mathbf{A}_0 \exp(-i\omega t + i\mathbf{k} \cdot \mathbf{x})$ where $\mathbf{A}_0$ is a constant vector.

Equations (7.166) to (7.170) constitute five equations in five unknowns. From (7.168), $B_1 = (kc/\omega)E_y$; therefore, (7.169) and (7.170) yield for V_x, V_y,

$$V_x = \frac{-i\omega}{4\pi n_0 e} E_x \tag{7.171}$$

$$V_y = \left(\frac{ik^2c^2}{4\pi n_0 e\,\omega} + \frac{-i\omega}{4\pi n_0 e}\right) E_y \tag{7.172}$$

Inserting (7.171), (7.172) into (7.166), (7.167), we obtain

$$\begin{bmatrix} -i\omega m_e\left(\dfrac{-i\omega}{4\pi n_0 e}\right) + e & \dfrac{e}{c} B_0\left(\dfrac{ik^2c^2}{4\pi n_0 e\omega} + \dfrac{-i\omega}{4\pi n_0 e}\right) \\ \dfrac{-eB_0}{c}\left(\dfrac{-i\omega}{4\pi n_0 e}\right) & -i\omega m_e\left(\dfrac{ik^2c^2}{4\pi n_0 e\omega} + \dfrac{-i\omega}{4\pi n_0 e}\right) + e \end{bmatrix}\begin{pmatrix} E_x \\ E_y \end{pmatrix} = \begin{pmatrix} 0 \\ 0 \end{pmatrix} \tag{7.173}$$

The determinant of the coefficients must vanish. Dividing each term by e, we find

$$\left(1 - \frac{\omega^2}{\omega_e^2}\right)\left(1 + \frac{k^2c^2}{\omega_e^2} - \frac{\omega^2}{\omega_e^2}\right) + \left(\frac{\omega\Omega_e}{\omega_e^2}\right)\Omega_e\left(\frac{k^2c^2}{\omega_e^2\omega} - \frac{\omega}{\omega_e^2}\right) = 0 \tag{7.174}$$

or

$$\boxed{\left(1 - \frac{\omega^2}{\omega_e^2}\right)\left(1 + \frac{k^2c^2}{\omega_e^2} - \frac{\omega^2}{\omega_e^2}\right) + \frac{\Omega_e^2k^2c^2}{\omega_e^4} - \frac{\omega^2\Omega_e^2}{\omega_e^4} = 0} \tag{7.175}$$

which is the rather imposing dispersion relation for the *extraordinary mode*.

Separating a factor k^2c^2/ω_e^2 from (7.175) yields

$$\frac{k^2c^2}{\omega_e^2} = \frac{\dfrac{\omega^2\Omega_e^2}{\omega_e^4} - \left(1 - \dfrac{\omega^2}{\omega_e^2}\right)^2}{1 - \dfrac{\omega^2}{\omega_e^2} + \dfrac{\Omega_e^2}{\omega_e^2}} \tag{7.176}$$

Multiplying (7.176) by ω_e^2/ω^2 and recalling $\omega_{UH}^2 = \omega_e^2 + \Omega_e^2$ yields

$$\begin{aligned} n^2 \equiv \frac{k^2c^2}{\omega^2} &= \frac{\omega_e^4 - 2\omega^2\omega_e^2 + \omega^4 - \omega^2\Omega_e^2}{\omega^2(\omega^2 - \omega_{UH}^2)} \\ &= \frac{\omega_e^4 - \omega^2\omega_e^2 + \omega^2(\omega^2 - \omega_{UH}^2)}{\omega^2(\omega^2 - \omega_{UH}^2)} \end{aligned} \tag{7.177}$$

or

$$\boxed{n^2 = \frac{k^2c^2}{\omega^2} = 1 - \frac{\omega_e^2}{\omega^2}\,\frac{\omega^2 - \omega_e^2}{\omega^2 - \omega_{UH}^2}} \tag{7.178}$$

for the extraordinary mode, where n is the index of refraction.

Equation (7.178) is the dispersion relation for the extraordinary mode, which is a perpendicular mode, partially transverse and partially longitudinal. It could be shown, by solving (7.178) for ω and inserting in (7.173), that the components E_x and E_y are out of phase with each other, so that at a given point in space, the electric field vector performs an elliptical rotation as a function of time, as shown in Fig. 7.12.

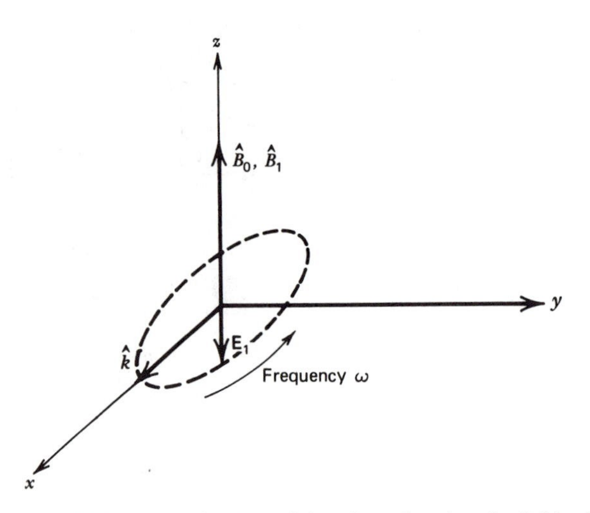

Fig. 7.12 At a fixed spatial point, the electric field of an extraordinary wave rotates elliptically.

It is useful to define two properties of the X-mode (short for extraordinary; O-mode means ordinary mode). These are the *cutoffs* and the *resonances*. A *cutoff* is any frequency ω at which $k \to 0$; thus, the cutoff frequencies are given by the zeros of the index of refraction (7.178). A *resonance* is any frequency for which the wave number $k \to \pm\infty$; thus, we have a resonance whenever the denominator of (7.178) vanishes. [Some people remember the difference between cutoff ($k \to 0$) and resonance ($k \to \infty$) by the fact that $c(k \to 0)$ comes before $r(k \to \infty)$ in the alphabet.]

The resonances are easy; by (7.178) they happen at $\omega = 0$ and $\omega = \omega_{UH}$. The cutoffs are obtained by setting (7.178) equal to zero. We find

$$\omega^2 = \omega_e^2 \frac{\omega^2 - \omega_e^2}{\omega^2 - \omega_{UH}^2} \tag{7.179}$$

or

$$\omega^4 - \omega^2\omega_{UH}^2 - \omega_e^2\omega^2 + \omega_e^4 = 0 \tag{7.180}$$

or

$$\omega = \left[\frac{\omega_{UH}^2 + \omega_e^2}{2} \pm \frac{1}{2}\sqrt{(\omega_{UH}^2 + \omega_e^2)^2 - 4\omega_e^4}\right]^{1/2} \tag{7.181}$$

Recalling $\omega_{UH}^2 = \omega_e^2 + \Omega_e^2$, we see that the inner radical is $4\omega_e^2\Omega_e^2 + \Omega_e^4$; thus,

$$\omega = \left[\omega_e^2 + \frac{\Omega_e^2}{2} \pm \Omega_e\sqrt{\omega_e^2 + \Omega_e^2/4}\right]^{1/2} \tag{7.182}$$

or

$$\boxed{\omega_{\binom{L}{R}} = \pm\frac{\Omega_e}{2} + \sqrt{\omega_e^2 + \Omega_e^2/4}} \tag{7.183}$$

where L, R refer to *l*eft and *r*ight, for reasons that will become clear in Section 7.10. (Recall that $\Omega_e < 0$.)

EXERCISE Demonstrate the equivalence of (7.182) and (7.183).

We see that ω_L is somewhat below ω_e, and ω_R is somewhat above ω_e. With this knowledge of cutoffs and resonances, we are able to draw the dispersion diagram for the extraordinary mode, using (7.178) to tell us if a resonance is for $k^2 \to +\infty$ or $k^2 \to -\infty$, and if a cutoff is for $k^2 \to 0^+$ or $k^2 \to 0^-$. For completeness, our diagram can include the ordinary mode, which by (7.165) is

$$n^2 = \frac{k^2c^2}{\omega^2} = 1 - \frac{\omega_e^2}{\omega^2} \tag{7.184}$$

for the ordinary mode, which has a cutoff at $\omega = \omega_e$ and a resonance at $\omega = 0$. Both dispersion diagrams are sketched in Fig. 7.13. Unlike previous dispersion diagrams, which are frequency ω vs. wave number k, Fig. 7.13 is a sketch of the square of the index of refraction $n^2 = c^2k^2/\omega^2$ vs. frequency ω.

EXERCISE Is Fig. 7.13 accurate for low frequencies $\omega << \omega_L$? Why not?

We could also solve (7.178) and (7.184) for the usual dispersion function $\omega = \omega(k)$. A sketch of this function for both modes is shown in Fig. 7.14.

We note that Fig. 7.13 shows a frequency for every $n^2 = k^2c^2/\omega^2$, while in Fig. 7.14 there are regions of frequency where there are no waves. Why is this? It is because in certain regions, $n^2 = k^2c^2/\omega^2 < 0$ for real ω; thus k is imaginary. These waves are therefore evanescent, and they do not appear in Fig. 7.14, which is a sketch of real ω vs. real k. The bands where there are no propagating waves ($0 < \omega < \omega_e$ for the O-mode; $0 < \omega < \omega_L$ and $\omega_{UH} < \omega < \omega_R$ for the X-mode) are called *stop bands* from radio engineering, while the other bands are called *pass bands*.

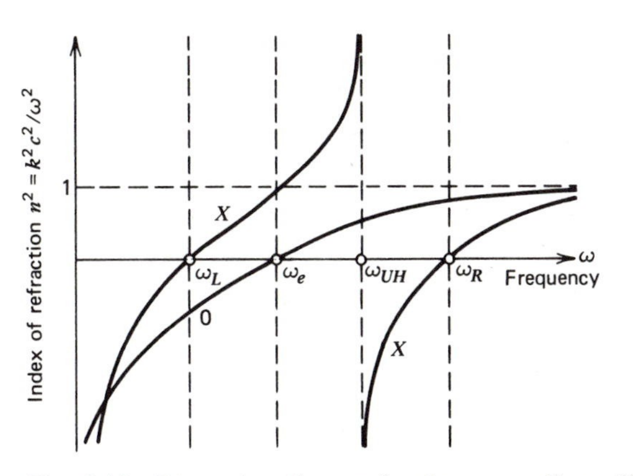

Fig. 7.13 Dispersion diagram for the extraordinary (X) mode and the ordinary (O) mode.

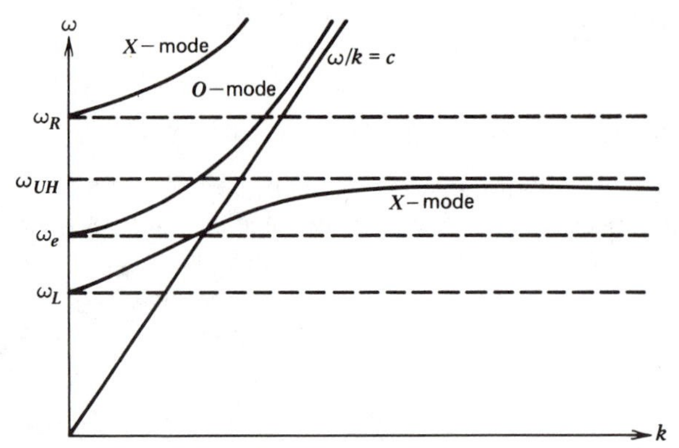

Fig. 7.14 Sketch of the dispersion relation $\omega = \omega(k)$ for the extraordinary (X) mode and the ordinary (O) mode.

7.10 ELECTROMAGNETIC WAVES ALONG $\mathbf{B}_0$

Continuing our discussion of electromagnetic waves in magnetized plasma, we next look for parallel waves, traveling along $\mathbf{B}_0$. We need only three basic equations: Ampere's law, Faraday's law, and the electron force equation (ignoring electron temperature and ion motion). These are, Fourier transforming and linearizing immediately,

$$i\mathbf{k} \times \mathbf{E}_1 = \frac{i\omega}{c}\mathbf{B}_1 \tag{7.185}$$

$$i\mathbf{k} \times \mathbf{B}_1 = \frac{-4\pi n_0 e}{c}\mathbf{V} - \frac{i\omega}{c}\mathbf{E}_1 \tag{7.186}$$

and

$$-i\omega m_e \mathbf{V} = -e\mathbf{E}_1 - \frac{e}{c}\mathbf{V} \times \mathbf{B}_0 \tag{7.187}$$

where $\mathbf{V} \equiv \mathbf{V}_e$. Referring to Fig. 7.15, we see that a consistent solution to (7.185)–(7.187) is one that has $\mathbf{V}$, $\mathbf{E}_1$, and $\mathbf{B}_1$ all in the x-y plane with $\mathbf{k} = k\hat{z}$ along $\hat{\mathbf{B}}_0$. When we take $\mathbf{E}_1 = (E_x, E_y, 0)$, $\mathbf{B}_1 = (B_x, B_y, 0)$, and $\mathbf{V} = (v_x, v_y, 0)$, Eqs. (7.185) to (7.187) yield

$$-ikE_y = \frac{i\omega}{c}B_x \tag{7.188}$$

$$ikE_x = \frac{i\omega}{c}B_y \tag{7.189}$$

$$-ikB_y = \frac{-4\pi n_0 e}{c}v_x - \frac{i\omega}{c}E_x \tag{7.190}$$

$$ikB_x = \frac{-4\pi n_0 e}{c}v_y - \frac{i\omega}{c}E_y \tag{7.191}$$

$$-i\omega m_e v_x = -eE_x - \frac{e}{c}B_0 v_y \tag{7.192}$$

and

$$-i\omega m_e v_y = -eE_y + \frac{e}{c} B_0 v_x \tag{7.193}$$

Inserting (7.188) and (7.189) for B_x, B_y, in (7.190) and (7.191), we find

$$v_x = \frac{\left(\dfrac{-ik^2c}{\omega} + \dfrac{i\omega}{c}\right)}{-4\pi n_0 e/c} E_x \tag{7.194}$$

and

$$v_y = \frac{\left(\dfrac{-ik^2c}{\omega} + \dfrac{i\omega}{c}\right)}{-4\pi n_0 e/c} E_y \tag{7.195}$$

Inserting (7.194), (7.195) in (7.192), (7.193), we obtain for E_x, E_y, the matrix equation

$$\begin{bmatrix} -i\omega m_e \dfrac{\left(\dfrac{-ik^2c}{\omega} + \dfrac{i\omega}{c}\right)}{-4\pi n_0 e/c} + e & \dfrac{eB_0}{c} \dfrac{\left(\dfrac{-ik^2c}{\omega} + \dfrac{i\omega}{c}\right)}{-4\pi n_0 e/c} \\ \dfrac{-eB_0}{c} \dfrac{\left(\dfrac{-ik^2c}{\omega} + \dfrac{i\omega}{c}\right)}{-4\pi n_0 e/c} & -i\omega m_e \dfrac{\left(\dfrac{-ik^2c}{\omega} + \dfrac{i\omega}{c}\right)}{-4\pi n_0 e/c} + e \end{bmatrix} \begin{pmatrix} E_x \\ E_y \end{pmatrix} = \begin{pmatrix} 0 \\ 0 \end{pmatrix} \tag{7.196}$$

Setting the determinant of the coefficients equal to zero, we find

$$\left(1 + \frac{k^2c^2}{\omega_e^2} - \frac{\omega^2}{\omega_e^2}\right)^2 = \frac{\Omega_e^2}{\omega_e^4} \left(\omega - \frac{k^2c^2}{\omega}\right)^2 \tag{7.197}$$

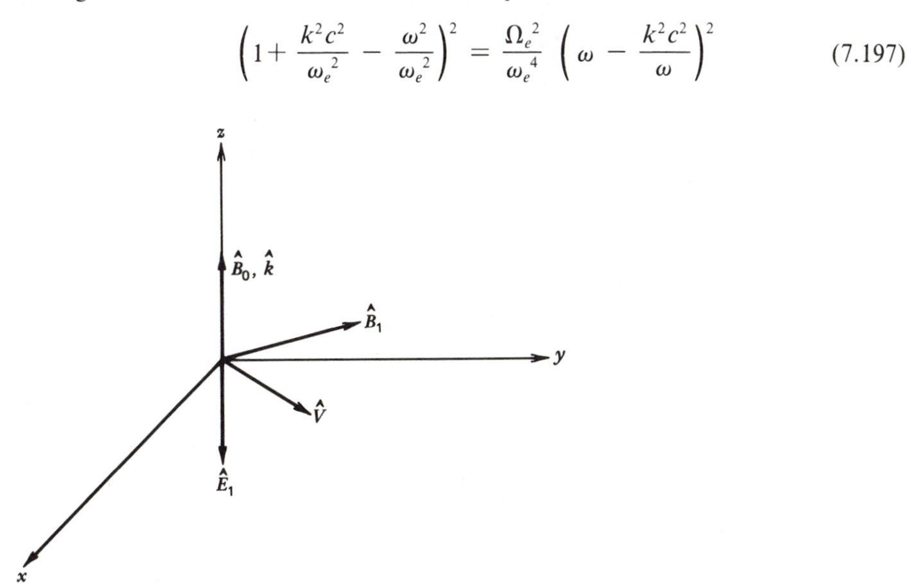

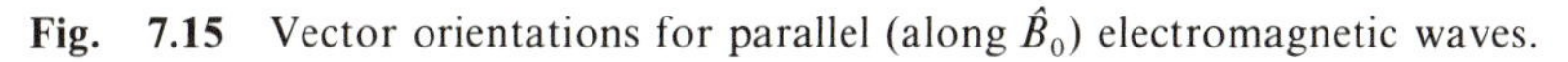

Fig. 7.15 Vector orientations for parallel (along $\hat{B}_0$) electromagnetic waves.

We take the square root of (7.197), retaining both signs, to obtain

$$1 - \frac{\omega}{\omega_e^2}\left(\omega - \frac{k^2c^2}{\omega}\right) = \pm\frac{\Omega_e}{\omega_e^2}\left(\omega - \frac{k^2c^2}{\omega}\right) \tag{7.198}$$

or

$$1 = \left(\frac{\omega}{\omega_e^2} \pm \frac{\Omega_e}{\omega_e^2}\right)\left(\omega - \frac{k^2c^2}{\omega}\right) \tag{7.199}$$

or

$$\boxed{n^2 \equiv \frac{k^2c^2}{\omega^2} = 1 - \frac{\omega_e^2/\omega^2}{1 \pm \Omega_e/\omega}} \tag{7.200}$$

which is the index of refraction for electromagnetic waves traveling along the magnetic field.

EXERCISE Verify all steps leading to (7.200).

Recalling $\Omega_e < 0$, the top sign in (7.200) is called the *R-wave*, meaning *right circularly polarized*, while the bottom sign is called the *L-wave*, meaning *left circularly polarized.* These terms come from the rotation of the electric field vector as the wave propagates (the right-hand rule places the thumb along **k** and the fingers in the direction of the **E** rotation for the R-wave; and opposite for the left wave; see Fig. 7.16). Because this situation is cylindrically symmetric about $\hat{B}_0$, the $\mathbf{E}_1$ vector describes a circle, rather than an ellipse as in the X-mode case.

Note that the direction of rotation of the R-wave corresponds to the direction of gyration of electrons. Further note that when $\omega = |\Omega_e|$, the R-wave has $1 + (\Omega_e/\omega) = 0$, which by (7.200) is a resonance, $k \to \infty$. Thus we see a physical

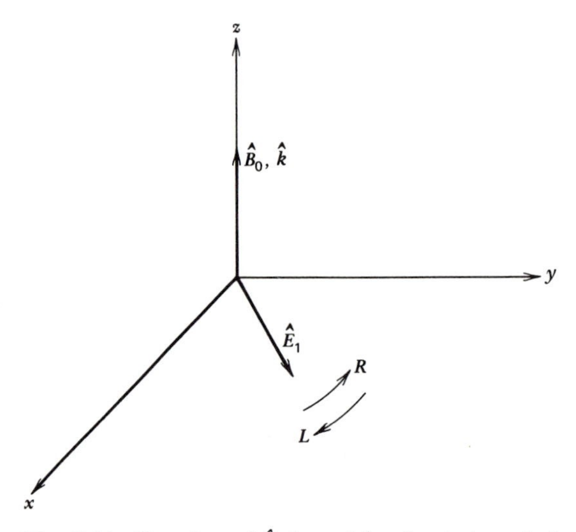

Fig. 7.16 Rotation of $\hat{E}_1$ in a right circularly polarized wave (R) and in a left circularly polarized wave (L).

connection between a resonance ($k \to \infty$), and a resonance between the wave and the electrons; the electric field of the R-wave will continuously accelerate electrons when $\omega = -\Omega_e$.

From (7.200) we see that the L-wave has no resonances; this makes physical sense because the L-wave rotates in the direction opposite to the gyration of electrons. The cutoffs are obtained by setting $k = 0$ in (7.200); we find

$$\omega_{\binom{R}{L}} = \mp \frac{\Omega_e}{2} + \sqrt{\omega_e^2 + (\Omega_e^2/4)} \tag{7.201}$$

as the cutoffs for the R-wave and the L-wave. These are precisely the cutoffs found for the extraordinary mode, Eq. (7.183), and we now understand why we called them the left and right cutoffs. We note from (7.201) that one always has $\omega_R > |\Omega_e|$ and $\omega_R > \omega_L$. However, for ω_L and $|\Omega_e|$, there are two possibilities: $\omega_L > |\Omega_e|$, and $\omega_L < |\Omega_e|$.

EXERCISE Show that $\omega_L = |\Omega_e|$ when $\omega_e = 2^{1/2}|\Omega_e|$.

The dispersion diagrams are different in the two cases. These are found in Ref. [3], p. 195. The dispersion relation $\omega = \omega(k)$ is shown in Fig. 7.17 for the case $\omega_L > |\Omega_e|$. Note that the R-wave has two "pass bands," $0 < \omega < |\Omega_e|$, and $\omega > \omega_R$, separated by a "stop band." The L-wave exists only for $\omega > \omega_L$. Both high frequency branches asymptote to $\omega \approx kc$ at high frequencies. The locations of the pass and stop bands can be seen more clearly by drawing $n^2 = k^2c^2/\omega^2$ vs. ω, as shown in Fig. 7.18 (see Ref. [3], p. 194). The low frequency branch of the R-wave is often called the *electron-cyclotron wave*. Once again, the "stop bands" occur where $n^2 < 0$ and the "pass bands" occur when $n^2 > 0$. (Do we trust this theory for low frequencies $\omega < \Omega_i$?)

For the low density plasma, $\omega_L < |\Omega_e|$, the character of the dispersion relation $\omega = \omega(k)$ changes, as shown in Fig. 7.19 (see Ref. [3], p. 195). The low frequency branch of the R-wave is again called the *electron-cyclotron wave*.

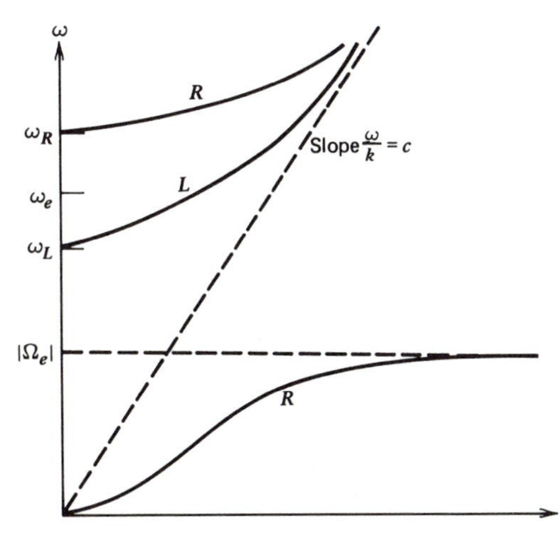

Fig. 7.17 Dispersion diagram for parallel electromagnetic waves for the case $\omega_L > |\Omega_e|$.

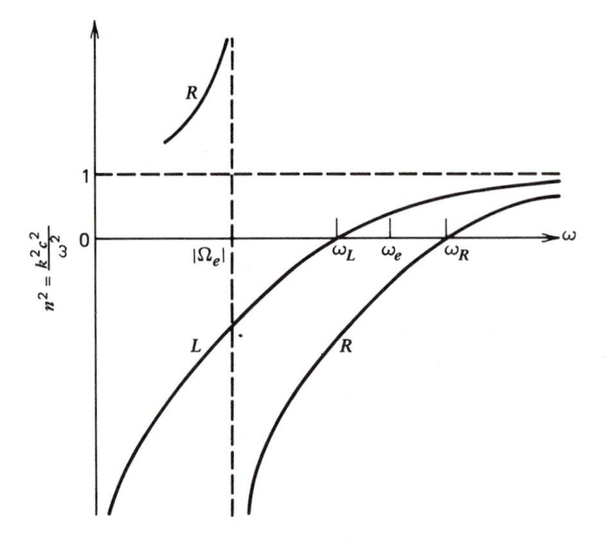

Fig. 7.18 Dispersion diagram for parallel electromagnetic waves for the case $\omega_L > |\Omega_e|$.

From Fig. 7.19 we can see that the electron-cyclotron wave has a portion where $V_g \equiv d\omega/dk$ increases as ω increases. This is called the *whistler* wave, because the high frequency components of a wave packet travel faster than its low frequency components. An observer some distance away from a source (a lightning stroke, for example) will then hear a whistle starting at high frequencies and descending to lower frequencies.

In both of our dispersion diagrams, the R-wave at very high frequencies is seen to have a higher phase speed than the L-wave. Thus, if a plane wave is incident on

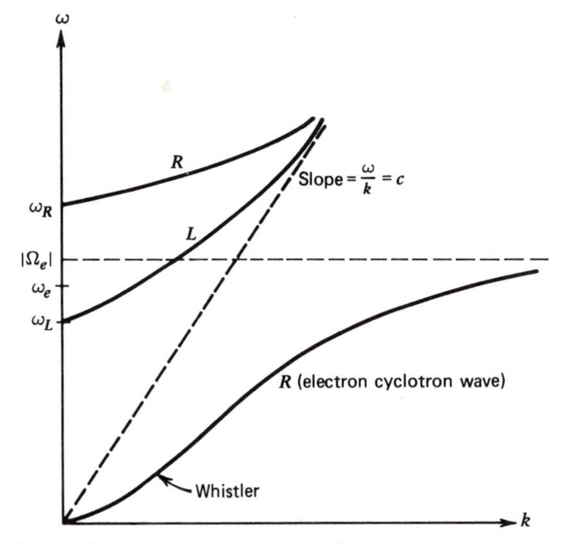

Fig. 7.19 Dispersion diagram for parallel electromagnetic waves for the case $\omega_L < |\Omega_e|$.

a plasma along $\hat{B}_0$, its two normal mode components, R and L, travel at different speeds, and the plane of polarization of the plane wave rotates as it travels. This is known as *Faraday rotation*, and is useful in measuring plasma densities in laboratory plasma and in interstellar space.

This completes our discussion of high frequency electromagnetic waves (ignoring ion motion) traveling in a magnetized plasma. We have discussed only waves traveling across $\hat{B}_0$ (O-mode and X-mode) and along $\hat{B}_0$ (R-wave and L-wave). Of course, waves can travel at any angle to $\hat{B}_0$. When they do, there will be two modes for any angle of propagation, and their properties will be some combination of the properties of the O, X, R, and L waves.

7.11 ALFVÉN WAVES

Up until this point, we have considered electromagnetic waves ignoring ion motion, and ion waves that were electrostatic and thus ignored electromagnetic effects. Let us next combine ion motion with electromagnetic effects; we shall find parallel Alfvén waves ($\mathbf{k} \parallel \hat{B}_0$) and perpendicular magnetosonic waves ($\mathbf{k} \perp \hat{B}_0$).

First we look for low frequency waves traveling along $\hat{B}_0$. For simplicity we take a cold plasma, $T_i = T_e = 0$. We can also ignore electron inertia ($m_e \to 0$). Just as in the case of R-waves and L-waves, we look for waves with $\mathbf{k} = k\hat{z}$, $\mathbf{B}_0 = B_0\hat{z}$, and $\mathbf{V}_e$, $\mathbf{V}_i$, $\mathbf{E}_1$, $\mathbf{B}_1$ all in the x-y plane (Fig. 7.20). Unlike the R-, L-wave case, we do not look for a rotating $\mathbf{E}_1$, $\mathbf{B}_1$; rather, we take $\mathbf{E}_1 = E_x\hat{x}$ and $\mathbf{B}_1 = B_y\hat{y}$. As we shall see, this form for $\mathbf{E}_1$ and $\mathbf{B}_1$ is not entirely self-consistent. The relevant fluid equations are then (linearizing and Fourier transforming immediately):

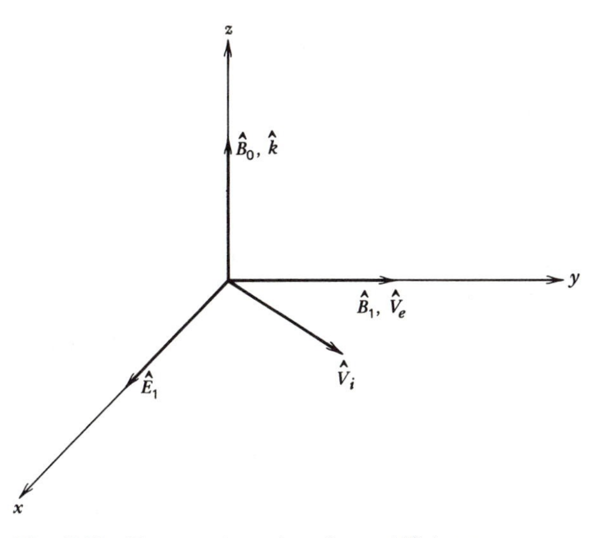

Fig. 7.20 Vector orientations in an Alfvén wave.

$$ik \times \mathbf{E}_1 = \frac{i\omega}{c} \mathbf{B}_1 \tag{7.202}$$

$$ik \times \mathbf{B}_1 = \frac{4\pi}{c} n_0 e(\mathbf{V}_i - \mathbf{V}_e) - \frac{i\omega}{c} \mathbf{E}_1 \tag{7.203}$$

$$0 = -en_0\mathbf{E}_1 - \frac{en_0}{c} \mathbf{V}_e \times \mathbf{B}_0 \tag{7.204}$$

$$-i\omega m_i n_0 \mathbf{V}_i = en_0\mathbf{E}_1 + \frac{en_0}{c} \mathbf{V}_i \times \mathbf{B}_0 \tag{7.205}$$

Considering the motions of individual electrons, we are keeping the electron $\mathbf{E}_1 \times \mathbf{B}_0$ drift in the $\hat{y}$-direction, while ignoring the polarization drift in the $\hat{x}$-direction. We recall from Chapter 2 that the polarization drift speed is proportional to mass. Thus, for the ions, we have an $\mathbf{E}_1 \times \mathbf{B}_0$ drift in the $\hat{y}$-direction, which approximately equals the electron drift and prevents any current in the $\hat{y}$-direction. We also have for the ions a component of velocity in the $\hat{x}$-direction (a polarization drift) that provides the current $n_0 e(\mathbf{V}_i - \mathbf{V}_e)$ in (7.203) to drive the magnetic field $\mathbf{B}_1$. The approximation in this derivation is to ignore that portion of V_{ey} due to the $V_{ex}\hat{x} \times \hat{B}_0$ force.

With this introduction, we write the relevant components of (7.202) to (7.205) as

$$\frac{-ik^2c}{\omega} E_x = \frac{4\pi n_0 e}{c} V_{ix} - \frac{i\omega}{c} E_x \tag{7.206}$$

$$-i\omega m_i V_{ix} = eE_x + \frac{e}{c} V_{iy} B_0 \tag{7.207}$$

and

$$-i\omega m_i V_{iy} = -\frac{e}{c} V_{ix} B_0 \tag{7.208}$$

Solving (7.208) for V_{iy}, we find

$$V_{iy} = -\frac{i\Omega_i}{\omega} V_{ix} \tag{7.209}$$

We insert (7.209) in (7.207) to obtain

$$V_{ix} = \left(\frac{e/m_i}{-i\omega + i\Omega_i^2/\omega}\right) E_x \tag{7.210}$$

Combining (7.210) and (7.206) then yields the dispersion relation

$$1 - \frac{k^2c^2}{\omega^2} + \frac{\omega_i^2}{\Omega_i^2 - \omega^2} = 0 \tag{7.211}$$

Enforcing the assumption $\omega << \Omega_i$, implicit in the above discussion, we ignore $\omega^2 << \Omega_i^2$ in the second denominator, finding

$$\omega^2 = \frac{k^2c^2}{1 + \omega_i^2/\Omega_i^2} = \frac{k^2c^2}{1 + 4\pi\rho_m c^2/B_0^2} \tag{7.212}$$

where $\rho_m \equiv n_0 m_i$ is the ion mass density. If we define the *Alfvén speed* $V_A \equiv (B_0^2/4\pi\rho_m)^{1/2}$, (7.212) is

$$\omega^2 = \frac{k^2c^2}{1 + (c^2/V_A{}^2)} \tag{7.213}$$

Multiplying top and bottom by $V_A{}^2/c^2$ we finally obtain

$$\boxed{\omega^2 = \frac{k^2V_A{}^2}{1 + (V_A{}^2/c^2)}} \tag{7.214}$$

as the dispersion relation for *Alfvén waves*. Note that for $V_A << c$, this is $\omega = kV_A$; therefore we have an acoustic dispersion relation. Recall that acoustic waves in air have an acoustic speed $(P/\rho_m)^{1/2}$ where P is pressure; here we have a speed $V_A = (B_0{}^2/4\pi\rho_m)^{1/2}$, which is of the same form if we relate $B_0{}^2/4\pi$ to a magnetic pressure.

The physical interpretation of Alfvén waves is very interesting. We have seen that electrons and ions are $\mathbf{E}_1 \times \mathbf{B}_0$ drifting together in the $\hat{y}$-direction, with speed $-(E_x/B_0)c$. Thus, both plasma fluids move together in the $\hat{y}$-direction. Now what is happening to magnetic field lines? They are being distorted by the addition of $\mathbf{B}_1 = B_y\hat{y}$ to the background magnetic field $\mathbf{B}_0 = B_0\hat{z}$, as shown in Fig. 7.21. The position function of a magnetic field line can be defined as

$$Y_B(z,t) = \int^z \frac{B_y(z',t)}{B_0}\, dz' \tag{7.215}$$

Then the $\hat{y}$ velocity V_B of a magnetic field line is the time derivative of Y_B, or [with $B_y \sim \exp(-i\omega t + ikz)$]

$$V_B = -\, i\omega \int^z \frac{B_y}{B_0}\, dz' = \frac{-i\omega}{ik}\,\frac{B_y}{B_0} \tag{7.216}$$

Now from (7.202) and Fig. 7.20 we have

$$B_y = \frac{ck}{\omega}\, E_x \tag{7.217}$$

or

$$V_B = -(E_x/B_0)c \tag{7.218}$$

which is precisely the $\hat{y}$-velocity of fluid flow. Thus, in the $\hat{y}$-direction, we say that the particles are *frozen to the field lines*. This concept will prove useful later

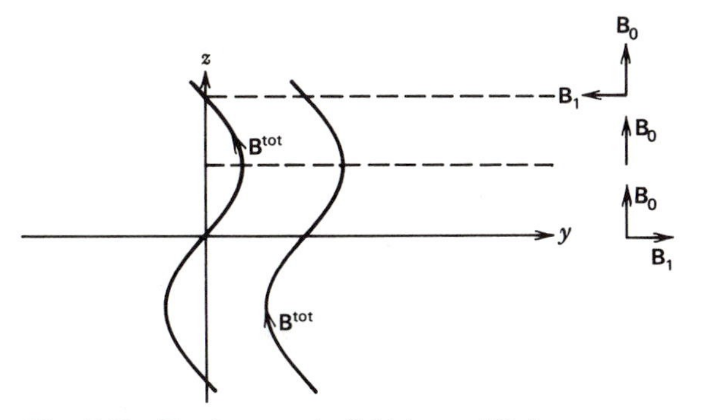

Fig. 7.21 Total magnetic field in an Alfvén wave.

(Chapter 8) for other low frequency plasma motions. Note that in the $\hat{z}$-direction, we can take the fluid speed and the field line speed both to be zero, satisfying this concept. However, in the $\hat{x}$-direction, we have seen that $V_{ex} \approx 0$ while $V_{ix} \neq 0$. Thus, we cannot have both kinds of particles frozen to the field lines in the $\hat{x}$-direction.

One may recall that the wave equation of a stretched string is $\omega = kc_T$ with $c_T = \sqrt{T/\rho_m}$, where T is the tension on the string and ρ_m is the mass per unit length. If we identify $B^2/4\pi$ as a tension per unit area and ρ_m as a mass per unit volume, then the Alfvén wave dispersion relation $\omega = kV_A$ can be thought of as representing the wave that propagates when a field line, loaded with plasma, is plucked in the transverse direction.

7.12 FAST MAGNETOSONIC WAVE

The Alfvén wave of the previous section is a low frequency parallel electromagnetic wave, traveling along $\hat{B}_0$. Let us now look for a low frequency perpendicular electromagnetic wave, traveling across $\hat{B}_0$; this is the *fast magnetosonic wave.*

For simplicity, consider a cold plasma, $T_e = T_i = 0$, and ignore electron inertia, $m_e \to 0$. We look for a wave with $\mathbf{k} \cdot \mathbf{B}_0 = 0, \mathbf{k} \cdot \mathbf{E}_1 = 0$, and $\mathbf{E}_1 \cdot \mathbf{B}_0 = 0$, as in Fig. 7.22.

EXERCISE Why don't we look for low frequency waves with $\mathbf{E}_1$ along $\mathbf{B}_0$? Have we ever looked for a wave with $\mathbf{E}_1$ along $\mathbf{B}_0$? What did we find?

We choose $\mathbf{k} = k\hat{y}$, $\mathbf{E}_1 = E_1\hat{x}$, and $\mathbf{B}_1 = B_1\hat{z}$; thus $\mathbf{B}_1$ is along $\mathbf{B}_0$, and the relevant fluid equations, linearized and Fourier transformed, are

$$i\mathbf{k} \times \mathbf{E}_1 = \frac{i\omega}{c} \mathbf{B}_1 \tag{7.219}$$

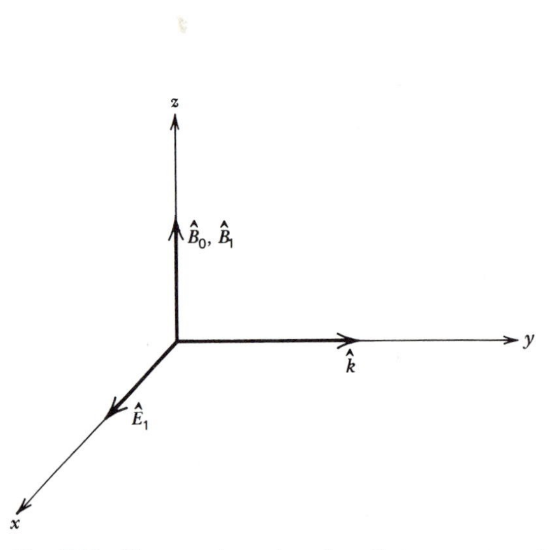

Fig. 7.22 Vector orientations in a fast magnetosonic wave.

$$ik \times \mathbf{B}_1 = \frac{4\pi}{c} n_0 e(\mathbf{V}_i - \mathbf{V}_e) - \frac{i\omega}{c} \mathbf{E}_1 \tag{7.220}$$

$$0 = - en_0\mathbf{E}_1 - \frac{en_0}{c} \mathbf{V}_e \times \mathbf{B}_0 \tag{7.221}$$

$$-i\omega m_i n_0 \mathbf{V}_i = en_0\mathbf{E}_1 + \frac{en_0}{c} \mathbf{V}_i \times \mathbf{B}_0 \tag{7.222}$$

From (7.221) we see that the electrons will have only an $\mathbf{E}_1 \times \mathbf{B}_0$ drift in the $\hat{k}$-direction, while the ions, for very small ω, will have approximately the same $\mathbf{E}_1 \times \mathbf{B}_0$ drift in the $\hat{k}$-direction. The ions, however, will have an extra component of velocity in the $\hat{x}$-direction, along $\hat{E}_1$, because of the polarization drift, which produces a current in the $\hat{x}$-direction that produces the perturbed magnetic field in the $\hat{z}$-direction. The relevant components of (7.219) to (7.222) are then

$$-ikE_1 = \frac{i\omega}{c} B_1 \tag{7.223}$$

$$ikB_1 = \frac{4\pi n_0 e}{c} V_{ix} - \frac{i\omega}{c} E_1 \tag{7.224}$$

$$-i\omega m_i V_{ix} = eE_1 + \frac{e}{c} V_{iy} B_0 \tag{7.225}$$

$$-i\omega m_i V_{iy} = \frac{-e}{c} V_{ix} B_0 \tag{7.226}$$

These equations are identical to (7.206) to (7.209) for the Alfvén waves, and we can immediately write the dispersion relation (7.214) for small frequencies; this is

$$\boxed{\omega^2 = \frac{k^2 V_A{}^2}{1 + V_A{}^2/c^2}} \tag{7.227}$$

for the fast magnetosonic wave traveling across $\hat{B}_0$.

This completes our discussion of linear wave equations in infinite uniform plasma. We have often looked at parallel and perpendicular waves; in practice, waves can propagate at any angle to the magnetic field. Waves propagating at an arbitrary angle usually have some combination of the properties of the corresponding parallel wave and the corresponding perpendicular wave. Because of the complexity of these waves, we shall not derive all of their properties here. However, a useful qualitative device exists for thinking about these waves. This is called the CMA diagram, after its inventors Clemmow, Mullaly, and Allis [4, 5]. The diagram is valid only for cold plasmas, $T_i = T_e = 0$. It shows all of the waves that can propagate at a given angle to the magnetic field for any combination of density and magnetic field intensity. This useful diagram is discussed in Refs. [6] and [7].

In the next section, we turn our attention from linear *waves* characterized by a real frequency and a real wave number, to linear *instabilities* characterized by a complex frequency and a real wave number.

7.13 TWO-STREAM INSTABILITY

Previous sections of this chapter have treated examples of linear waves that arise from the fluid theory of plasma. These waves are characterized by a real frequency and a real wave number, and would all be excited if a magnetized Maxwellian plasma were perturbed. When a plasma does not consist of Maxwellian electrons and Maxwellian ions, some of the waves (normal modes) of the system can become unstable. This subject is treated within the Vlasov theory in Section 6.9. Within the fluid theory, unstable normal modes can arise whenever the zero order electron and ion velocities are different, or whenever one species consists of two or more components each with different zero order velocities. Such instabilities are called *streaming instabilities.*

As an example of a streaming instability, consider a plasma in which the ions are stationary, while the electrons are traveling with speed V_0. The linearized fluid equations are then

$$\partial_t n_{e1} + n_0\, \partial_x V_{e1} + V_0\, \partial_x n_{e1} = 0 \tag{7.228}$$

$$m_e n_0\, \partial_t V_{e1} + m_e n_0 V_0\, \partial_x V_{e1} = -\, e n_0 E \tag{7.229}$$

$$\partial_t n_{i1} + n_0\, \partial_x V_{i1} = 0 \tag{7.230}$$

$$m_i n_0\, \partial_t\, V_{i1} = e n_0 E \tag{7.231}$$

and

$$\partial_x E = 4\pi e(n_{i1} - n_{e1}) \tag{7.232}$$

where we have assumed one-dimensional motions and $T_e = T_i = 0$, and the zeroth order speed V_0 contributes an extra term in (7.228) and in (7.229). Because we have no reason to suspect that the oscillations found here will be low frequency, we keep Poisson's equation (7.232) and we do not assume quasineutrality. In fact, if we allow $m_i \to \infty$ in (7.231), we would simply obtain the drifting cold plasma waves discussed in Problem 7.4; these are high frequency waves that become Langmuir waves in the limit that the drift speed $V_0 \to 0$. Here, we keep m_i large but finite and show that the drifting cold plasma waves are unstable; that is, when the frequency ω is obtained from (7.228) to (7.232), one finds $\text{Im}(\omega) > 0$; thus $\exp(-i\omega t) \sim \exp[\text{Im}(\omega)t]$, which grows exponentially with time. Since no instability is found in Problem 7.4, it must be the case that $\text{Im}(\omega) \to 0$ as $m_i \to \infty$

Fourier transforming (7.228) to (7.232), Eq. (7.229) yields

$$(-i\omega + ikV_0)V_{e1} = -\, eE/m_e \tag{7.233}$$

which when inserted in (7.228) yields

$$(-i\omega + ikV_0)n_{e1} - \frac{ikn_0 eE/m_e}{-i\omega + ikV_0} = 0 \tag{7.234}$$

while (7.231) in (7.230) yields

$$-i\omega n_{i1} = \frac{-ikn_0 eE/m_i}{-i\omega} \tag{7.235}$$

Then using (7.234) and (7.235) in Poisson's equation (7.232), we find the dispersion relation

$$ik = 4\pi e \left(\frac{ikn_0 e}{m_i \omega^2} + \frac{ikn_0 e}{m_e(\omega - kV_0)^2} \right) \tag{7.236}$$

or

$$\boxed{\epsilon(k,\omega) = 1 - \frac{\omega_i^2}{\omega^2} - \frac{\omega_e^2}{(\omega - kV_0)^2} = 0} \tag{7.237}$$

where we have identified the dielectric function $\epsilon(k,\omega)$ (see Section 7.4). In the limit $m_i \to \infty$, $\omega_i \to 0$, we find

$$\omega = kV_0 \pm \omega_e \tag{7.238}$$

in agreement with Problem 7.4.

With m_i finite, Eq. (7.237) is a quartic equation in ω, with four roots. Since (7.237) is a real equation, the complex conjuate of any root is also a root. (Why?) Thus, if we find any complex roots, either that root or its complex conjugate will have $\text{Im}(\omega) > 0$, and there will be an instability.

Let us use enlightened guessing to solve (7.237). Since the ions are important, we look for a wave such that the frequency is low in the laboratory frame [e.g., $kV_0 \approx \omega_e$ and the lower sign in (7.238)]. However, low frequency means only $|\omega| << \omega_e$; a vigorous instability could well lead to $|\omega| >> \omega_i$. Let us then look for a solution (possibly complex) to (7.237) that satisfies $\omega_i << |\omega| << \omega_e$. Then, because the second term in (7.237) is much less than unity, in order to cancel the first term we must have the third term close to unity; this leads us to look at wave numbers k such that $kV_0 = \omega_e$. Then (7.237) yields

$$\begin{aligned} 0 &= 1 - \frac{\omega_i^2}{\omega^2} - \frac{\omega_e^2}{(\omega - \omega_e)^2} \\ &= 1 - \frac{\omega_i^2}{\omega^2} - \frac{1}{(1 - \omega/\omega_e)^2} \\ &\approx 1 - \frac{\omega_i^2}{\omega^2} - \left(1 + \frac{2\omega}{\omega_e}\right) \\ &= -\frac{\omega_i^2}{\omega^2} - \frac{2\omega}{\omega_e} \end{aligned} \tag{7.239}$$

or

$$\omega^3 = -\frac{1}{2}\,\omega_i^2\omega_e \tag{7.240}$$

or

$$\boxed{\frac{\omega}{\omega_e} = \left(-\frac{1}{2}\right)^{1/3} \left(\frac{m_e}{m_i}\right)^{1/3}} \tag{7.241}$$

which represents instability since one of the three values of $(-1)^{1/3}$ is $(1/2) + i(3^{1/2}/2)$. In the frame moving with the electrons, the Doppler shifted frequency is $\omega' = \omega - kV_0$; since $kV_0 = \omega_e$ and $|\omega| << \omega_e$ this is roughly $\omega' \approx -\omega_e$, so that the electrons see an oscillation at nearly their natural frequency of oscillation.

There is another useful way to determine that (7.237) yields instability. From (7.237) we define

$$F(k,\omega) \equiv \frac{\omega_i^2}{\omega^2} + \frac{\omega_e^2}{(\omega - kV_0)^2} \tag{7.242}$$

We can plot this function versus real frequency ω at fixed wave number k, as sketched in Fig. 7.23. From (7.242) and the illustration we see that when the line at unity intersects the graph of $F(k,\omega)$ at four different points, there are four real roots and no instability for the chosen value of k. However, suppose the central minimum of $F(k,\omega)$ occurs at a value greater than unity; then there are only two real roots, as shown in Fig. 7.24. To determine when this happens, we determine when

$$F_{\min}(k,\omega) > 1 \tag{7.243}$$

where $F_{\min}$ is determined by $\partial F/\partial\omega = 0$ from (7.242). We find

$$\omega_{\min} \approx \left(\frac{m_e}{m_i}\right)^{1/3} kV_0 \tag{7.244}$$

and

$$F_{\min} \cong \frac{\omega_i^2}{(m_e/m_i)^{2/3}\, k^2V_0^2} + \frac{\omega_e^2}{k^2V_0^2} \tag{7.245}$$

which satisfies (7.243) and predicts instability whenever

$$|kV_0| \lesssim \omega_e \tag{7.246}$$

Thus, there is a broad range $-\omega_e/V_0 < k < \omega_e/V_0$ of unstable wave numbers.

Two-stream instabilities are very common in plasma physics. They happen whenever one fairly cold plasma component has a relative velocity with respect to another plasma component. These components need not be of different species; a cold electron beam impinging on an existing electron–ion plasma will produce its own instability. These instabilities are nature's way of saying that Maxwellians are desirable, and any configuration that is too far from Maxwellian will not last forever, even in the absence of collisions.

The linear theories of streaming instabilities for both cold components (fluid theory) and warm components (Vlasov theory) are very well understood. The

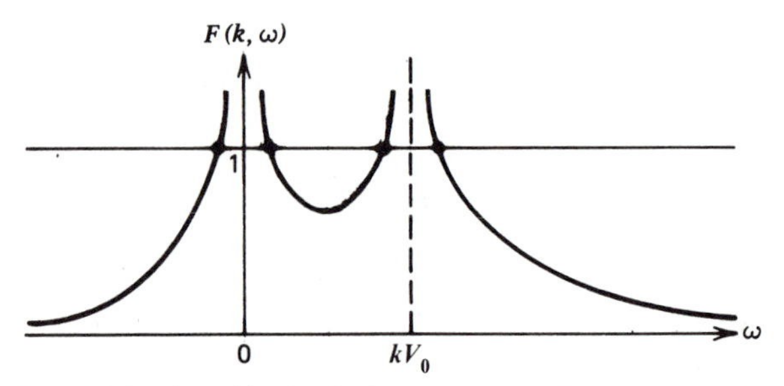

Fig. 7.23 Graphical solution of (7.237), for wave numbers k that yield four real roots.

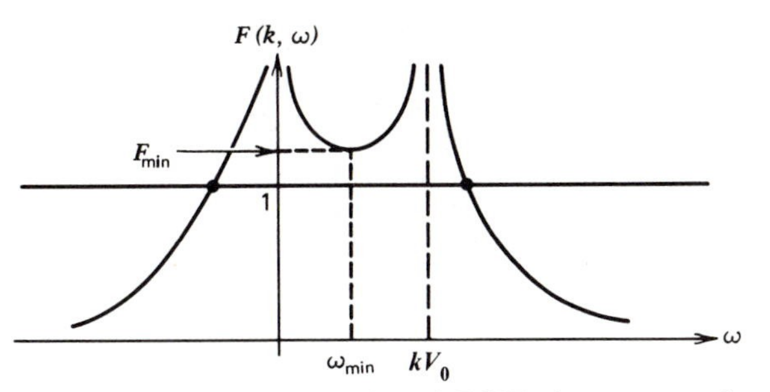

Fig. 7.24 Graphical solution of (7.237), for wave numbers k that yield only two real roots.

nonlinear saturation of these instabilities, involving such concepts as particle orbit modification, nonlinear wave-wave interactions, and strong turbulence, are not so well understood, and are the subject of considerable current research.

Up to this point in our study of the fluid theory, the waves and instabilities have propagated in a spatially homogeneous plasma. In the next section, we consider waves that propagate in a spatially inhomogeneous plasma.

7.14 DRIFT WAVES

Spatial inhomogeneities can give rise to their own wave motions. Consider an electrostatic wave, with frequency high enough that ions are unperturbed, but low enough that electrons perform an $\mathbf{E} \times \mathbf{B}_0$ drift in the wave field (Fig. 7.25). The wave number is predominantly in the $\hat{y}$-direction, but has a small $\hat{z}$ component to allow electrons to flow freely along the field lines. With $\mathbf{E}_1 = E_y\hat{y} + E_z\hat{z}$, the $E_y\hat{y} \times B_0\hat{z}$ drift is in the $\hat{x}$-direction, causing a charge separation. The continuity equation is

$$\partial_t n + \nabla \cdot (n\mathbf{V}) = 0 \tag{7.247}$$

or

$$-i\omega n_1 + \partial_x n_0 V_{1x} = 0 \tag{7.248}$$

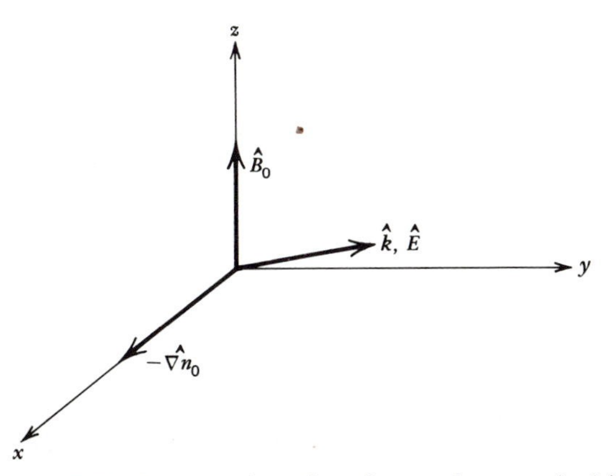

Fig. 7.25 Vector orientations for an electrostatic drift wave.

where we assume V_{1x} is not a function of x, we ignore the $k_z V_{1z}$ term because k_z is small, and we ignore the $k_y V_{1y}$ term because V_{1y} is small, being mostly a result of a polarization drift. Since V_{1x} results primarily from the $\mathbf{E} \times \mathbf{B}$ drift, we have

$$V_{1x} = \frac{E_1 c}{B_0} \tag{7.249}$$

since $E_y \approx E_1$; thus (7.248) yields

$$n_1 = \frac{1}{i\omega} \frac{\partial n_0}{\partial x} \frac{E_1}{B_0} c \tag{7.250}$$

Now the force equation in the $\hat{z}$-direction, ignoring $-i\omega m_e n_0 V_{1z}$ because of the smallness of ω, is

$$\begin{aligned} 0 &\approx - T_e i k_z n_1 - e n_0 E_z \\ &\approx - T_e i k_z n_1 - e n_0 E_1 (k_z/k_y) \end{aligned} \tag{7.251}$$

or

$$n_1 = \frac{i e n_0 E_1}{T_e k_y} \tag{7.252}$$

or equating (7.250) and (7.252) and eliminating E_1,

$$\omega = - \frac{T_e k_y}{e n_0} \frac{\partial n_0}{\partial x} \frac{c}{B_0} = \frac{v_e^2}{|\Omega_e| L_n} k_y \tag{7.253}$$

where the *density scale length*

$$L_n \equiv - \left(\frac{1}{n_0} \frac{dn_0}{dx} \right)^{-1} > 0 \tag{7.254}$$

Defining the *electron diamagnetic drift speed* (see Chapter 2)

$$v_{\mathrm{De}} = \frac{v_e^2}{|\Omega_e| L_n} \tag{7.255}$$

we can write (7.253) in the form

$$\boxed{\omega = k_y v_{\mathrm{De}}} \tag{7.256}$$

which is the dispersion relation for *electrostatic drift waves.*

There is a whole zoo of drift waves, matching in diversity all of the waves in homogeneous magnetized plasma. Drift waves are very important in magnetic confinement devices for controlled fusion such as the tokomak and mirror machine, and in the study of planetary ionospheres and magnetospheres. They are discussed in greater detail in Refs. [3], [6], and [8] to [19].

This brings us to the end of our study of *linear* fluid waves in magnetized and unmagnetized, homogeneous and inhomogeneous plasma. In the next two sections, we introduce the important subject of *nonlinear* fluid waves by adding one nonlinear term to two of the most important waves in plasma physics: ion-acoustic waves and Langmuir waves.

7.15 NONLINEAR ION-ACOUSTIC WAVES—KORTEWEG-DeVRIES EQUATION

Up to this point in the fluid theory we have considered only *linear* waves. We must always remember that the theory of linear waves restricts us to very small amplitudes. A wave with a finite amplitude will be susceptible to nonlinear effects, which show up mathematically as products of first order terms. This section and the next section are intended to introduce the concept of nonlinear wave equations and their corresponding solutions, which often take the form of solitons and shock waves.

Here we consider an example of one nonlinear wave equation, the Korteweg-deVries equation [20]:

$$\partial_t v + v\, \partial_z v + \alpha \partial_z^{\,3} v = 0 \tag{7.257}$$

This equation is obtained by adding one nonlinear term in the derivation of the ion-acoustic wave equation.

Although it is possible to give a rigorous derivation of (7.257), we give here only a heuristic derivation that indicates how one might arrive at (7.257). The origin of the terms in (7.257) is fairly easy to see. The first two terms might arise from the total time derivative of the ion fluid velocity. The third term can be seen in the ion-acoustic dispersion relation (7.104), which upon taking $T_i = 0$, $\gamma_e = 1$, is

$$\omega^2 = \frac{k^2 c_s^{\,2}}{1 + k^2 \lambda_e^{\,2}} \tag{7.258}$$

The square root of (7.258) is, for small $k\lambda_e$,

$$\omega = \frac{k c_s}{(1 + k^2 \lambda_e^{\,2})^{1/2}} = k c_s \left(1 - \frac{k^2 \lambda_e^{\,2}}{2}\right) \tag{7.259}$$

If we now multiply (7.259) on the right by the ion fluid velocity v, and identify $-i\omega$ with ∂_t and ik with ∂_x, we obtain

$$\frac{\partial v}{\partial t} = -\, c_s \frac{\partial v}{\partial x} - \frac{c_s \lambda_e^{\,2}}{2} \frac{\partial^3 v}{\partial x^3} \tag{7.260}$$

In a frame $x' = x - c_s t$ moving with the velocity c_s, and defining $\alpha = c_s \lambda_e^{\,2}/2$, we obtain

$$\partial_t v + \alpha \partial_{x'}^{\,3} v = 0 \tag{7.261}$$

which are the linear terms in (7.257). The nonlinear term is obtained by replacing the partial time derivative ∂_t with the convective time derivative $\partial_t + v\partial_{x'}$.

We begin our heuristic derivation with the five fluid equations. Taking $T_i \to 0$ so that we can neglect ion pressure in the ion force equation, and taking $m_e \to 0$ so that we can neglect electron inertia in the electron force equation, we find

$$\partial_t n_e + \partial_x (n_e V_e) = 0 \tag{7.262}$$

$$0 = -\, T_e\, \partial_x n_e - e n_e E \tag{7.263}$$

$$\partial_t n_i + \partial_x (n_i V_i) = 0 \tag{7.264}$$

$$m_i n_i\, \partial_t V_i + m_i n_i V_i\, \partial_x V_i = e n_i E \tag{7.265}$$

and

$$\partial_x E = 4\pi e (n_i - n_e) \tag{7.266}$$

where we have chosen $\gamma_e = 1$. We next linearize (7.262) to (7.266) everywhere except one place: we keep one nonlinear term, the $m_i n_0 V_i \partial_x V_i$ term on the left side of (7.265). We have then

$$\partial_t n_{e1} + n_0 \partial_x V_e = 0 \tag{7.267}$$

$$0 = -T_e \partial_x n_{e1} - e n_0 E \tag{7.268}$$

$$\partial_t n_{i1} + n_0 \partial_x V_i = 0 \tag{7.269}$$

$$m_i n_0 \partial_t V_i + m_i n_0 V_i \partial_x V_i = e n_0 E \tag{7.270}$$

$$\partial_x E = 4\pi e(n_{i1} - n_{e1}) \tag{7.271}$$

EXERCISE Can you find seven other nonlinear terms neglected in going from (7.262)–(7.266) to (7.267)–(7.271)?

A more rigorous derivation would show us the regime of validity implied by our neglect of seven other nonlinear terms while retaining only one nonlinear term. It turns out that this regime is reasonably large.

We next assume a plane wave solution, everywhere *except in (7.270)*. [What would happen if we tried to assume a plane wave solution $\sim \exp(-i\omega t + ikx)$ in (7.270)?] We also take $v \equiv V_e \approx V_i$, which means that (7.267) and (7.269) have the same information; we retain the difference between n_{e1} and n_{i1} so that (7.271) can be used. Solving (7.268) for n_{e1}, we find

$$n_{e1} = -\frac{e n_0 E}{ikT_e} \tag{7.272}$$

which inserted in Poisson's equation (7.271) yields

$$E = \frac{4\pi e n_{i1}}{ik - (\omega_i^2 m_i / ikT_e)} \tag{7.273}$$

n_{i1} is from (7.269)

$$n_{i1} = \frac{k n_0}{\omega} v \tag{7.274}$$

Both (7.274) in (7.273) and the result in (7.270) yield

$$\partial_t v + v \partial_x v = -\frac{ik^2 c_s^2}{\omega}(1 + k^2\lambda_e^2)^{-1} v \tag{7.275}$$

Here, we are still treating the right side as linear; therefore ω and k have their meanings as differential operators, while the left side is nonlinear. It proves convenient to eliminate ω on the right side; we do this by using the linear ion-acoustic dispersion relation (7.258), which is obtained from (7.275) by ignoring the nonlinear term and replacing the left side with $-i\omega v$. Solving for ω and substituting in the right side of (7.275), we have

$$\partial_t v + v \partial_x v = -ikc_s(1 + k^2\lambda_e^2)^{-1/2} v \tag{7.276}$$

For small $k\lambda_e$, we can expand the right side of (7.276) to obtain

$$\partial_t v + v \partial_x v = -ikc_s\left(1 - \frac{1}{2} k^2\lambda_e^2\right) v \tag{7.277}$$

Reinterpreting ik as ∂_x, this becomes

$$\boxed{\partial_t v + (c_s + v)\,\partial_x v + \alpha\,\partial_x{}^3 v = 0} \tag{7.278}$$

where $\alpha = \lambda_e{}^2 c_s/2$. In the frame $z = x - c_s t$, this is the Korteweg–deVries equation (7.257).

EXERCISE Show the above relationship.

Recall that $v(x,t)$ represents fluid velocity in the laboratory frame; this identification of $v(x,t)$ remains true even if we transform to a moving frame. What physics do the various terms in (7.278) represent? The first two terms by themselves,

$$\partial_t v + c_s\,\partial_x v = 0 \tag{7.279}$$

merely represent our old ion-acoustic waves in the limit $k\lambda_e \to 0$. The solution of (7.279) is simply a dispersionless wave, $\omega = kc_s$, with phase velocity $V_\varphi \equiv \omega/k = c_s$, and group velocity $d\omega/dk = c_s$ a constant independent of k. Suppose we add the nonlinear term to obtain

$$\partial_t v + (c_s + v)\,\partial_x v = 0 \tag{7.280}$$

The effect of the nonlinear term is as follows. Consider an initial waveform as shown in Fig. 7.26. As the wave moves, the part with larger v moves faster, so that it overtakes the part with smaller v. Eventually, at $t = t_2$, there is an infinite slope, and at $t = t_3$, the wave has broken. Now suppose we had included the dispersive term in (7.278); the term $\alpha\,\partial_x{}^3 v$ is called dispersive because it contributes a term k^3 to the linear dispersion relation $\omega = kc_s - \alpha k^3$; then $V_g \equiv d\omega/dk = c_s - 3\alpha k^2$, which depends on k, making this a dispersive wave. We know the effect of dispersion on a wave; it makes a wave packet spread out as it travels. This is just opposite to the steepening observed in the figure. Consider the time between $t = t_1$ and $t = t_2$. Here, the slope is becoming very large. A large slope corresponds to a large x-derivative, which makes the $\alpha\,\partial_x{}^3 v$ term in (7.278) become large. Since we know that the effect of this $\alpha\,\partial_x{}^3 v$ term is to spread out the wave, we might expect that there could be a balance between the nonlinear steepening and the linear dispersion. Indeed this is the case. One can obtain nonlinear wave packets, known as *solitons*, which travel without change of shape (Fig. 7.27). The physical basis for these solitons involves a balance between dispersion and nonlinearity.

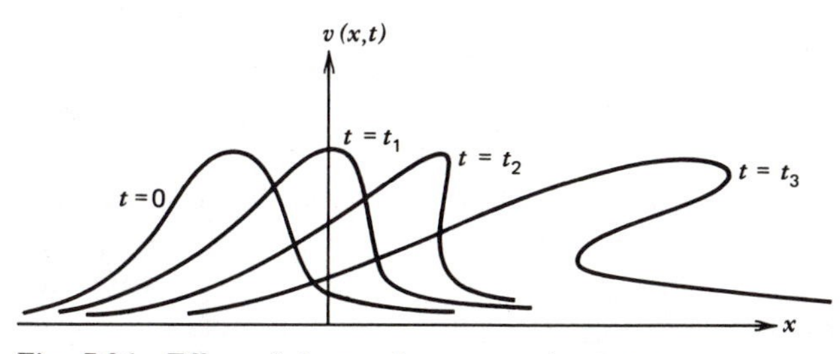

Fig. 7.26 Effect of the nonlinear term in (7.282).

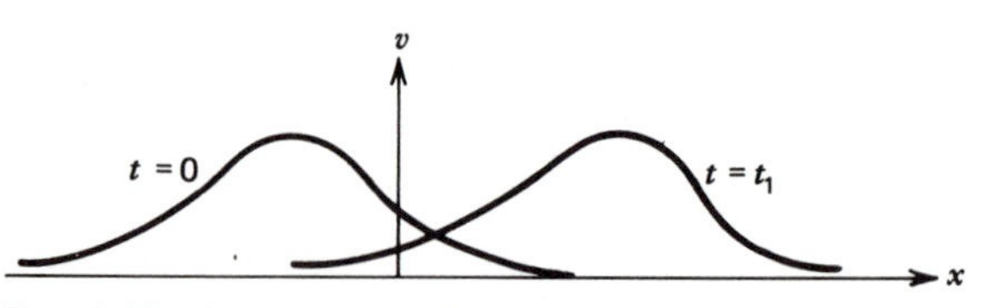

Fig. 7.27 Sketch of a soliton solution.

Let us proceed to find a soliton solution to the Korteweg–deVries equation (7.278). We look for stationary solutions in a moving frame,

$$x' = x - v_0 t \tag{7.281}$$

$$t' = t \tag{7.282}$$

so that

$$\partial_x = (\partial x'/\partial x)\partial_{x'} + (\partial t'/\partial x)\partial_{t'} = \partial_{x'} \tag{7.283}$$

and

$$\partial_t = (\partial x'/\partial t)\partial_{x'} + (\partial t'/\partial t)\partial_{t'} = \partial_{t'} - v_0 \partial_{x'} \tag{7.284}$$

Since stationary implies $\partial_{t'} = 0$, the Korteweg–deVries equation (7.278) becomes

$$(-v_0 + c_s + v)\,\partial_{x'} v + \alpha\,\partial_{x'}{}^3 v = 0 \tag{7.285}$$

Remember that $v(x',t')$ is still that function of space and time which represents the fluid velocity in the lab frame. Equation (7.285) can be integrated once immediately, to give

$$(c_s - v_0)v + \frac{v^2}{2} + \alpha v'' = 0 \tag{7.286}$$

where $(\;)' \equiv \partial_{x'}(\;)$ and we have taken the integration constant to vanish. Equation (7.286) is in the form

$$\alpha v'' = (v_0 - c_s)v - \frac{v^2}{2} \tag{7.287}$$

which has the same mathematical form as Newton's law of motion,

$$m\ddot{x} = F(x) = -\,\partial_x V(x) \tag{7.288}$$

where $V(x)$ is the potential energy. Thus, (7.287) has the form

$$\alpha v'' = -\,\partial_v \left[(c_s - v_0)\,\frac{v^2}{2} + \frac{v^3}{6}\right] \tag{7.289}$$

Equation (7.289) has the same mathematical form as a force equation for a particle of mass α moving under the influence of a potential field given by the quantity in brackets. We call the quantity in brackets the pseudopotential,

$$\Phi(v) = (c_s - v_0)\,\frac{v^2}{2} + \frac{v^3}{6} \tag{7.290}$$

A graph of $\phi(v)$ is shown for $c_s - v_0 > 0$ in Fig. 7.28. A similar graph of the pseudopotential $\Phi(v)$ for $(c_s - v_0) < 0$ is shown in Fig. 7.29. Only the second form is suitable for our purposes. This is because we desire a localized wave form, $v(x' \to \pm\infty) \to 0$. This will occur in Fig. 7.29 when the pseudoparticle leaves $v = 0$ when the pseudotime $x' \to -\infty$, falling once through the well to reach $v_{\max}$

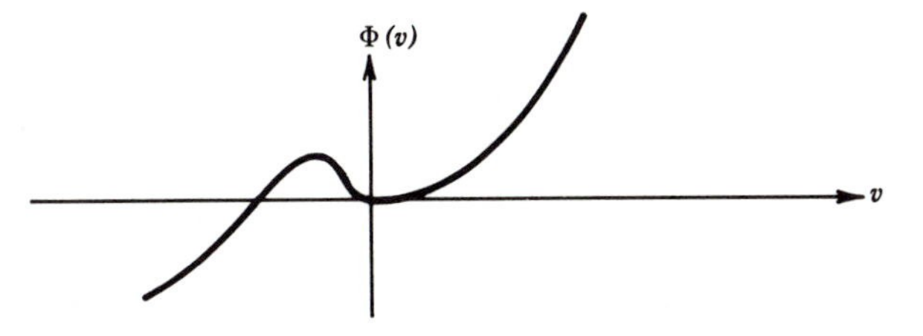

Fig. 7.28 Sketch of pseudopotential when $c_s > v_0$.

at $x' = 0$, and taking an infinite amount of pseudotime x' to fall back through the well to reach $v = 0$ as the pseudotime $x' \to +\infty$. We thus obtain the shape shown in Fig. 7.30. The pseudopotential in Fig. 7.28 would not allow $v(x' \to \pm\infty) \to 0$.

Let us now solve (7.287) exactly, with $c_s - v_0 < 0$ or $v_0 > c_s$. We all know how to solve force equations of the form (7.287). Multiply (7.287) by v' and integrate, to obtain,

$$\frac{\alpha}{2}(v')^2 = (v_0 - c_s)\frac{v^2}{2} - \frac{v^3}{6} \tag{7.291}$$

where we have chosen the constant of integration to be zero because we want $v' = 0$ when $v = 0$ (Fig. 7.30). Then

$$\frac{dv}{dx'} = v' = \left(\frac{2}{\alpha}\right)^{1/2}\left[(v_0 - c_s)\frac{v^2}{2} - \frac{v^3}{6}\right]^{1/2} \tag{7.292}$$

or

$$\frac{dv}{\left[\dfrac{(v_0 - c_s)}{2}v^2 - \dfrac{v^3}{6}\right]^{1/2}} = \left(\frac{2}{\alpha}\right)^{1/2} dx' \tag{7.293}$$

Each side of (7.293) can be integrated. The left side is of the form

$$I = \int \frac{dv}{\sqrt{v^2 - \beta v^3}} = \int \frac{dv}{v\sqrt{1 - \beta v}} \tag{7.294}$$

where $\beta = 1/[3(v_0 - c_s)]$. Let $u = \sqrt{1 - \beta v}$, then $v = (1 - u^2)/\beta$ and $du = (-\beta\, dv/2)/\sqrt{1 - \beta v}$. We find

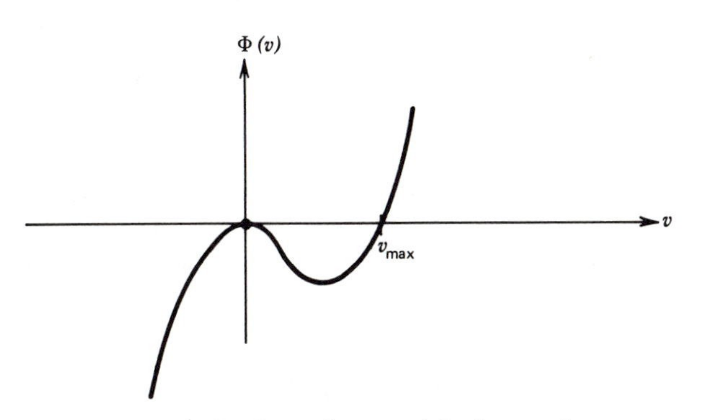

Fig. 7.29 Sketch of pseudopotential when $c_s < v_0$.

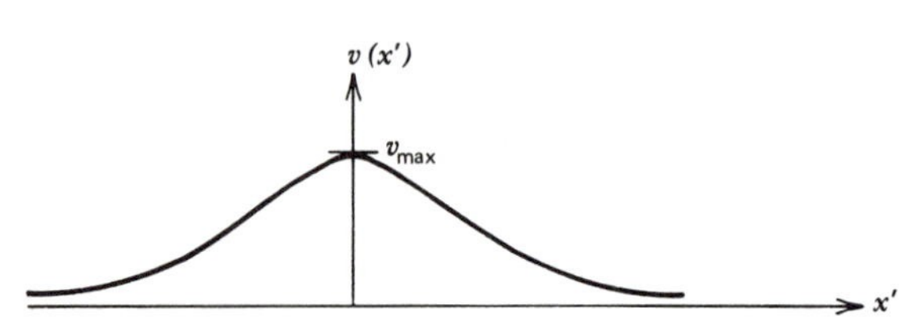

Fig. 7.30 Sketch of soliton solution to the Korteweg–deVries equation.

$$I = -2\int \frac{du}{1-u^2} = -\int du\left(\frac{1}{1-u}+\frac{1}{1+u}\right) = \ln\left(\frac{1-u}{1+u}\right) \tag{7.295}$$

Then

$$\left(\frac{2}{v_0 - c_s}\right)^{1/2} \ln\left(\frac{1-u}{1+u}\right) = \left(\frac{2}{\alpha}\right)^{1/2} x' \tag{7.296}$$

from (7.293). With $\gamma \equiv [(v_0 - c_s)/\alpha]^{1/2}$, and exponentiating both sides, we get

$$\frac{1-u}{1+u} = e^{\gamma x'} \tag{7.297}$$

Then $1 - u = (1 + u)e^{\gamma x'}$, implying that

$$u = \frac{1 - e^{\gamma x'}}{1 + e^{\gamma x'}} \tag{7.298}$$

and $v = (1 - u^2)/\beta$ is

$$v = \frac{1}{\beta}\left[\frac{(1+e^{\gamma x'})^2 - (1-e^{\gamma x'})^2}{(1+e^{\gamma x'})^2}\right] = \frac{1}{\beta}\left[\frac{4e^{\gamma x'}}{(1+e^{\gamma x'})^2}\right] \tag{7.299}$$

or

$$v = \frac{1}{\beta}\frac{4}{(e^{\gamma x'/2} + e^{-\gamma x'/2})^2} = \frac{1}{\beta}\operatorname{sech}^2(\gamma x'/2) \tag{7.300}$$

which is

$$\boxed{v = 3(v_0 - c_s)\operatorname{sech}^2\left[\left(\frac{v_0 - c_s}{4\alpha}\right)^{1/2} x'\right]} \tag{7.301}$$

In fact, this solution has only been derived for $x' < 0$ since we chose the $v' > 0$ branch in (7.292); nevertheless, it would be easy to obtain the part of (7.301) for $x' > 0$ by choosing the $v' < 0$ branch in (7.292); therefore, (7.301) applies to all x' and is the soliton solution. Note that the larger amplitude solitons are more sharply peaked, having a smaller scale length. This behavior is in accordance with our picture of nonlinearity $v\,\partial_x v$ which balances dispersion $\partial_x{}^3 v$ (Fig. 7.31). Back in the lab frame, where $x = x' + v_0 t$, this solution is

$$\boxed{v(x,t) = 3(v_0 - c_s)\operatorname{sech}^2\left[\left(\frac{v_0 - c_s}{4\alpha}\right)^{1/2}(x - v_0 t)\right]} \tag{7.302}$$

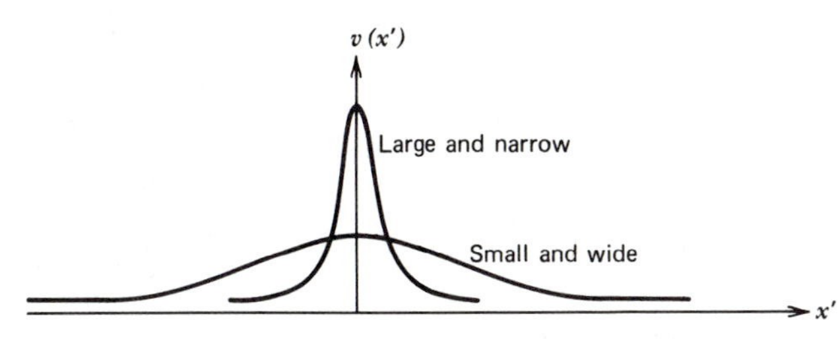

Fig. 7.31 Soliton solutions exhibit the balance between dispersion and nonlinearity.

EXERCISE Write out $\text{sech}^2 (x')$ in terms of exponentials, and show that it reproduces the solition behavior shown in Fig. 7.31.

7.16 NONLINEAR LANGMUIR WAVES—ZAKHAROV EQUATIONS

In the previous section, the addition of one nonlinear term to the equation for ion-acoustic waves led to a nonlinear wave equation with soliton solutions. In this section, the addition of a different nonlinear term, representing the ponderomotive force, to the equation for ion-acoustic waves leads to a set of coupled nonlinear wave equations that describe the nonlinear interaction between high frequency Langmuir waves and low frequency ion-acoustic waves.

Consider a collection of linear Langmuir waves in one spatial dimension whose electric field (the subscript h stands for *h*igh frequency) can be written

$$\mathrm{E}_h(x,t) = \frac{1}{2} \tilde{E}(x,t) \exp(-i\omega_e t) + \text{c.c.} \tag{7.303}$$

The amplitude $\tilde{E}(x,t)$ contains the $\frac{3}{2}k^2\lambda_e^2\omega_e$ frequency dependence of the Langmuir waves. Since $k^2\lambda_e^2 << 1$ for Langmuir waves, the function $\tilde{E}(x,t)$ varies slowly in time compared to the rapidly varying $\exp(-i\omega_e t)$. Thus, in the ponderomotive force equation (2.76) the constant field E_0 can be replaced by the slowly varying amplitude $\tilde{E}(x,t)$ so that the low frequency ponderomotive force acting on electrons is

$$F_p = \frac{-e^2}{4m_e\omega_e^2} \frac{d}{dx} |\tilde{E}|^2 \tag{7.304}$$

where the plasma frequency ω_e appears in the denominator because all components of the Langmuir wave field E_h have frequency near the plasma frequency.

Our goal is to rederive the ion-acoustic wave equation including the ponderomotive force (7.304). This couples the low frequency ion-acoustic waves to the high frequency Langmuir waves. If we then rederive the high frequency Langmuir-wave equation including the change in the background density due to the presence of ion-acoustic waves, we will have two coupled nonlinear equations in the two unknowns representing Langmuir-wave electric field and ion-acoustic wave density perturbation.

The derivation of these equations makes explicit use of the fact that the problem has two time scales. Thus, we shall often encounter equations of the general form

$$a(t) + b(t) \exp(-i\omega_e t) = c(t) + d(t) \exp(-i\omega_e t) \tag{7.305}$$

where a, b, c, and d vary slowly compared to the time scale ω_e^{-1}, that is,

$$\left|\frac{1}{a}\frac{da}{dt}\right| \frac{1}{\omega_e} << 1 \tag{7.306}$$

and likewise for b, c, and d. Then to a good approximation we can hold a, b, c, and d constant over the short time interval $2\pi/\omega_e$ and integrate (7.305) from any time t to $t + 2\pi/\omega_e$; the exponential terms vanish leaving

$$a(t) = c(t) \tag{7.307}$$

This procedure is called *averaging over the fast time scale*. Similarly, one can first multiply (7.305) by exp $(i\omega_e t)$; averaging over the fast time scale then yields

$$b(t) = d(t) \tag{7.308}$$

In this manner, one can pick out all of the terms in a given equation that have either a fast or a slow time dependence. Similar considerations apply to terms with different wave numbers.

With these preliminaries, let us derive a unified set of fluid equations that describes Langmuir-wave physics, ion-acoustic wave physics, and the nonlinear coupling between them. The discussion is heuristic; we keep only the nonlinear terms we are looking for and throw away many others without rigorous justification. All quantities are separated into high frequency (subscript h) and low frequency (subscript l) components,

$$n_e(x,t) = n_0 + n_{el}(x,t) + n_{eh}(x,t) \tag{7.309}$$

$$n_i(x,t) = n_0 + n_{il}(x,t) \tag{7.310}$$

$$V_e(x,t) = V_{el}(x,t) + V_{eh}(x,t) \tag{7.311}$$

$$V_i(x,t) = V_{il}(x,t) \tag{7.312}$$

$$E(x,t) = E_l(x,t) + E_h(x,t) \tag{7.313}$$

where we ignore the high frequency portions of ion quantities because of their large mass. The density perturbations n_{el}, n_{eh}, and n_{il} are all considered to be much smaller than n_0.

First, we repeat the derivation of the Langmuir-wave equation including the density perturbation $n_{el}(x,t)$ due to the low frequency waves. The high frequency components of Poisson's equation, the electron continuity equation, and the electron force equation are

$$\partial_x E_h = -4\pi e n_{eh} \tag{7.314}$$

$$\partial_t n_{eh} + \partial_x[(n_0 + n_{el})V_{eh}] = 0 \tag{7.315}$$

$$m_e n_0 \, \partial_t V_{eh} = -3T_e \, \partial_x n_{eh} - e n_0 E_h \tag{7.316}$$

where we note that the product of a high frequency term and a low frequency term is a high frequency term. The total low frequency electron density $n_0 + n_{el}$ has been replaced by n_0 in several places in (7.315) and (7.316), and the term $\partial_x(n_{eh}V_{el})$ has been ignored in (7.315). Taking the time derivative of only the high frequency terms in (7.315), eliminating $\partial_t V_{eh}$ using (7.316), and eliminating n_{eh} using (7.314) yield

$$\partial_t^2 E_h + \omega_e^2 E_h - 3v_e^2 \partial_x^2 E_h = -\omega_e^2 \frac{n_{el}}{n_0} E_h \qquad (7.317)$$

where $\omega_e^2 \equiv 4\pi n_0 e^2/m_e$. The left side is easily recognizable as the linear Langmuir-wave equation, while the right side gives the change in the effective plasma frequency due to the fact that the low frequency electron density is $n_0 + n_{el}$ rather than n_0.

Inserting the form (7.303) into (7.317) and keeping only terms with time dependence exp $(-i\omega_e t)$, we find

$$\boxed{i\,\partial_t \tilde{E} + \frac{3}{2}\frac{v_e^2}{\omega_e}\partial_x^2 \tilde{E} = \frac{\omega_e}{2}\frac{n_{el}}{n_0}\tilde{E}} \qquad (7.318)$$

where the term $|\partial_t^2 \tilde{E}| << |\omega_e \partial_t \tilde{E}|$ has been discarded. Equation (7.318) is now a low frequency equation describing the time evolution of the slowly varying *envelope* $\tilde{E}(x,t)$ of the rapidly varying electric field $E_h(x,t)$.

Next, we repeat the derivation of the ion-acoustic wave equation including the ponderomotive force (7.304) in the electron force equation. Assuming quasineutrality $n_{el} \approx n_{il}$ and $V_{el} \approx V_{il}$, the low frequency part of the electron continuity equation is

$$\partial_t n_{el} + n_0 \partial_x V_{el} = 0 \qquad (7.319)$$

where the term $\partial_x(n_{el}V_{el})$ has been ignored; the low frequency part of the electron force equation, ignoring $m_e\partial_t V_{el}$ because of the small electron mass, is

$$0 = -\frac{T_e\gamma_e}{n_0}\partial_x n_{el} - eE_l - \frac{e^2}{4m_e\omega_e^2}\partial_x|\tilde{E}|^2 \qquad (7.320)$$

and the ion force equation yields

$$m_i \partial_t V_{el} = -\frac{T_i\gamma_i}{n_0}\partial_x n_{el} + eE_l \qquad (7.321)$$

Here, γ_e and γ_i are the usual factors relating pressure change to density change, and $n_0 + n_{el}$ has been replaced by n_0 in several places in (7.320) and (7.321). Solving (7.321) for E_l, substituting the result in (7.320), taking the spatial derivative, and eliminating V_{el} using (7.319), yield

$$\boxed{\partial_t^2 n_{el} - c_s^2 \partial_x^2 n_{el} = \frac{1}{16\pi m_i}\partial_x^2|\tilde{E}|^2} \qquad (7.322)$$

where the sound speed is defined by

$$c_s^2 = \frac{\gamma_e T_e + \gamma_i T_i}{m_i} \qquad (7.323)$$

as usual. The coupled equations (7.318) and (7.322) were first derived by Zakharov [21] and are known as the *Zakharov equations.*

We wish to study the consequences of the nonlinear equations (7.318) and (7.322), including soliton solutions and *parametric instabilities*. It is convenient to define dimensionless variables as

$$\eta \equiv \frac{\gamma_e T_e + \gamma_i T_i}{T_e} \tag{7.324}$$

$$\tau \equiv \left(\frac{2\eta}{3}\right)\left(\frac{m_e}{m_i}\right)(\omega_e t) \tag{7.325}$$

$$z \equiv \left(\frac{2}{3}\right)\left(\frac{\eta m_e}{m_i}\right)^{1/2}\left(\frac{x}{\lambda_e}\right) \tag{7.326}$$

$$E \equiv \left(\frac{1}{\eta}\right)\left(\frac{m_i}{m_e}\right)^{1/2}\left(\frac{3\tilde{E}^2}{64\pi n_0 T_e}\right)^{1/2} \tag{7.327}$$

$$n = \left(\frac{3m_i}{4\eta m_e}\right)\left(\frac{n_{el}}{n_0}\right) \tag{7.328}$$

whereupon (7.318) and (7.322) become

$$i\,\partial_\tau E + \partial_z^2 E = nE \tag{7.329}$$

$$\partial_\tau^2 n - \partial_z^2 n = \partial_z^2 |E|^2 \tag{7.330}$$

Let us look for soliton solutions to (7.329) and (7.330). The simplest soliton solution [22] is one that is stationary in the laboratory frame; it is a bump of electric field intensity that exists self-consistently with the hole in ion density dug out by the ponderomotive force. The first term on the left of (7.330) vanishes; integrating twice and setting the constants of integration equal to zero yield

$$n = -\,|E|^2 \tag{7.331}$$

so that (7.329) becomes

$$\boxed{i\,\partial_\tau E + \partial_z^2 E + |E|^2 E = 0} \tag{7.332}$$

which is called the nonlinear Schrödinger equation because it resembles the quantum mechanical Schrödinger equation.

Looking for a solution of the form

$$E(z,\tau) = \exp(i\Omega\tau)\,f(z) \tag{7.333}$$

we find that Eq. (7.332) becomes [with $(\ )' \equiv d(\ )/dz$]

$$f'' = \Omega f - f^3 \tag{7.334}$$

This can be solved by the same pseudopotential method used in the previous section to solve the Korteweg–deVries equation. We write (7.334) in the form

$$f'' = -\frac{d}{df}\left[\frac{1}{4}f^4 - \frac{1}{2}\Omega f^2\right] \tag{7.335}$$

Multiplying both sides by f' and integrating yield

$$(f')^2 = \Omega f^2 - \frac{1}{2}f^4 \tag{7.336}$$

or

$$\frac{df}{dz} = f\left(\Omega - \frac{1}{2} f^2\right)^{1/2} \tag{7.337}$$

or

$$\frac{df}{f\left(\Omega - \frac{1}{2} f^2\right)^{1/2}} = dz \tag{7.338}$$

or

$$\int \frac{df}{f\left(\Omega - \frac{1}{2} f^2\right)^{1/2}} = z \tag{7.339}$$

This integral can be performed with the substitution $u = (1 - f^2/2\Omega)^{1/2}$. With $f > 0$ everywhere, (7.339) can be integrated to yield

$$\frac{1}{2} \ln \left(\frac{1 - u}{1 + u}\right) = \Omega^{1/2} z \tag{7.340}$$

Solving for u and converting back to f, one finds

$$f = (2\Omega)^{1/2} \operatorname{sech} (\Omega^{1/2} z) \tag{7.341}$$

so that the total field, as given by (7.333), is

$$\boxed{E(z,\tau) = (2\Omega)^{1/2} \exp (i\Omega\tau) \operatorname{sech} (\Omega^{1/2} z)} \tag{7.342}$$

which can be called a *Langmuir soliton*.

EXERCISE Sketch the solution (7.342) and show that it is indeed a localized "bump." Sketch the density perturbation (7.331).

A more general class of solitons exists [23], moving at any speed the absolute value of which is less than the sound speed.

In the next section, we turn our attention to another important subject that can be studied within the context of the Zakharov equations: *parametric instabilities*. The study of solitons and parametric instabilities is one of the most active areas of research in plasma physics [24–26].

7.17 PARAMETRIC INSTABILITIES

Consider a plasma that contains a single plane wave of finite amplitude. Within the fluid theory, the system of plasma plus wave can be thought of as a time-dependent equilibrium state. We can then ask the question: Is such an equilibrium stable or unstable? This is the same question we asked about time-independent equilibria in Chapter 6 on Vlasov theory and in Section 7.13 on the two-stream instability. The answer to the question often indicates instability, and such instabilities are called *parametric instabilities*, the "parameter" being the amplitude of the single wave.

One can look for such instabilities with any of the waves studied in this book. For example, we shall use Langmuir waves, the stability of which can be studied within the context of the Zakharov equations of the previous section. It turns out that the most general instability in this case involves the single finite-amplitude Langmuir wave, two other Langmuir waves, and one low frequency wave. The stability analysis proceeds by assuming that the amplitudes of the two other Langmuir waves and the low frequency wave are infinitesimal. We choose

$$E(z,\tau) = E_0 \exp(-i\omega_0\tau + ik_0 z) + E_+ \exp[-i(\omega_0 + \omega)\tau + i(k_0 + k)z] + E_- \exp[-i(\omega_0 - \omega^*)\tau + i(k_0 - k)z] \tag{7.343}$$

and

$$n = \tilde{n} \exp(-i\omega\tau + ikz) + \text{complex conjugate} \tag{7.344}$$

where $\tilde{n}$, E_-, and E_+ are all much smaller than E_0. The equilibrium solution $E(z,\tau) = E_0 \exp(-i\omega_0\tau + ik_0 z)$, $n(z,\tau) = 0$, is chosen to satisfy the Zakharov equations with E_0 real.

EXERCISE Show that this solution implies $\omega_0 = k_0^2$.

Inserting the forms (7.343) and (7.344) into the first Zakharov equation (7.329), and keeping only those terms with spatial dependence $\sim \exp[i(k_0 + k)z]$, we find

$$(\omega_0 + \omega)E_+ - (k_0 + k)^2 E_+ = \tilde{n} E_0 \tag{7.345}$$

Likewise, the terms with spatial dependence $\sim \exp[i(k_0 - k)z]$ yield

$$(\omega_0 - \omega^*)E_- - (k_0 - k)^2 E_- = \tilde{n}^* E_0 \tag{7.346}$$

Solving (7.345) and (7.346) for E_+ and E_-, inserting these into the second Zakharov equation (7.330), keeping only terms with spatial variation $\sim \exp(ikz)$, and eliminating $\tilde{n}$ from each term yield the dispersion relation

$$\omega^2 - k^2 = k^2 E_0^2 \left(\frac{1}{\omega - k^2 - 2k_0 k} + \frac{1}{-\omega - k^2 + 2k_0 k} \right) \tag{7.347}$$

There are several types of solutions. With $k_0 > 0$, we first look for an instability with $k < 0$. If $|\omega|$ is small, the second denominator on the right is larger than the first, so we ignore the second. This is equivalent to ignoring the term E_- in the electric field (7.343), so this instability involves only E_0, E_+, and $\tilde{n}$ and is thus known as a *three-wave interaction*. The dispersion relation is now

$$(\omega^2 - k^2)(\omega - 2kk_0 - k^2) - k^2 E_0^2 = 0 \tag{7.348}$$

Looking for a solution with $\omega = k + \delta$ where $|\delta| << |k|$, we write $(\omega^2 - k^2) = (\omega + k)(\omega - k) = (2k + \delta)\delta \approx 2k\delta$; Eq. (7.348) then yields

$$\delta^2 + \delta(k - 2kk_0 - k^2) - \frac{kE_0^2}{2} = 0 \tag{7.349}$$

At the particular negative wave number that satisfies $k - 2kk_0 - k^2 = 0$ or $k = -2k_0 + 1$, this is

$$\delta = \pm \left(\frac{kE_0^2}{2}\right)^{1/2} \tag{7.350}$$

which indicates instability since $k < 0$. If $k_0 >> 1$, this becomes

$$\delta = ik_0^{1/2}E_0 \tag{7.351}$$

EXERCISE Show that in physical units denoted by a tilde, $k_0 >> 1$ means $\tilde{k}_0\lambda_e >> (m_e/m_i)^{1/2}$.

EXERCISE What does $|\delta| << |k|$ mean in physical units?

Since the electric field E_+ has a wave number $k_0 + k = k_0 - 2k_0 + 1 = -k_0 + 1$, E_+ has a negative wave number and travels in the opposite direction to E_0. It is known as a *backscatter instability*, and is one example of a *parametric decay instability*. The physical growth rate $\tilde{\gamma}$ is

$$\frac{\tilde{\gamma}}{\omega_e} = \left(\frac{m_e}{\eta m_i}\right)^{1/4} (\tilde{k}_0\lambda_e)^{1/2} \frac{\tilde{E}_0}{(32\pi n_0 T_e)^{1/2}} \tag{7.352}$$

where physical quantities are denoted by a tilde.

EXERCISE Demonstrate (7.352) from (7.351).

The dispersion relation (7.347) also yields an instability that involves all of the terms and thus is known as a *four-wave interaction*. The simplest case is when $\omega_0 = k_0 = 0$; that is, the physical field represented by E_0 is oscillating exactly at the plasma frequency ω_e and has zero wave number (a so-called *dipole field*). Looking for a purely growing instability $\omega = i\gamma$, we see that the dispersion relation (7.347) becomes

$$(\gamma^2 + k^2)(\gamma^2 + k^4) - 2k^4 E_0^2 = 0 \tag{7.353}$$

the solution of which is

$$\gamma^2 = -\frac{1}{2}(k^2 + k^4) + \frac{1}{2}[(k^2 - k^4)^2 + 8k^4 E_0^2]^{1/2} \tag{7.354}$$

With both $k << 1$ and $E_0 << 1$ the k^8 term within the bracket can be discarded, and the square root can be expanded to yield

$$\gamma = k(2E_0^2 - k^2)^{1/2} \tag{7.355}$$

which is the growth rate of the four-wave interaction known as the *oscillating two-stream instability* [27]. The growth rate versus wave number is sketched in Fig. 7.32. The maximum growth rate $\gamma = E_0^2$ occurs at $k = \pm E_0$.

EXERCISE Show that in physical units these are

$$\tilde{\gamma}/\omega_e = \frac{\tilde{E}_0^2}{32\pi n_0 T_e} \tag{7.356}$$

and

$$\tilde{k}\lambda_e = \left(\frac{\tilde{E}_0^2}{48\eta\pi n_0 T_e}\right)^{1/2} \tag{7.357}$$

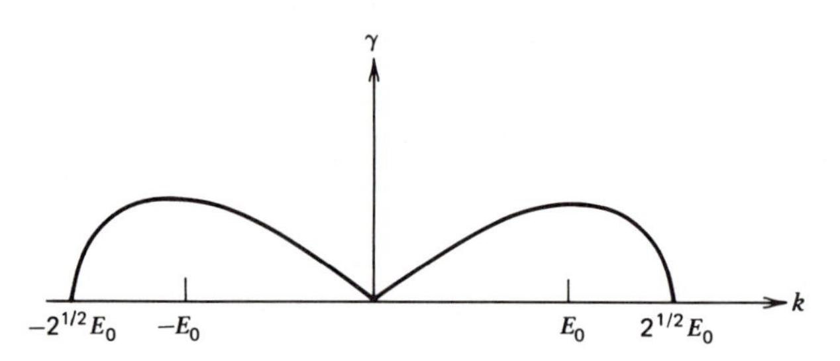

Fig. 7.32 Growth rate versus wave number for the oscillating two-stream instability.

The study of parametric instabilities is very important for such fields as laser fusion, particle beam fusion, radio-frequency heating of the ionosphere and of magnetic confinement devices, and solar radio physics.

This brings us to the end of our study of the fluid equations of plasma physics. In the next chapter, the fluid equations for each species are combined to yield the equations of *magnetohydrodynamics*.

REFERENCES

[1] K. R. Symon, *Mechanics*, Addison-Wesley, Reading, Mass., 1960.

[2] A. N. Kaufman, in *Plasma Physics in Theory and Application*, edited by W. B. Kunkel, McGraw-Hill, New York, 1966, p. 91.

[3] N. A. Krall and A. W. Trivelpiece, *Principles of Plasma Physics*, McGraw-Hill, New York, 1973.

[4] P. C. Clemmow and R. F. Mullaly, in *Physics of the Ionosphere: Report of Phys. Soc. Conf. Cavendish Lab.*, Physical Society, London, 1955, p. 340.

[5] W. P. Allis, in *Sherwood Conf. Contr. Fusion*, Gatlinburg, Tennessee, April 27–28, 1959, TID-7582, p. 32.

[6] F. F. Chen, *Introduction to Plasma Physics*, Plenum, New York, 1974.

[7] T. H. Stix, *The Theory of Plasma Waves*, McGraw-Hill, New York, 1962.

[8] W. M. Manheimer, *An Introduction to Trapped-Particle Instability in Tokamaks*, TID-27157, National Technical Information Service, Springfield, Virginia, 1977.

[9] N. A. Krall, in *Advances in Plasma Physics*, Vol. 1, edited by A. Simon and W. B. Thompson, Wiley-Interscience, New York, 1968, p. 153.

[10] T. K. Chu, B. Coppi, H. W. Hendel, and F. W. Perkins, *Phys. Fluids*, *12*, 203 (1969).

[11] B. Coppi, *Revista del Nuovo Cimento*, *1*, 357 (1969).

[12] P. Rutherford and E. Frieman, *Phys. Fluids*, *11*, 569 (1965).

[13] A. A. Rukhadze and V. P. Silin, *Sov. Phys. Uspekhi*, *11*, 659 (1969).

[14] J. D. Jukes, *Rep. Prog. Phys.*, *31*, 305 (1968).

[15] P. H. Rebut, *Plasma Phys.*, *9*, 671 (1967).

[16] A. B. Mikhailovskii, *Rev. Plasma Phys.*, *3*, 159 (1967).

[17] B. B. Kadomtsev and O. P. Pogutse, *Rev. Plasma Phys.*, *5*, 249 (1970).

[18] A. B. Mikhailovskii, *Theory of Plasma Instabilities*, Vol. 2: *Instabilities of an Inhomogeneous Plasma*, Consultant Bureau, New York, 1974.

[19] F. F. Chen, *Sci. Am.*, *217*, 76 (1967).

[20] D. J. Korteweg and G. deVries, *Phil. Mag.*, *39*, 422 (1895).

[21] V. E. Zakharov, *Zh. Eksp. Teor. Fiz.*, *62*, 1745 (1972) [*Sov. Phys.-JETP*, *35*, 908 (1972)].

[22] P. J. Hansen and D. R. Nicholson, *Am. J. Phys.*, *47*, 769 (1979).

[23] G. Schmidt, *Phys. Rev. Lett.*, *34*, 724 (1975).

[24] A. C. Scott, F. Y. F. Chu, and D. W. McLaughlin, *Proc. IEEE*, *61*, 1443 (1973).

[25] G. B. Whitham, *Linear and Nonlinear Waves*, Wiley, New York, 1974.

[26] K. E. Lonngren and A. C. Scott, eds., *Solitons in Action*, Academic, New York, 1978.

[27] K. Nishikawa, *J. Phys. Soc. Jpn.*, *24*, 916, 1152 (1968).

PROBLEMS

7.1 Energy Transport Equation

Obtain an equation for the fluid transport of particle kinetic energy by multiplying the Vlasov equation by $\frac{1}{2}m_s v^2$ and integrating over all velocity space. Simplify your result in any convenient fashion.

7.2 Fluid Conservation Properties

Suppose we have an electron-proton plasma that is finite in extent in all three dimensions. Suppose there is no magnetic field. Using the fluid equations, prove that

(a) For each species, total particles are conserved.
(b) Momentum is not necessarily conserved for each species.
(c) Total momentum, summed over species, is conserved.

7.3 Langmuir Waves

One does not need to look for individual sinusoidal wave solutions to solve wave equations. Consider the electron fluid equations in the differential form (7.38) to (7.40).

(a) Combine these equations, and linearize, to obtain the linear partial differential wave equation

$$(\partial_t^2 - \omega_e^2 - 3v_e^2\,\partial_x^2)E(x,t) = 0$$

(b) Suppose the initial conditions for $E(x,t)$ are

$$E(x,t = 0) = f(x)$$
$$\dot{E}(x,t = 0) = 0$$

where an overdot indicates a time derivative. Using Fourier and Laplace transform techniques, find an exact explicit solution for the time evolution of $E(x,t)$.

(c) Suppose $f(x)$ represents a standard wave packet, a sinusoidal variation with space accompanied by a Gaussian envelope

$$E(x,t = 0) = f(x) = E_0 e^{-x^2/2L^2} \sin k_0 x$$

where we assume $k_0 >> L^{-1}$. By using an appropriate approximation, if necessary, in the exact solution from (b), show that the wave packet travels with the group speed

$$|V_g| = \left|\frac{d\omega}{dk}\right|_{k=k_0} = |3(k_0\lambda_e)v_e|$$

Show also that the packet spreads as it propagates, and the rate of spreading (dispersion) is proportional to $|dV_g/dk|_{k=k_0}$. Does the packet move to the right, to the left, or does it split into right- and left-going pieces?

7.4 Negative Energy Waves

Suppose a plasma has cold electrons drifting with velocity v_0 with respect to cold ions. Derive the wave dispersion relation corresponding to high frequency electron plasma waves. Show that in the frame moving with the electrons, these are just our old cold plasma waves. Plot the two branches of $\omega(k)$ vs. k. Use the wave energy formula (6.72) to evaluate the wave energy. Indicate the regions of your dispersion diagram where the energy is negative.

7.5 Upper Hybrid vs. Right Cutoff Frequency

Prove that the right cutoff frequency

$$\omega_R = \frac{|\Omega_e|}{2} + \left[\omega_e^2 + \left(\Omega_e^2/4\right)\right]^{1/2}$$

is always greater than or equal to the upper hybrid frequency

$$\omega_{UH} = (\omega_e^2 + \Omega_e^2)^{1/2}$$

7.6 Upper Hybrid Wave

Compare the derivations of the upper hybrid wave (a perpendicular, electrostatic wave) and the extraordinary wave (a perpendicular, partially electromagnetic and partially electrostatic wave). Does the assumed form of the upper hybrid wave satisfy Maxwell's equations? Why not? In which parameter regime does it approximately satisfy Maxwell's equations? In this parameter regime, is there any difference between the upper hybrid wave and the extraordinary wave? In the X-mode derivation, show that Faraday's law contains the same information as would be contained by Poisson's equation plus the electron continuity equation. Reproduce the graph of $n \equiv ck/\omega$ for the extraordinary wave and draw the dispersion diagram for the upper hybrid wave. For which portion of each diagram do we trust the derivation? Which portion of the X-mode diagram corresponds to the upper hybrid wave?

7.7 Model of Collisions

We wish to derive a simple model of collisional effects in a plasma. Suppose that a typical particle suffers collisions at a rate ν. Then the particle when oscillating in the electric field of a wave will occasionally suffer a collision (assuming $\nu << \omega$) and to the first approximation can be thought to lose all of its directed energy. An electron fluid element with velocity $\mathbf{v}$ will thus lose momentum at a rate $-\nu n_0 \mathbf{v} m_e$, where we have assumed that each colliding electron loses momentum $-m_e\mathbf{v}$. Thus, we can add a term in the electron force equation to represent collisions,

$$n_e m_e \, \partial_t \mathbf{V}_e + n_e m_e \mathbf{V}_e \cdot \nabla \mathbf{V}_e = - \nabla P_e - e n_e \mathbf{E} - \nu n_e m_e \mathbf{V}_e$$

With this extra term, rederive the Langmuir wave dispersion relation, assuming $\nu << \omega$. At what rate does the electric field damp away? At what rate does the wave energy damp away?

7.8 Low Frequency Dielectric Constant

Recall that in the theory of dielectrics, one likes to include the currents in

$$\nabla \times \mathbf{B} = \frac{4\pi}{c} \mathbf{J} + \frac{1}{c} \partial_t \mathbf{E}$$

in the dielectric function ϵ; thus

$$\nabla \times \mathbf{B} = \frac{1}{c} \partial_t \mathbf{D}$$

where

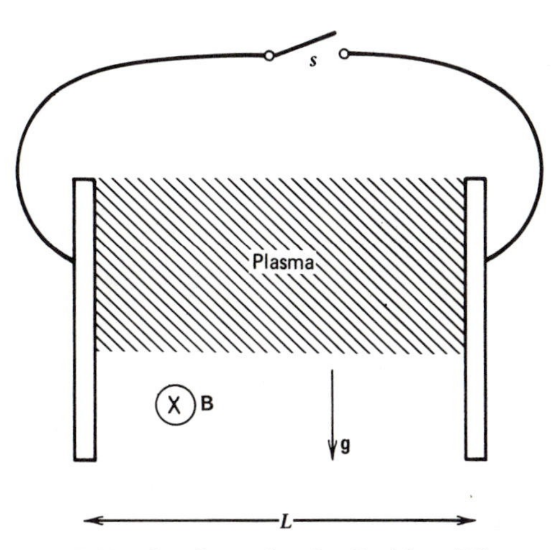

Fig. 7.33 Configuration for Problem 7.9.

$$\mathbf{D} = \epsilon\mathbf{E}$$

Suppose that a slowly varying sinusoidal electric field is applied across a magnetic field. Derive an expression for ϵ by considering the polarization current produced by the electric field. What do you suppose "slowly varying" means?

7.9 Kunkel's Problem

A plasma of mass density $\rho = n_0(m_i + m_e)$ is bounded by two parallel conducting plates separated by a distance L. A gravitational acceleration $\mathbf{g}$ is applied at right angles to a uniform magnetic field $\mathbf{B}$, and both of these are parallel to the plates as shown in Fig. 7.33. Show by means of a careful particle drift analysis that the plasma can *accelerate* freely downward only if switch S is open, and if the low frequency dielectric function from the previous problem is $\epsilon >> 1$. What is the voltage between the two plates in that case? If S is closed, what is the current density between the two plates?

7.10 Laser Fusion

In order to obtain controlled thermonuclear fusion using deuterium and tritium, one needs to satisfy the Lawson criterion $n\tau > 10^{14}$ (c.g.s.) at a temperature $T \sim 10$ keV, where n is the number of particles per cm^3, and τ is the confinement time in sec. Use the Lawson criterion to derive the corresponding requirement for laser pellet fusion, $\rho r > 1$ (c.g.s.) where ρ is the density of the compressed pellet in g/cm^3 and r is the radius of the compressed pellet in cm. (*Hint*: How does one define "confinement time" for inertial "confinement"? How can one estimate this physically?)

CHAPTER 8

Magnetohydrodynamics

8.1 INTRODUCTION

Chapter 7 is concerned with a set of equations, the fluid equations, which were derived from the Vlasov equation, which in turn was derived from the Klimontovich equation by neglecting all collision effects. Thus, all of the phenomena discussed in Chapter 7 will occur only when collisions are not important. A rough criterion for the importance of collisions is obtained by comparing the collision frequency ν_{ei} to the frequency ω of the phenomenon under consideration; the fluid treatment is valid when $\nu_{ei} << \omega$. Since we have seen (Section 1.6) that the collision frequency $\nu_{ei} \approx \omega_e/\Lambda$, there is a huge range of frequencies where the fluid treatment applies. However, there is a significant range of frequencies, $0 \leq \omega \leq \nu_{ei}$, where the fluid treatment does not apply. In particular, one often wants to find an equilibrium plasma configuration; for example, in tokamaks, mirror machines, planetary magnetospheres, pulsar magnetospheres, and stellar winds. An equilibrium is equivalent to $\omega = 0$ ($\partial_t = 0$), and collisions must be included in such considerations.

Let us then develop a set of equations that are valid for low frequencies. [Students who have not studied Chapters 3 and 6 can skip directly to (8.3) and (8.7) with the understanding that the extra terms represent the physics of collisions.] Recall the plasma kinetic equation (3.26),

$$\partial_t f_s + \mathbf{v} \cdot \nabla_{\mathbf{x}} f_s + \left(\frac{q_s}{m_s} \mathbf{E} + \frac{q_s}{m_s c} \mathbf{v} \times \mathbf{B} \right) \cdot \nabla_{\mathbf{v}} f_s$$

$$= - \frac{q_s}{m_s} \left\langle \left(\delta\mathbf{E} + \frac{\mathbf{v}}{c} \times \delta\mathbf{B} \right) \cdot \nabla_{\mathbf{v}} \, \delta N_s \right\rangle \qquad (8.1)$$

Here, $f_s(\mathbf{x},\mathbf{v},t)$ is a smooth function obtained by averaging the Klimontovich density $N_s(\mathbf{x},\mathbf{v},t)$ over an appropriate volume, while δN_s is the difference between the smooth function f_s and N_s, which is the sum of delta functions. In Chapter 3 we argued that the right side represents discrete particle effects, including collisions. We can therefore invent a symbol for the right side of (8.1), $(\partial f/\partial t)_c$, and write (8.1) as

$$\partial_t f_s + \mathbf{v}\cdot\nabla_{\mathbf{x}} f_s + \frac{q_s}{m_s}\left(\mathbf{E} + \frac{\mathbf{v}}{c}\times\mathbf{B}\right)\cdot\nabla_{\mathbf{v}} f_s = \left(\frac{\partial f_s}{\partial t}\right)_c \tag{8.2}$$

Thus, we are thinking of f_s as the number of particles in a small volume of six-dimensional phase space $(\mathbf{x},\mathbf{v})$, divided by that volume. (See the discussion in Section 6.1.) Then $(\partial f_s/\partial t)_c$ is the rate that particles are gained or lost by that small volume because of collisions. Recall that **E** and **B** in (8.2) are also averaged quantities, so that they do not include the fields due to individual particles. This identification of $(\partial f_s/\partial t)_c$ is admittedly crude, but we shall not attempt to do better here. There does exist a large body of more exact literature that starts from the formally exact expression (8.1). Here, we use (8.2) to try to obtain the most significant effects of collisions.

Having identified $(\partial f_s/\partial t)_c$ as the change in $f_s(\mathbf{x},\mathbf{v},t)$ due to collisions, we expect it to have a much stronger influence on the velocity dependence of f_s than on the spatial dependence of f_s. This is because a collision can cause a huge change in a particle's velocity, but does not cause much change at all in a particle's position.

In Section 7.2 we obtained the fluid equations by integrating the Vlasov equation (6.5) over velocity space after multiplying by an appropriate power of velocity. Let us repeat that procedure with (8.2). Multiplying by unity and integrating over all velocity space, we obtain

$$\partial_t n_s(\mathbf{x},t) + \nabla\cdot(\mathbf{V}_s n_s) = \int d\mathbf{v}\left(\frac{\partial f_s}{\partial t}\right)_c \tag{8.3}$$

where the left side is as in (7.15). The right side represents the change in the number of particles in a small volume of real space due to collisions; we have just argued that this is very small since collisions do not cause large changes in particle positions; therefore we set this to zero and obtain

$$\partial_t n_s(\mathbf{x},t) + \nabla\cdot(n_s\mathbf{V}_s) = 0 \tag{8.4}$$

which is just the continuity equation (7.15).

Next, multiply (8.2) by $\mathbf{v}$ and integrate over all velocity space; we obtain the force equation (7.28) with the addition of one term. This is

$$m_s n_s\,\partial_t\mathbf{V}_s + m_s n_s(\mathbf{V}_s\cdot\nabla)\mathbf{V}_s = -\nabla P_s + q_s n_s\left(\mathbf{E} + \frac{1}{c}\mathbf{V}_s\times\mathbf{B}\right) + m_s\int d\mathbf{v}\,\mathbf{v}\left(\frac{\partial f_s}{\partial t}\right)_c \tag{8.5}$$

The term

$$\mathbf{K}_s \equiv m_s\int d\mathbf{v}\,\mathbf{v}\left(\frac{\partial f_s}{\partial t}\right)_c \tag{8.6}$$

represents the change in the momentum of species s at position $\mathbf{x}$ due to collisions. A species cannot change its own momentum by colliding with itself; the center of mass of two electrons, for example, is not accelerated during a collision of the two electrons (see Section 2.9). However, the momentum of electrons in a certain volume of space can certainly be changed by collisions with the ions. For example, a beam of electrons incident on a plasma will slow down due to collisions with the ions; the ions begin to move in the initial direction of the electron beam because they have taken up the electron momentum. Thus, we expect $\mathbf{K}_e(\mathbf{x}) = -\mathbf{K}_i(\mathbf{x})$. It would be possible to develop simple but crude models for $\mathbf{K}_s(\mathbf{x})$, but here we shall leave $\mathbf{K}_s$ in general form. Our force equation then is written as

$$m_s n_s \partial_t \mathbf{V}_s + m_s n_s (\mathbf{V}_s \cdot \nabla)\mathbf{V}_s = -\nabla P_s + q_s n_s \left(\mathbf{E} + \frac{1}{c}\mathbf{V}_s \times \mathbf{B}\right) + \mathbf{K}_s(\mathbf{x}) \quad (8.7)$$

The fluid equations (8.4) and (8.7) are the same as the fluid equations in Chapter 7, without the term $\mathbf{K}_s(\mathbf{x})$ in (8.7). When written for both electrons and for ions, this set is called the *two-fluid model.* We now wish to combine the electron equations with the ion equations to obtain a *one-fluid model,* also known as the equations of *magnetohydrodynamics* (MHD). We thus wish to think of a single fluid characterized by a mass density

$$\rho_M(\mathbf{x}) \equiv m_e n_e(\mathbf{x}) + m_i n_i(\mathbf{x}) \approx m_i n_i(\mathbf{x}) \quad (8.8)$$

a charge density

$$\rho_c(\mathbf{x}) \equiv q_e n_e(\mathbf{x}) + q_i n_i(\mathbf{x}) = e(n_i - n_e) \quad (8.9)$$

a center of mass fluid flow velocity

$$\mathbf{V} \equiv \frac{1}{\rho_M}(m_i n_i \mathbf{V}_i + m_e n_e \mathbf{V}_e) \quad (8.10)$$

a current density

$$\mathbf{J} \equiv q_i n_i \mathbf{V}_i + q_e n_e \mathbf{V}_e \quad (8.11)$$

and a total pressure

$$P \equiv P_e + P_i \quad (8.12)$$

We wish to derive four equations relating these quantities: a *mass conservation* equation, a *charge conservation* equation, a *momentum* equation, and a *generalized Ohm's law.*

First we derive a mass conservation law. Multiply the ion continuity equation (8.4) by m_i, the electron continuity equation (8.4) by m_e, and add to obtain

$$\boxed{\frac{\partial \rho_M}{\partial t} + \nabla \cdot (\rho_M \mathbf{V}) = 0} \quad (8.13)$$

which is the *mass conservation law.*

Next, multiply the ion continuity equation (8.4) by q_i, the electron continuity equation (8.4) by q_e, and add to obtain

$$\boxed{\frac{\partial \rho_c}{\partial t} + \nabla \cdot \mathbf{J} = 0} \tag{8.14}$$

which is the charge continuity equation or *charge conservation law.*

Consider next the force equation (8.7). Regarding V_s and $\partial_t n_s$ as small quantities, neglecting the products of small quantities, and recalling $\mathbf{K}_e = -\mathbf{K}_i$, we add (8.7) for electrons and ions to obtain

$$\boxed{\rho_M \frac{\partial \mathbf{V}}{\partial t} = -\nabla P + \rho_c \mathbf{E} + \frac{1}{c}\mathbf{J} \times \mathbf{B}} \tag{8.15}$$

which is the one-fluid *force equation*, or momentum equation.

Finally, we desire an equation for the time derivative of the current, called a *generalized Ohm's law*. Multiplying the force equation (8.7) by q_s/m_s, adding the ion version to the electron version, neglecting quadratic terms in the small quantities $\partial_t n_s$ and $\mathbf{V}_s$, and using $q_i = -q_e = e$, we find

$$\begin{aligned}\frac{\partial \mathbf{J}}{\partial t} = \frac{-e}{m_i}\nabla P_i + \frac{e}{m_e}\nabla P_e + \left(\frac{e^2 n_e}{m_e} + \frac{e^2 n_i}{m_i}\right)\mathbf{E} + \frac{e^2 n_e}{m_e c}\mathbf{V}_e \times \mathbf{B} \\ + \frac{e^2 n_i}{m_i c}\mathbf{V}_i \times \mathbf{B} + \left(\frac{e}{m_i} + \frac{e}{m_e}\right)\mathbf{K}_i\end{aligned} \tag{8.16}$$

We notice that

$$\begin{aligned}\frac{n_e e^2}{m_e c}\mathbf{V}_e &= \frac{e}{m_e c}(n_e e \mathbf{V}_e - n_i e \mathbf{V}_i) + \frac{e^2}{m_e m_i c}(m_i n_i \mathbf{V}_i) \\ &= \frac{-e}{m_e c}\mathbf{J} + \frac{e^2}{m_e m_i c}(m_i n_i \mathbf{V}_i + m_e n_e \mathbf{V}_e) \\ &= \frac{-e}{m_e c}\mathbf{J} + \frac{e^2}{m_e m_i c}(\rho_M \mathbf{V})\end{aligned} \tag{8.17}$$

where the first line merely adds and subtracts the same quantity, and the second line adds the tiny quantity $(m_e/m_i)\mathbf{V}_e$, which is negligible compared to $\mathbf{V}_e$ as already incorporated into $\mathbf{J}$.

Using (8.17) in (8.16), neglecting $m_i^{-1} << m_e^{-1}$ wherever possible, and assuming that $P_e \approx P_i \approx \frac{1}{2}P$ and $n_i \approx n_e$, we find

$$\frac{\partial \mathbf{J}}{\partial t} = \frac{e}{2m_e}\nabla P + \frac{e^2 \rho_M}{m_e m_i}\left(\mathbf{E} + \frac{1}{c}\mathbf{V} \times \mathbf{B}\right) - \frac{e}{m_e c}\mathbf{J} \times \mathbf{B} + \frac{e}{m_e}\mathbf{K}_i \tag{8.18}$$

Recall that $\mathbf{K}_i$ represents the change in ion momentum due to collisions with electrons. It is reasonable then to assume that $\mathbf{K}_i$ is a function of the relative velocity $\mathbf{V}_i - \mathbf{V}_e$ between the two species; keeping only the first term in a Taylor expansion of $\mathbf{K}_i$, we find

$$\begin{aligned}\mathbf{K}_i &= C_1(\mathbf{V}_i - \mathbf{V}_e) \\ &= C_2 \mathbf{J}\end{aligned}$$

$$= -\frac{\rho_M e}{m_i \sigma}\,\mathbf{J} \tag{8.19}$$

where the constant C_2 has been put in a form such that σ can be identified as a conductivity, as we shall see. The minus sign has been chosen because we expect collisions to decrease the current caused by relative species velocity. When we multiply by $m_e m_i/\rho_M e^2$, (8.18) becomes

$$\frac{m_e m_i}{\rho_M e^2}\frac{\partial \mathbf{J}}{\partial t} = \frac{m_i}{2\rho_M e}\nabla P + \mathbf{E} + \frac{1}{c}\mathbf{V}\times\mathbf{B} - \frac{m_i}{\rho_M e c}\mathbf{J}\times\mathbf{B} - \frac{\mathbf{J}}{\sigma} \tag{8.20}$$

which is the *generalized Ohm's law*. The name comes from the fact that if the only important terms are the second and the fifth terms on the right side, we have

$$\mathbf{J} = \sigma\mathbf{E} \tag{8.21}$$

which is Ohm's law and in which σ is clearly the conductivity.

This completes our derivation of the MHD equations. Collecting these equations, we have

$$\frac{\partial \rho_M}{\partial t} + \nabla\cdot(\rho_M \mathbf{V}) = 0 \tag{8.13}$$

$$\frac{\partial \rho_c}{\partial t} + \nabla\cdot\mathbf{J} = 0 \tag{8.14}$$

$$\rho_M\frac{\partial \mathbf{V}}{\partial t} = -\nabla P + \rho_c\mathbf{E} + \frac{1}{c}\mathbf{J}\times\mathbf{B} \tag{8.15}$$

$$\frac{m_e m_i}{\rho_M e^2}\frac{\partial \mathbf{J}}{\partial t} = \frac{m_i}{2\rho_M e}\nabla P + \mathbf{E} + \frac{1}{c}\mathbf{V}\times\mathbf{B} - \frac{m_i}{\rho_M e c}\mathbf{J}\times\mathbf{B} - \frac{\mathbf{J}}{\sigma} \tag{8.20}$$

When coupled to Maxwell's equations

$$\nabla\times\mathbf{E} = -\frac{1}{c}\frac{\partial \mathbf{B}}{\partial t} \tag{8.22}$$

and

$$\nabla\times\mathbf{B} = \frac{4\pi}{c}\mathbf{J} + \frac{1}{c}\frac{\partial \mathbf{E}}{\partial t} \tag{8.23}$$

we have 14 equations in the 14 unknowns ρ_M, ρ_c, $\mathbf{V}$, $\mathbf{J}$, $\mathbf{E}$, and $\mathbf{B}$; this assumes that the pressure P can be expressed in terms of the mass density ρ_M.

For very low frequencies, one can ignore the $\partial_t\mathbf{J}$ term in the generalized Ohm's law, whereas for low temperatures the ∇P term can be ignored. In addition, when the current is small we can neglect the $\mathbf{J}\times\mathbf{B}$ term (known as the Hall term) compared to the $\mathbf{V}\times\mathbf{B}$ term; under all these assumptions, Ohm's law (8.20) becomes

$$0 = \mathbf{E} + \frac{1}{c}\mathbf{V}\times\mathbf{B} - \frac{\mathbf{J}}{\sigma} \tag{8.24}$$

or

$$\mathbf{J} = \sigma \left(\mathbf{E} + \frac{1}{c} \mathbf{V} \times \mathbf{B} \right) \tag{8.25}$$

When collisions vanish, the conductivity becomes infinite and, in order to have only finite currents, we must have

$$\mathbf{E} + \frac{1}{c} \mathbf{V} \times \mathbf{B} = 0 \tag{8.26}$$

or

$$\mathbf{E} = -\frac{1}{c} \mathbf{V} \times \mathbf{B} \tag{8.27}$$

Under this infinite conductivity, low frequency condition, no charge imbalances are allowed and we have $\rho_c = 0$. Under these *ideal MHD* conditions, our basic equations (8.13), (8.14), (8.15), (8.20), (8.22), and (8.23) become

$$\partial_t \rho_M + \nabla \cdot (\rho_M \mathbf{V}) = 0 \tag{8.28}$$

$$\rho_M \partial_t \mathbf{V} = -\nabla P + \frac{1}{c} \mathbf{J} \times \mathbf{B} \tag{8.29}$$

$$\nabla \times (\mathbf{V} \times \mathbf{B}) = \partial_t \mathbf{B} \tag{8.30}$$

$$\nabla \times \mathbf{B} = \frac{4\pi}{c} \mathbf{J} \tag{8.31}$$

where the low frequency assumption is used to ignore $(1/c)(\partial \mathbf{E}/\partial t)$ on the right of (8.31).

We shall not attempt to further justify (8.28) to (8.31); instead, we shall take for granted that these equations can be justified in useful physical situations. In the next section, we shall use these equations to consider the equilibrium and stability of various plasma configurations.

8.2 MHD EQUILIBRIUM

In many cases one is interested in the *equilibrium* configurations of a plasma. Once the equilibrium is found, one then asks whether or not the equilibrium is *stable*. For example, the problem of the earth's magnetosphere and its interaction with the solar wind can be approached by first looking for MHD equilibria. (Since equilibrium implies that all zeroth order quantities have no time derivatives, we can feel some confidence in using the ideal MHD equations to look for equilibria.) Once we find the zeroth order quantities $\mathbf{B}^0$, $\mathbf{V}^0$, $\mathbf{J}^0$, and $\rho_M{}^0$ (and thus P^0) satisfying the ideal MHD equations (8.28)–(8.31) with $\partial_t \to 0$, we can then linearize about the equilibrium to determine whether it is stable. In other words, we let $\rho_M = \rho_M{}^0 + \rho_M{}^1$, etc., and $\partial_t \to -i\omega$, and we solve for all possible values of ω. If one of these values of ω has $\text{Im}(\omega) > 0$, we have instability and any tiny perturbation of the equilibrium will grow as $\exp[\text{Im}(\omega)t]$ until the equilibrium is destroyed. If no unstable values of ω are found, then we have MHD stability, but this does not imply overall stability. Much of the high frequency physics has been lost in first going from the Vlasov equation to the two-fluid equations, and in next adding the two-fluid equations to obtain the MHD equations. A system that is MHD stable may well be unstable when two-fluid effects or Vlasov effects are considered. Thus, MHD stability is a necessary but not a sufficient condition for overall stability.

Before proceeding to questions of MHD equilibrium and stability, let us try to gain a measure of intuition regarding our MHD equations. From Ohm's law (8.25) and Maxwell's equations (neglecting the displacement current)

$$\nabla \times \mathbf{B} = \frac{4\pi}{c}\,\mathbf{J} \tag{8.32}$$

and

$$\nabla \times \mathbf{E} = -\frac{1}{c}\,\partial_t \mathbf{B} \tag{8.33}$$

we obtain

$$\nabla \times (\nabla \times \mathbf{B}) = \frac{4\pi\sigma}{c}\left[-\frac{1}{c}\,\partial_t \mathbf{B} + \frac{1}{c}\,\nabla \times (\mathbf{V} \times \mathbf{B})\right] \tag{8.34}$$

or

$$\partial_t \mathbf{B} = \nabla \times (\mathbf{V} \times \mathbf{B}) + \frac{c^2}{4\pi\sigma}\,\nabla^2 \mathbf{B} \tag{8.35}$$

Thus, the magnetic field at a point in a plasma can be changed by the fluid convection, the first term on the right side of (8.35), or by diffusion due to the second term on the right side of (8.35). When $\mathbf{V} = 0$, (8.35) is

$$\partial_t \mathbf{B} = \frac{c^2}{4\pi\sigma}\,\nabla^2 \mathbf{B} \tag{8.36}$$

which is the standard form of a diffusion equation; the constant $c^2/4\pi\sigma$ is known as the *magnetic diffusivity*. Dimensional analysis of (8.36) shows that the field can diffuse away with a diffusion time τ_D given by

$$\tau_D \sim \frac{4\pi\sigma L^2}{c^2} \tag{8.37}$$

where L is the scale length. Since the conductivity σ is inversely proportional to the collision rate, the physics of the diffusion must involve the disruption of particle orbits due to collisions, which in turn disrupts the current that gives rise to the magnetic fields; the disruption of the current allows the magnetic field to diffuse.

We consider next a very important concept, that of *frozen-in field lines*. Consider a surface ΔS drawn perpendicular to the field lines, and suppose that the boundary of this surface is moving with some specified velocity field (Fig. 8.1). What is the time rate of change of the total magnetic flux

$$\Phi \equiv \int_{\Delta S} d\mathbf{A} \cdot \mathbf{B} \tag{8.38}$$

through the surface ΔS? It is

$$\dot{\Phi} = \int_{\Delta S} d\mathbf{A} \cdot \dot{\mathbf{B}} + \int_{\Delta S} d\dot{\mathbf{A}} \cdot \mathbf{B} \tag{8.39}$$

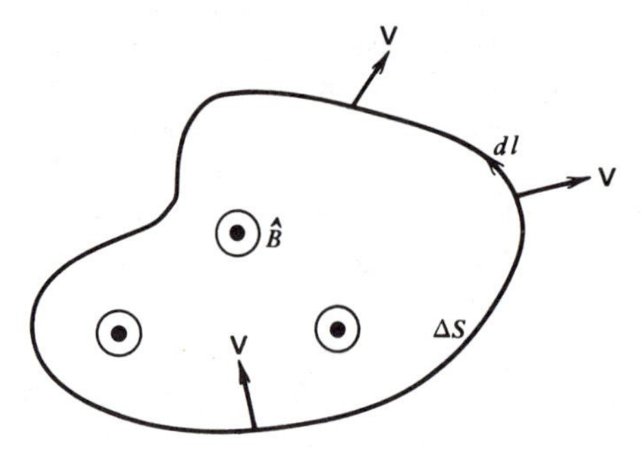

Fig. 8.1 Surface ΔS perpendicular to the magnetic field lines.

where the first term gives the contribution from the time rate of change of **B**, and the second term gives the contribution from the movement of the boundary of the surface ΔS. At every point on the boundary of ΔS, the rate of change of the area enclosed by the boundary is proportional to the perpendicular component of **V** at that point, and therefore it is given by $\mathbf{V} \times d\mathbf{l}$. Thus, (8.39) is

$$\dot{\Phi} = \int_{\Delta S} d\mathbf{A} \cdot \frac{\partial \mathbf{B}}{\partial t} + \oint \mathbf{B} \cdot (\mathbf{V} \times d\mathbf{l}) \tag{8.40}$$

In ideal MHD ($\sigma \to \infty$) we define the surface ΔS to be attached to the fluid, and we evaluate (8.40) for $\dot{\Phi}$. We first integrate (8.30) over the surface ΔS, obtaining

$$\int_{\Delta S} d\mathbf{A} \cdot \partial_t \mathbf{B} = \int_{\Delta S} d\mathbf{A} \cdot [\nabla \times (\mathbf{V} \times \mathbf{B})] \tag{8.41}$$

If we recall Stoke's theorem,

$$\int_{\Delta S} (\nabla \times \mathbf{C}) \cdot d\mathbf{A} = \oint \mathbf{C} \cdot d\mathbf{l} \tag{8.42}$$

(8.41) becomes

$$\int_{\Delta S} d\mathbf{A} \cdot \partial_t \mathbf{B} = \oint d\mathbf{l} \cdot (\mathbf{V} \times \mathbf{B}) \tag{8.43}$$

Next recalling the vector identity

$$\mathbf{A} \cdot (\mathbf{B} \times \mathbf{C}) = (\mathbf{A} \times \mathbf{B}) \cdot \mathbf{C} \tag{8.44}$$

we find

$$\int_{\Delta S} d\mathbf{A} \cdot \partial_t \mathbf{B} + \oint \mathbf{B} \cdot (\mathbf{V} \times d\mathbf{l}) = 0 \tag{8.45}$$

But by (8.40), the left side of (8.45) is just the total time rate of change of Φ through the surface ΔS; therefore

$$\dot{\Phi} = 0 \tag{8.46}$$

as long as the surface ΔS moves with the fluid.

Suppose that at $t = 0$ we draw two surfaces ΔS_1 and ΔS_2, and we then sweep the surfaces along the instantaneous field lines to form two flux tubes, as shown in Fig. 8.2. If the two surfaces ΔS_1 and ΔS_2 intersect at one point, then there is one

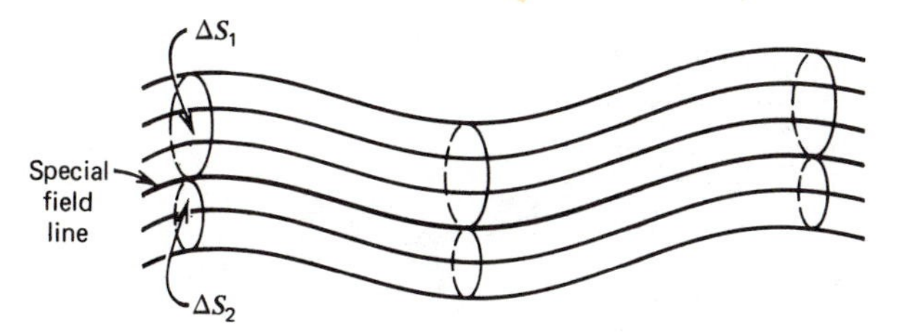

Fig. 8.2 Two flux tubes.

special field line that is the line at which the two flux tubes touch. By coloring the fluid in one flux tube red and the fluid in the other tube blue, we can follow the two flux tubes forever. For any reasonable flow, the tubes will always touch, and the line of touching identifies that particular magnetic field line forever. Thus, in ideal MHD, we can similarly label each and every field line, and we can say that the plasma is *frozen to the field lines*.

In nonideal MHD ($\sigma \neq \infty$), it becomes much more difficult to label field lines. This is true partially because lines can disappear because of resistivity.

Let us now return to consider the requirements for an MHD equilibrium. Looking for equilibrium solutions to (8.28)–(8.31), with no fluid flow, we require that

$$0 = -\nabla P + \frac{1}{c}\mathbf{J} \times \mathbf{B} \tag{8.47}$$

and

$$\nabla \times \mathbf{B} = \frac{4\pi}{c}\mathbf{J} \tag{8.48}$$

which yield

$$\nabla P - \frac{1}{4\pi}(\nabla \times \mathbf{B}) \times \mathbf{B} = 0 \tag{8.49}$$

Recalling

$$(\nabla \times \mathbf{B}) \times \mathbf{B} = (\mathbf{B} \cdot \nabla)\mathbf{B} - \frac{1}{2}\nabla B^2 \tag{8.50}$$

we have

$$\nabla P + \frac{1}{8\pi}\nabla B^2 = \frac{1}{4\pi}(\mathbf{B} \cdot \nabla)\mathbf{B} \tag{8.51}$$

When $(\mathbf{B} \cdot \nabla)\mathbf{B} = 0$, this is

$$\nabla\left(P + \frac{B^2}{8\pi}\right) = 0 \tag{8.52}$$

which leads us to *define* the magnetic pressure $B^2/8\pi$, and to state that in equilibrium, magnetic pressure must balance plasma pressure (Fig. 8.3),

$$P + \frac{B^2}{8\pi} = \text{constant} \tag{8.53}$$

Returning to (8.47), we see that in equilibrium,

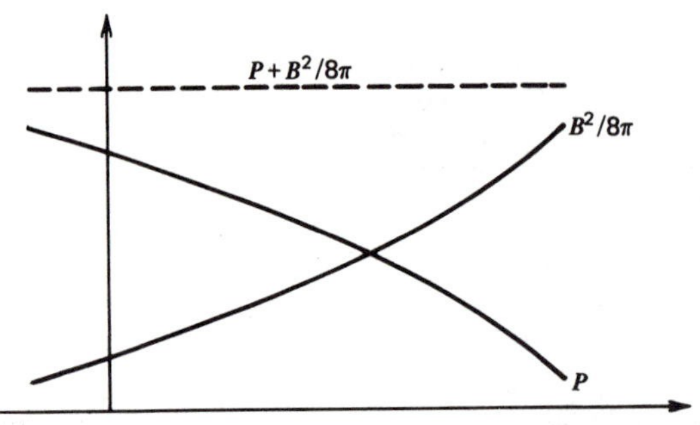

Fig. 8.3 Magnetic field pressure balances plasma pressure in ideal magnetohydrodynamic equilibrium.

$$\nabla P = \frac{1}{c} \mathbf{J} \times \mathbf{B} \tag{8.54}$$

so that $\mathbf{B} \cdot \nabla P = \mathbf{J} \cdot \nabla P = 0$; in other words, **B** and **J** lie along surfaces of constant pressure.

Let us consider two simple cases of plasma equilibria, the *theta pinch* and the *z-pinch*. In the theta pinch, capacitor plates are discharged about a cylindrical conductor, as shown in Fig. 8.4. The azimuthal $\dot{\mathbf{J}}$ makes a $\dot{\mathbf{B}}$ into the paper; the $\dot{\mathbf{B}}$ into the paper induces an azimuthal **E** in the direction opposite to **J**, which in turn produces an internal current in the plasma in the direction opposite to J. A hypothetical final state could be as shown in Fig. 8.5 where the central plasma region has $\mathbf{B} = 0$. The pressures are then as shown in Fig. 8.6 so that $P + B^2/8\pi$ = constant, and there is an azimuthal current sheet at r_0 such that

$$\nabla P = \frac{1}{c} \mathbf{J} \times \mathbf{B} \tag{8.55}$$

is satisfied (Fig. 8.7).

A second possible equilibria is the *z-pinch*, where a current flows along a plasma, and the plasma is confined by its own magnetic field and its own current

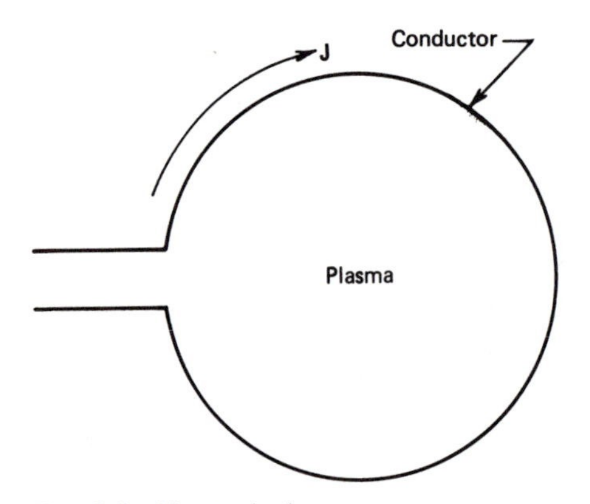

Fig. 8.4 Theta pinch.

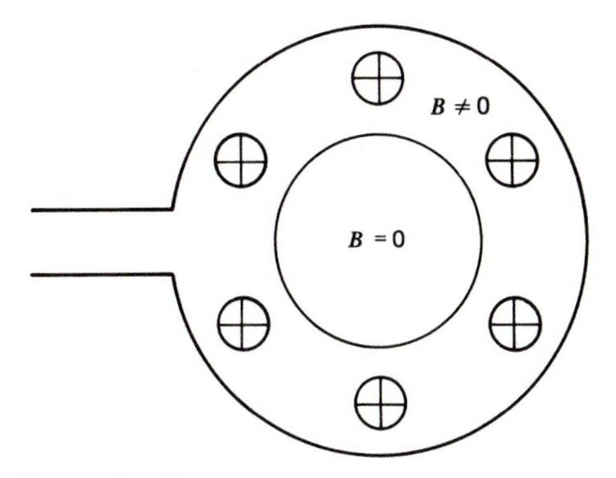

Fig. 8.5 Hypothetical final state of a theta pinch.

through the $\mathbf{J} \times \mathbf{B}$ force, as shown in Fig. 8.8. Then in cyclindrical coordinates we have

$$\nabla P = \frac{1}{c}\, \mathbf{J} \times \mathbf{B} = \frac{1}{4\pi}\, (\nabla \times \mathbf{B}) \times \mathbf{B} \tag{8.56}$$

or

$$\frac{\partial P}{\partial r} = -\frac{B_\theta}{4\pi r}\, \frac{\partial}{\partial r}\, (rB_\theta) \tag{8.57}$$

The solution of (8.57) will be left for one of the problems; here we merely note that (8.57) does have well-balanced solutions for $B_\theta(r)$ and $P(r)$.

Before closing this discussion of equilibrium, we note two related points. First, because

$$P + \frac{B^2}{8\pi} = \text{constant} \tag{8.58}$$

the magnetic field is smaller inside a plasma ($P > 0$) than outside a plasma ($P = 0$). This is another illustration of the fact that a plasma is diamagnetic; we have seen this before in considering single particle motion.

Second, we notice from (8.48) and (8.49) that when $\mathbf{J}$ is parallel to $\mathbf{B}$, we have

$$\nabla P = 0 \tag{8.59}$$

and

$$(\nabla \times \mathbf{B}) \times \mathbf{B} = 0 \tag{8.60}$$

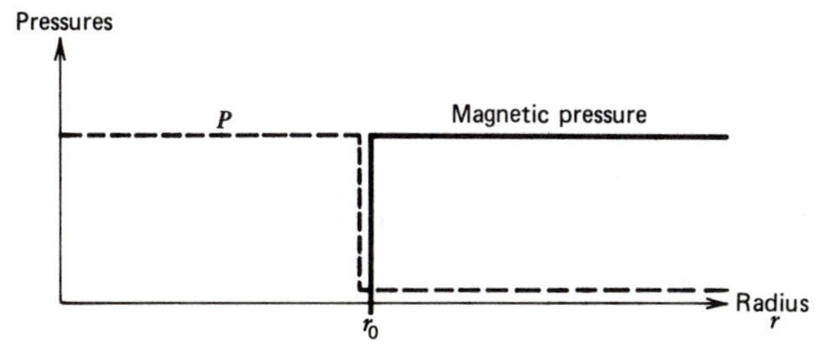

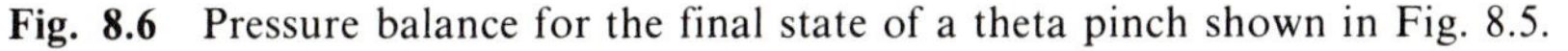
Fig. 8.6 Pressure balance for the final state of a theta pinch shown in Fig. 8.5.

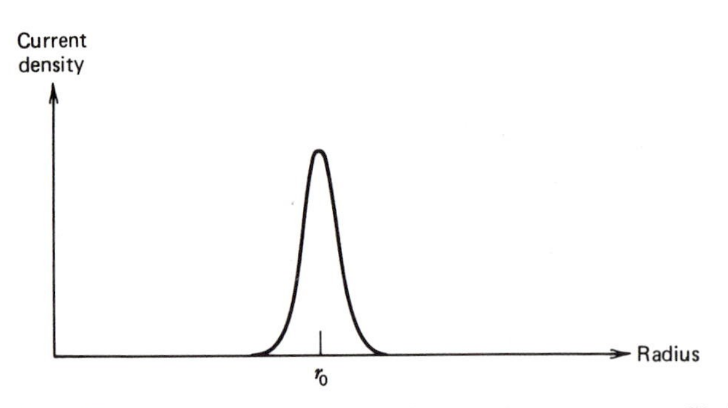

Fig. 8.7 Current profile that maintains the pressure profile in Fig. 8.6.

which is known as the *force free* situation. Any magnetic field for which

$$\nabla \times \mathbf{B}(\mathbf{x}) = f(\mathbf{x})\mathbf{B}(\mathbf{x}) \tag{8.61}$$

satisfies (8.60).

In the next section we proceed to discuss the stability of our equilibria. It is the unfortunate case that both the theta pinch and z-pinch are unstable, as well as the simple mirror machine. We shall find that stable equilibria are possible, but only when certain criteria are satisfied.

8.3 MHD STABILITY

In the last section we found various examples of MHD equilibria. We must now ask whether those equilibria are stable. There are two ways of doing this. The first is to linearize the equations of motion about the zero-order equilibrium, and solve for the frequencies in exactly the same way in which we previously found linear waves. If one of these frequencies has $\text{Im}(\omega) > 0$, then $\exp(-i\omega t) \sim \exp[\text{Im}(\omega)t]$ will grow with time, and the system is unstable. The second is to consider the total energy of a system, and to ask whether that energy increases or decreases under a perturbation. If the energy increases, the perturbation will not grow. If the energy decreases, the perturbation can happen and have energy left over to go into kinetic energy of expansion; this is the mark of instability.

The first method is to linearize the equations of motion. In ideal MHD, these are (8.28) to (8.31). For example, consider a plasma held up against the force of

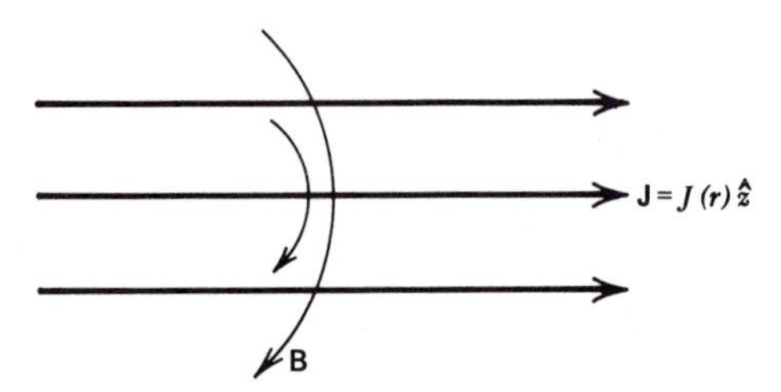

Fig. 8.8 Current and magnetic field configurations in a z-pinch.

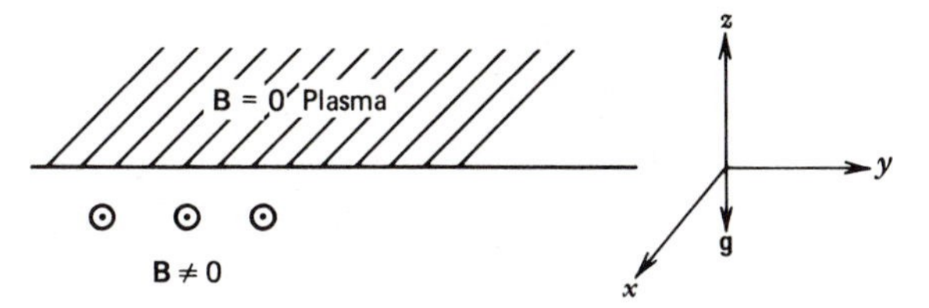

Fig. 8.9 Plasma held up against the force of gravity by a current sheet.

gravity by a magnetic field, as in Fig. 8.9. For this example, because we generalize the force equation (8.29) to include gravity, we have

$$\rho_M \partial_t \mathbf{V} = -\nabla P + \frac{1}{c}\mathbf{J} \times \mathbf{B} + \rho_M \mathbf{g} \tag{8.62}$$

Since $|\mathbf{B}|$ has a discontinuity at $z = 0$, (8.31) predicts a sheet current at $z = 0$ in the $(-)\hat{y}$-direction. The equilibrium quantities are, generalizing (8.51) and (8.52),

$$\frac{d}{dz}\left(P + \frac{B^2}{8\pi}\right) = -\rho_M g \tag{8.63}$$

We next consider a perturbed fluid velocity at the plasma-vacuum interface with sinusoidal variation only along y, an undetermined variation in z, and no variation in x,

$$\mathbf{V} = \mathbf{v}(z) \exp(-i\omega t + iky) \tag{8.64}$$

where $\mathbf{v} = (0, v_y, v_z)$. It can then be shown [1] that instability results, with

$$\omega^2 = -kg \tag{8.65}$$

or

$$\omega = \pm i(kg)^{1/2} \tag{8.66}$$

which implies instability. This is called the *Kruskal–Schwarzchild instability*, and is the MHD analog of the fluid *Rayleigh–Taylor instability.*

It is interesting to consider the microscopic physics of this instability. Recall the current in the $-\hat{y}$-direction. Microscopically, this current is due to the $\mathbf{g} \times \mathbf{B}$ drifts of the particles on the plasma surface. Since this drift is proportional to mass (why?), the ions are drifting much faster. Now consider the initial perturbation, as shown in Fig. 8.10. Since the ions drift to the left, and the electrons drift to the right, charges build up as shown. This creates an electric field as shown. The plasma on the surface then performs an $\mathbf{E} \times \mathbf{B}$ drift, down in the left section and

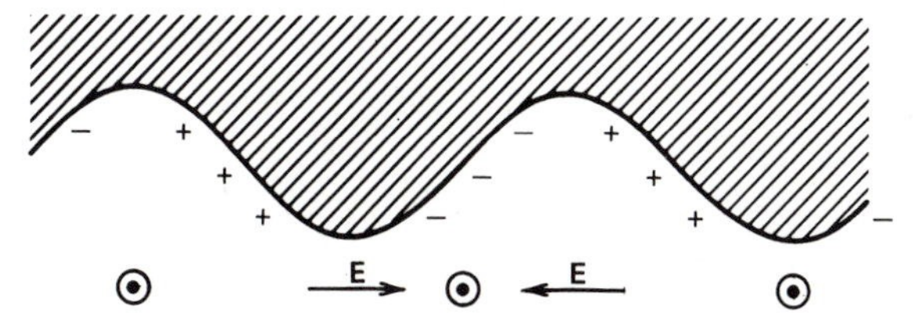

Fig. 8.10 Microscopic picture of the Kruskal–Schwarzchild instability.

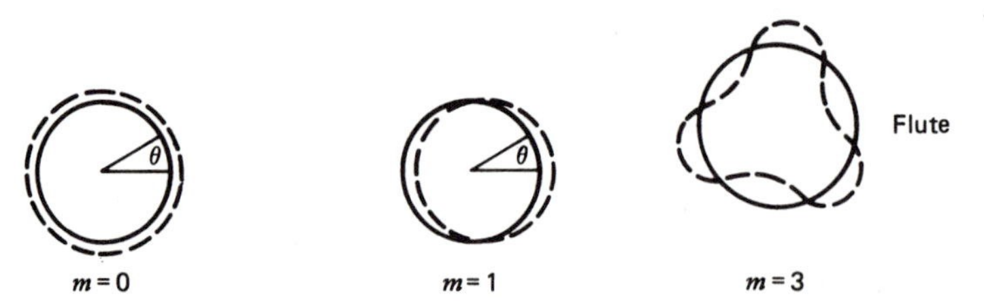

Fig. 8.11 Azimuthal variations of displacement for $m = 0$, $m = 1$, and $m = 3$.

up in the right section, thus intensifying the initial perturbation and leading to instability.

A similar analysis could be carried out for other equilibria, such as the theta pinch and z-pinch equilibria considered in the previous section. Consider the z-pinch equilibrium, and perturbations $\sim \exp(-i\omega t + im\theta + ikz)$, where θ is the azimuthal angle (Fig. 8.11). For $m = 0$, the instability is known as the *sausage* instability (Fig. 8.12). For $m = 1$, the instability is known as the *kink* instability. Higher values of m are known as *flute* instabilities, because their perturbations resemble fluted Greek columns.

We come to the second method of treating the question of stability in MHD systems, the *energy principle*. Consider a ball in a potential well, as shown in Fig. 8.13. In the unstable case, a small change in the particle's position leads to a decrease in the particle's potential energy; the difference in energy is available for kinetic energy and the implication is instability. On the other hand, the stable case is characterized by a positive change in potential energy for a small perturbation; thus, the perturbation is prohibited and the system is stable. It is interesting to contemplate the modification of these ideas when nonlinear effects are included.

It turns out that plasma systems behave in the same way. The energy of a plasma, which potentially could be turned into the kinetic energy of instability, is the integral over its volume of $B^2/8\pi$, the magnetic energy, plus its internal kinetic energy $\frac{3}{2}T$ per particle ($T_e = T_i = T$). Thus, the plasma energy W is

$$W = \int_V dV \, \frac{B^2}{8\pi} + \frac{3}{2} \int_V dV \, nT \tag{8.67}$$

or

$$W = \int_V dV \left(\frac{B^2}{8\pi} + \frac{3}{2} P \right) \tag{8.68}$$

If a hypothetical perturbation causes a decrease in W, the system is unstable. (We will not prove this here.) If W increases, the system is stable *to that perturbation*.

Fig. 8.12 Spatial variation of displacement for an $m = 0$ instability.

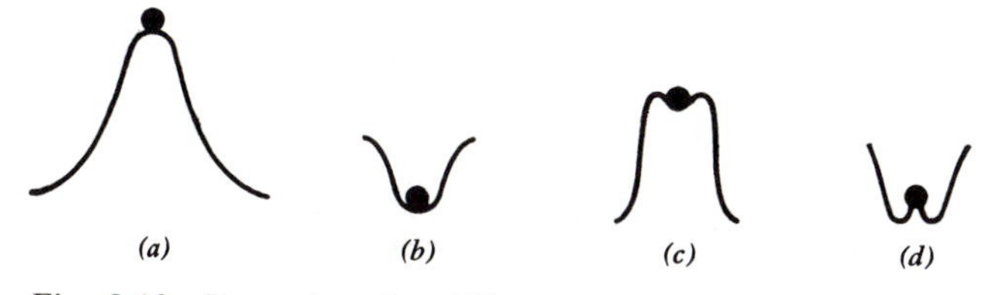

Fig. 8.13 Examples of stability: (*a*) unstable; (*b*) stable; (*c*) linearly stable, nonlinearly unstable; (*d*) linearly unstable, nonlinearly stable.

(Note Murphy's eighth law: To prove *instability*, one needs to find only one unstable perturbation. To prove *stability*, one needs to prove stability for each and every possible perturbation.)

One could apply this technique to each of the equilibria considered previously; each would yield a change $\delta W < 0$ for the types of perturbations considered before. Here we consider the more general problem of the hypothetical interchange of two neighboring tubes of magnetic flux. If this leads to $\delta W < 0$, we will call it an *interchange instability* (Fig. 8.14). We consider separately the change of magnetic energy W_m,

$$W_m \equiv \int dV \, \frac{B^2}{8\pi} \tag{8.69}$$

and the change of internal plasma energy W_p,

$$W_p \equiv \frac{3}{2} \int dV \, P \tag{8.70}$$

The interchange is accomplished by moving tube ② to where ① used to be. Then

$$W_m = \sum_{i=1,2} \int_0^{l_i} dl_i' A_i \, \frac{B_i^2}{8\pi} \tag{8.71}$$

where A_i is the cross-sectional area, and l_i is the length, of the ith flux tube. But in a flux tube the flux is constant; therefore $\varphi_i = B_i A_i$, or

$$W_m = \sum_{i=1,2} \frac{\varphi_i^2}{8\pi} \int_0^{l_i} \frac{dl_i'}{A_i} \tag{8.72}$$

The change δW_m in W_m by moving ① to ② is [① keeps its flux but finds a new l_i and A_i]

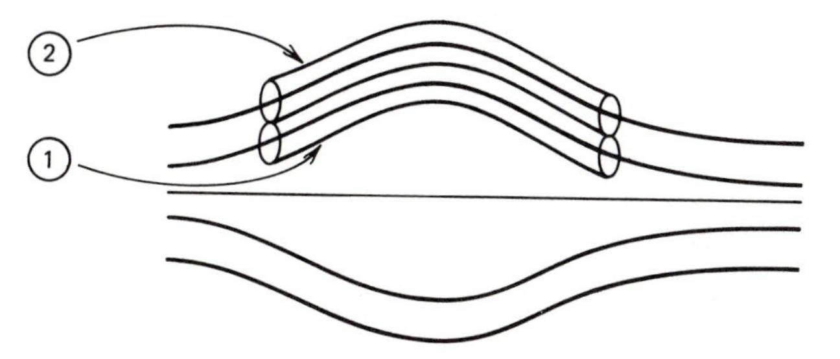

Fig. 8.14 Two neighboring flux tubes.

$$(\delta W_m)_{①\to②} = \frac{\varphi_1^2}{8\pi}\int_0^{l_2}\frac{dl_2'}{A_2} - \frac{\varphi_1^2}{8\pi}\int_0^{l_1}\frac{dl_1'}{A_1} \tag{8.73}$$

while the change for ② → ① is

$$(\delta W_m)_{②\to①} = \frac{\varphi_2^2}{8\pi}\int_0^{l_1}\frac{dl_1'}{A_1} - \frac{\varphi_2^2}{8\pi}\int_0^{l_2}\frac{dl_2'}{A_2} \tag{8.74}$$

The total change, adding (8.73) to (8.74), is then

$$\delta W_m = \left[\frac{\varphi_2^2 - \varphi_1^2}{8\pi}\right]\left[\int_0^{l_1}\frac{dl_1'}{A_1} - \int_0^{l_2}\frac{dl_2'}{A_2}\right] \tag{8.75}$$

or

$$\delta W_m = -\frac{\delta[\varphi^2]}{8\pi}\,\delta\left[\int\frac{dl}{A}\right] \tag{8.76}$$

If we pick two flux tubes with equal flux, we have

$$\delta W_m = 0 \tag{8.77}$$

With two flux tubes of equal flux, we next calculate the change in internal energy of the flux tubes, as the plasma expands or contracts to fit into a changing volume (Fig. 8.15).

The change in internal energy as we move the plasma in V_1 to the volume V_2 is (assuming $|V_2 - V_1|/V_1 << 1$)

$$\begin{aligned}\delta W_{p\,①\to②} &= \frac{3}{2}\,\delta(PV) = \frac{3}{2}\,\delta\left(\frac{PV^\gamma}{V^{\gamma-1}}\right)\\ &= \frac{3}{2}\,(PV^\gamma)_1\delta V^{1-\gamma}\\ &= \frac{3}{2}\,(PV^\gamma)_1(1-\gamma)V_1^{-\gamma}\delta V\\ &= \frac{3}{2}\,(PV^\gamma)_1(1-\gamma)V_1^{-\gamma}(V_2 - V_1)\end{aligned} \tag{8.78}$$

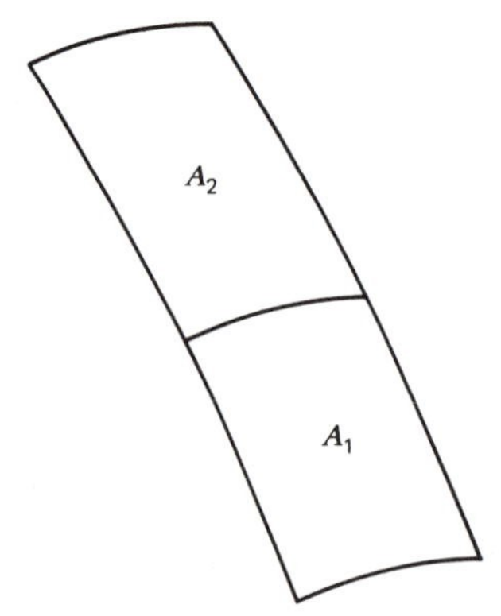

Fig. 8.15 Neighboring flux tubes with equal flux may have different areas and volumes.

where we have approximated $V \approx V_1 \approx V_2$ in the denominator only, and we have used the fact that PV^γ = constant in an adiabatic compression; assuming a three-dimensional compression we have $\gamma = 5/3$.

Next, the change $\delta W_{p\ ②\to①}$ is obtained from (8.78) by interchanging ① and ②; adding the two pieces yields

$$\delta W_p = -\frac{3}{2}(1-\gamma)\left[\frac{(PV^\gamma)_2}{V_2{}^\gamma} - \frac{(PV^\gamma)_1}{V_1{}^\gamma}\right](V_2 - V_1) \tag{8.79}$$

Approximating $V_1 \approx V_2$ in the denominator only, we have

$$\delta W_p = -\frac{3}{2}\frac{(1-\gamma)}{V_1{}^\gamma}\delta(PV^\gamma)\delta V \tag{8.80}$$

where $\delta(PV^\gamma)$ now means $(PV^\gamma)_2 - (PV^\gamma)_1$. With $\gamma = 5/3$ this is

$$\delta W_p = V_1{}^{-\gamma}\delta(PV^\gamma)\delta V \tag{8.81}$$

or

$$\delta W_p = V_1{}^{-\gamma}\delta V(\delta P V_1{}^\gamma + \gamma P_1 V^{\gamma-1}\delta V) \tag{8.82}$$

Now suppose we are in a low density part of the plasma, so that $P(\delta V/V) << \delta P$. (This is possible because δV will be very small when magnetic fields are large; recall that the flux of the two tubes is equal and we only need a small magnetic field gradient $\nabla B^2/8\pi \sim B\nabla B$ to balance the particle pressure gradient ∇P). Then

$$\delta W_p = \delta V\, \delta P \tag{8.83}$$

If the plasma density decreases from 1 to 2, $\delta P < 0$, there will be instability ($\delta W_p < 0$) if $\delta V > 0$ or

$$\delta V = \delta \int dlA = \varphi\, \delta \int \frac{dl}{B} > 0 \tag{8.84}$$

so that the condition for instability is

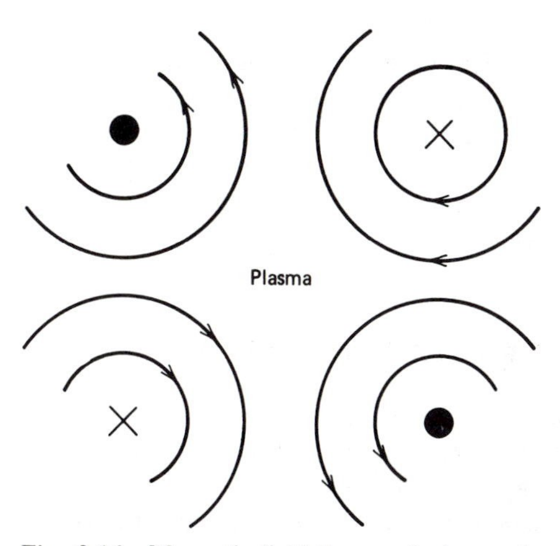

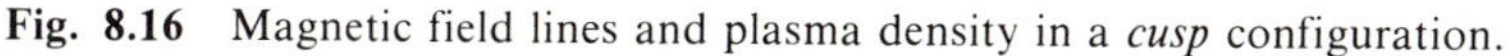

Fig. 8.16 Magnetic field lines and plasma density in a *cusp* configuration.

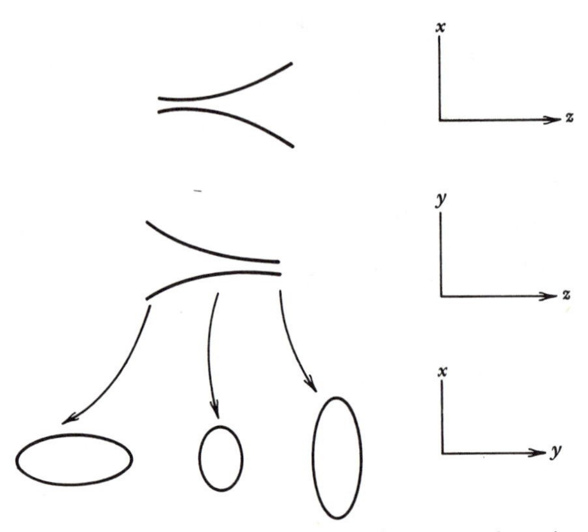

Fig. 8.17 Magnetic field curvature in present-day mirror machines.

$$\boxed{\delta \int \frac{dl}{B} > 0} \tag{8.85}$$

Consider the simple mirror configuration of Fig. 8.14. As we go from ① to ②, the length of $\int dl$ increases while $|B|$ decreases; (8.85) is easily satisfied and we find instability. The same result is easily obtained for both the θ-pinch and the z-pinch.

Equation (8.85) leads to a very important principle, which we shall not prove rigorously here. This principle states: Whenever the field lines curve toward the plasma, the plasma is unstable. Likewise, when field lines curve away from the plasma, the plasma is interchange stable. Thus, the simple mirror and pinches are unstable, while the *cusp* is stable, as shown in Fig. 8.16. Since the field lines curve away from the plasma the cusp is interchange stable. Furthermore, the magnetic field is a minimum in the center of the plasma, so that *minimum-B* is associated with interchange stability. The instability of the simple mirror has led mirror designers to look for a minimum-B configuration. They accomplish this by putting coils, known as Joffe bars, along the axis of the mirror. The net result is a field configuration that everywhere curves away from the plasma, has a minimum of $|B|$ in the center, and is MHD stable (Fig. 8.17).

8.4 MICROSCOPIC PICTURE OF MHD EQUILIBRIUM

Magnetohydrodynamics is based on a picture of the plasma as a single fluid. Yet we know that the plasma consists of two species of charged particles. Thus, any topic in MHD can also be understood by considering the detailed orbits of the charged particles. In this section, we look at the topic of MHD equilibrium from a microscopic point of view.

Consider a plasma whose density varies smoothly in the $\hat{x}$-direction, with magnetic field $\mathbf{B}_0 = B_0(x)\hat{z}$ (Fig. 8.18). Then the MHD equations (8.47) and (8.48) predict, in the steady state and in component form,

$$\frac{\partial P}{\partial x} = \frac{1}{c} J_y B_0 \tag{8.86}$$

and

$$-\partial_x B_0 = \frac{4\pi}{c} J_y \tag{8.87}$$

Suppose that the ions are cold, $T_i = 0$, and the electron temperature T_e is a constant in space. Then at $x = 0$, (8.86) yields

$$J_y = \frac{c}{B_0} T_e \frac{\partial n_0}{\partial x} < 0 \tag{8.88}$$

so that our macroscopic picture is one where the plasma pressure is being balanced by the $\mathbf{J} \times \mathbf{B}$ force. Suppose the ions are cold and very massive; then it is reasonable to suppose that the current is being contributed by the electrons, even though the one-fluid equations tell us nothing of the behavior of the individual species. If this is so, then there must be a mean electron flow speed in the $\hat{y}$-direction such that

$$J_y = -en_0 \langle v_y \rangle_{\text{electrons}} \tag{8.89}$$

or from (8.88)

$$v_{de} \equiv \langle v_y \rangle_{\text{electrons}} = -\frac{cT_e}{eB_0} \frac{1}{n_0} \frac{\partial n_0}{\partial x} \tag{8.90}$$

which with $L_n^{-1} \equiv (-1/n_0)(dn_0/dx) > 0$ is

$$\boxed{v_{de} = \frac{v_e^2}{|\Omega_e| L_n}} \tag{8.91}$$

where v_{de} is the *electron diamagnetic drift speed*, and v_e is, as usual, the electron thermal speed.

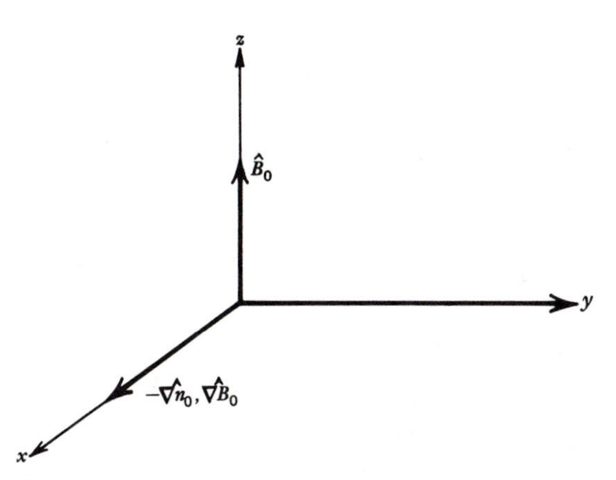

Fig. 8.18 Coordinate system for MHD equilibrium.

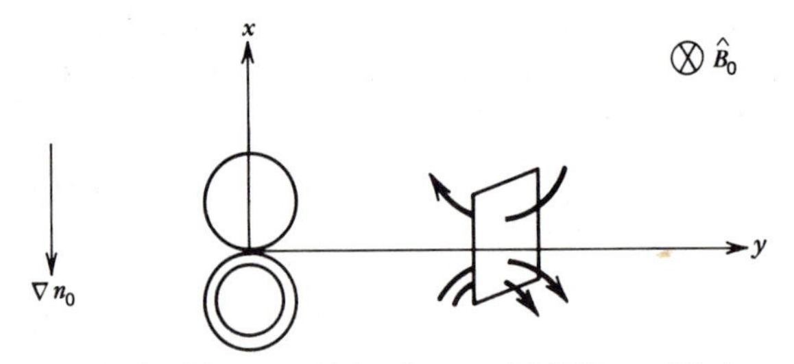

Fig. 8.19 Single particle picture of MHD equilibria.

EXERCISE Where does the "diamagnetic" come from?

Now let us look at the single particle picture. The electrons are gyrating about the magnetic field in the x-y plane. There is no electric field, hence no $\mathbf{E} \times \mathbf{B}_0$ drift. There is no curvature of the field lines, hence no curvature drift. There will be a ∇B drift but, if we assume a very strong magnetic field, then only a tiny ∇B is needed to make $\nabla(P + B^2/8\pi) = 0$, and we can ignore the ∇B drift. So where does the single particle drift come from to make v_{de} in (8.91)? The fact is that in terms of guiding centers, there is no drift, but in terms of fluid elements, there is a drift. Consider the x-y plane, as indicated in Fig. 8.19. Because there are more particles for $x < 0$ than for $x > 0$, a small area of the x-z plane will see more particles going through to the right than to the left; thus, there is a net flow of electrons to the right in every fluid element, and a net current to the left. This is true even though the guiding centers never move!

This concludes our introduction to magnetohydrodynamics. MHD is an extremely important approximation in plasma physics that is widely used in fusion plasma physics, solar physics, plasma astrophysics, and energy technology. For further discussion of MHD see Refs. [2]–[30].

REFERENCES

[1] G. Schmidt, *Physics of High Temperature Plasmas*, Academic, New York, 1966.

[2] T. G. Cowling, *Magnetohydrodynamics*, Interscience, New York, 1957.

[3] R. K. M. Landshoff, ed., *Magnetohydrodynamics*, Stanford University Press, Stanford, Calif., 1957.

[4] D. J. Rose and M. Clark, Jr., *Plasma and Controlled Fusion*, The M.I.T. Press, Cambridge, Mass., 1961.

[5] D. Bershader, ed., *Plasma Hydromagnetics*, Stanford University Press, Stanford, Calif., 1962.

[6] W. B. Thompson, *An Introduction to Plasma Physics*, Pergamon, New York, 1962.

[7] C. L. Longmire, *Elementary Plasma Physics*, Interscience, New York, 1963.

[8] S. Gartenhaus, *Elements of Plasma Physics*, Holt, Rinehart & Winston, New York, 1964.

[9] P. C. Kendall and C. Plumpton, *Magnetohydrodynamics with Hydrodynamics*, Vol. 1, Pergamon, New York, 1964.

[10] J. L. Delcroix, *Plasma Physics*, Wiley, New York, 1965.

[11] E. H. Holt and R. E. Haskell, *Foundations of Plasma Dynamics*, Macmillan, New York, 1965.

[12] J. A. Shercliff, *A Textbook of Magnetohydrodynamics*, Pergamon, Oxford, 1965.

[13] A. Jeffrey and T. Taniuti, *Magnetohydrodynamic Stability and Thermonuclear Containment*, Academic, New York, 1966.

[14] W. B. Kunkel, ed., *Plasma Physics in Theory and Application*, McGraw-Hill, New York, 1966.

[15] M. A. Leontovich, ed., *Reviews of Plasma Physics*, Vol. 2, Consultants Bureau, New York, 1966.

[16] A. Simon and W. B. Thompson, eds., *Advances in Plasma Physics*, Vol. 1, Interscience, New York, 1968.

[17] T. J. M. Boyd and J. J. Sanderson, *Plasma Dynamics*, Barnes & Nobel, New York, 1969.

[18] P. C. Clemmow and J. P. Dougherty, *Electrodynamics of Particles and Plasmas*, Addison-Wesley, Reading, Mass., 1969.

[19] H. Cabannes, *Theoretical Magnetofluiddynamics*, translated by M. Holt and A. A. Sfeir, Academic, New York, 1970.

[20] N. A. Krall and A. W. Trivelpiece, *Principles of Plasma Physics*, McGraw-Hill, New York, 1973.

[21] *Course on Instabilities and Confinement in Toroidal Plasmas*, Commission of the European Communities, Luxembourg, 1974.

[22] A. B. Mikhailovskii, *Theory of Plasma Instabilities*, Vol. 2: *Instabilities of an Inhomogeneous Plasma*, translated by J. B. Barbour, Consultants Bureau, New York, 1974.

[23] M. A. Leontovich, ed., *Reviews of Plasma Physics*, Vol. 6, Consultants Bureau, New York, 1975.

[24] D. E. Davis, ed., *Pulsed High Beta Plasmas*, Pergamon, Oxford, 1976.

[25] G. Bateman, *MHD Instabilities*, The MIT Press, Cambridge, Mass., 1978.

[26] *Plasma Transport, Heating and MHD Theory*, Proceedings of the Workshop, Varenna, Italy, 12–16 September 1977, Pergamon, Oxford, 1978.

[27] *Space Plasma Physics: The Study of Solar-System Plasmas*, Vol. 1: *Reports of the Study Committee and Advocacy Panels*, National Academy of Sciences, Washington, D.C., 1978.

[28] *Space Plasma Physics: The Study of Solar-System Plasmas*, Vol. 2: *Working Papers*, National Academy of Sciences, Washington, D.C., 1979.

[29] F. Krause and K.-H. Rädler, *Mean-Field Magnetohydrodynamics and Dynamo Theory*, Pergamon, New York, 1980.

[30] M. A. Leontovich, ed., *Reviews of Plasma Physics*, Vol. 8, Consultants Bureau, New York, 1980.

PROBLEMS

8.1 Alfvén Waves

Linearize the ideal MHD equations (8.28) to (8.31) with $\rho_M = \rho_{M0}$, $\mathbf{B} = B_0\hat{z} + B_1\hat{y}$, $\mathbf{V} = V_1\hat{y}$, $\mathbf{J} = J_x\hat{x}$, and $\mathbf{k} = k\hat{z}$ to obtain the dispersion relation for Alfvén waves. Compare your result to the result (7.214) from the two-fluid theory, and explain any differences.

8.2 Magnetosonic Waves

For a cold plasma, linearize the ideal MHD equations (8.28) to (8.31) with $\rho_M = \rho_{M0}$, $\mathbf{B} = B_0\hat{z} + B_1\hat{z}$, $\mathbf{V} = V_1\hat{y}$, $\mathbf{J} = J_1\hat{x}$, and $\mathbf{k} = k\hat{y}$ to obtain the dispersion relation for fast magnetosonic waves. Compare your result to the result (7.227) from two-fluid theory, and explain any differences.

CHAPTER 9

Discrete Particle Effects

9.1 INTRODUCTION

There are many effects in a plasma that are associated with the discrete nature of the plasma particles. One of these effects is that of collisions, which is studied in Chapters 1 and 3 to 5. A collision is the interaction of one discrete particle with another discrete particle. There are also discrete particle effects that are due to the interaction of one discrete particle with the plasma as a whole. For example, a fast electron moving through a plasma emits Langmuir waves. This phenomena depends on the fact that the fast electron is indeed a discrete particle, but it does not require that the rest of the plasma be made up of discrete particles; thus, the rest of the plasma can be treated through the Vlasov approximation. This leads to an extremely useful approach known as the *test-particle method.*

Since the effects to be studied in this chapter are discrete particle effects, it might have made more sense to study them together with the discrete particle collisional effects. However, it turns out that the test-particle method relies heavily on the *Vlasov dielectric function.* Chapter 6 on Vlasov theory taught us many of the properties of this dielectric function, so that we can now comprehend the results of the test-particle method more easily.

9.2 DEBYE SHIELDING

As a first application of the test-particle method, let us calculate the Debye shielding of a test charge q_T that moves through a uniform plasma with a constant speed $\mathbf{v}_0$, starting from position $\mathbf{x}_0 = 0$ at $t = 0$. For simplicity, we freeze the ions, and treat the plasma electrons via the Vlasov equation. The only discreteness in the problem is the test charge. Then Poisson's equation is

$$\nabla^2 \varphi(\mathbf{x},t) = -4\pi\rho$$

$$= -4\pi e[n_0 - \int d\mathbf{v}\, f_e(\mathbf{x},\mathbf{v},t)] - 4\pi q_T\, \delta(\mathbf{x} - \mathbf{v}_0 t)$$

$$= 4\pi e \int d\mathbf{v}\, f_1(\mathbf{x},\mathbf{v},t) - 4\pi q_T\, \delta(\mathbf{x} - \mathbf{v}_0 t) \tag{9.1}$$

where $f_e = f_0 + f_1$ and the f_0 term cancels the ion term. Assuming that the test charge makes only a small perturbation in the electron density, we can linearize the Vlasov equation to obtain

$$\partial_t f_1(\mathbf{x},\mathbf{v},t) + \mathbf{v} \cdot \nabla f_1 = -\frac{e}{m_e} \nabla\varphi \cdot \nabla_{\mathbf{v}} f_0(\mathbf{v}) \tag{9.2}$$

With these two equations, we use Laplace transform techniques to study the initial value problem. The test charge suddenly appears at $\mathbf{x} = 0$ at $t = 0$, and the distribution function is initially unperturbed,

$$f_1(\mathbf{x},\mathbf{v},t = 0) = 0 \tag{9.3}$$

We Fourier transform (9.1) and (9.2) in space, and Laplace transform in time. The Fourier and Laplace transform conventions are stated in Chapter 5. The spatial Fourier transform of Poisson's equation (9.1) is

$$-k^2\varphi(\mathbf{k},t) = 4\pi e \int d\mathbf{v}\, f_1(\mathbf{k},\mathbf{v},t) - \frac{q_T}{2\pi^2}\, e^{-i\mathbf{k}\cdot\mathbf{v}_0 t} \tag{9.4}$$

which has the Laplace transform

$$-k^2\varphi(\mathbf{k},\omega) = 4\pi e \int d\mathbf{v}\, f_1(\mathbf{k},\mathbf{v},\omega) + \frac{(2\pi^2)^{-1}\, q_T}{i\omega - i\mathbf{k} \cdot \mathbf{v}_0} \tag{9.5}$$

Because the initial value of f_1 is zero, the Fourier–Laplace transform of the linearized Vlasov equation (9.2) is

$$(-i\omega + i\mathbf{k} \cdot \mathbf{v}) f_1(\mathbf{k},\omega,\mathbf{v}) = -\frac{e}{m_e}\, i\mathbf{k} \cdot \nabla_{\mathbf{v}} f_0(\mathbf{v}) \varphi(\mathbf{k},\omega) \tag{9.6}$$

Solving (9.6) for f_1 and inserting in (9.5) yield

$$k^2\varphi(\mathbf{k},\omega) = -\frac{4\pi e^2}{m_e} \int d\mathbf{v}\, \frac{\mathbf{k} \cdot \nabla_{\mathbf{v}} f_0(\mathbf{v})}{\omega - \mathbf{k} \cdot \mathbf{v}}\, \varphi(\mathbf{k},\omega) + \frac{i(2\pi^2)^{-1}\, q_T}{\omega - \mathbf{k} \cdot \mathbf{v}_0} \tag{9.7}$$

The integrations of the two velocity directions perpendicular to $\mathbf{k}$ can be performed, yielding the form

$$k^2\varphi(\mathbf{k},\omega) \left[1 - \frac{\omega_e^2}{k^2} \int du\, \frac{d_u\, g(u)}{u - \omega/k}\right] = \frac{i(2\pi^2)^{-1}\, q_T}{\omega - \mathbf{k} \cdot \mathbf{v}_0} \tag{9.8}$$

where in square brackets we recognize our old friend the Vlasov dielectric function (6.34). Thus,

$$\varphi(\mathbf{k},\omega) = \frac{i(2\pi^2)^{-1}\, q_T}{k^2 \epsilon(\mathbf{k},\omega)(\omega - \mathbf{k} \cdot \mathbf{v}_0)} \tag{9.9}$$

Next, the Laplace transform is inverted to obtain

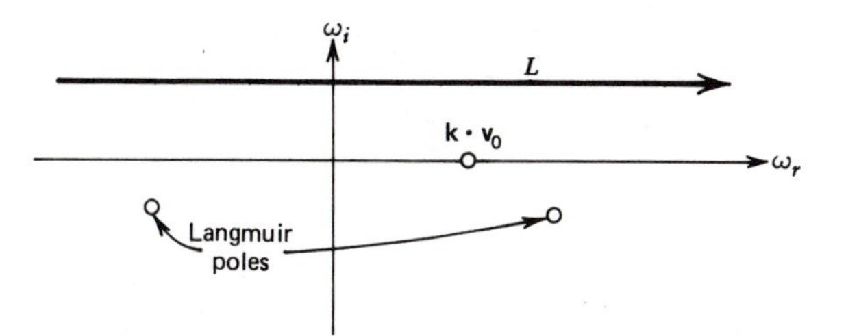

Fig. 9.1 Inverse Laplace contour used in calculating the Debye shielding of a test charge.

$$\varphi(\mathbf{k},t) = \int_L \frac{d\omega}{2\pi}\, e^{-i\omega t}\varphi(\mathbf{k},\omega)$$

$$= \int_L \frac{d\omega}{2\pi}\, e^{-i\omega t}\, \frac{i(2\pi^2)^{-1}\, q_T}{k^2\epsilon(\mathbf{k},\omega)(\omega - \mathbf{k}\cdot\mathbf{v}_0)} \tag{9.10}$$

The pole structure of the integrand is as shown in Fig. 9.1, where for Maxwellian electrons $\epsilon(\mathbf{k},\omega)$ has among others, the two zeros corresponding to Landau damped Langmuir waves. For this calculation, we can move the contour downward for $t > 0$ as in the calculation of Langmuir waves in Section 6.4. We ignore all of the transient contributions discussed in Section 6.4 (see Fig. 6.6). Furthermore, let us ignore the contribution from the Langmuir poles, which will damp away at large times. Then we pick up only the pole at $\omega = \mathbf{k}\cdot\mathbf{v}_0$, obtaining

$$\boxed{\varphi(\mathbf{k},t) = \frac{(2\pi^2)^{-1}\, q_T\, e^{-i\mathbf{k}\cdot\mathbf{v}_0 t}}{k^2\epsilon(\mathbf{k},\omega = \mathbf{k}\cdot\mathbf{v}_0)}} \tag{9.11}$$

Since ϵ is evaluated at the real frequency $\omega = \mathbf{k}\cdot\mathbf{v}_0$, we can use the exact formula (6.45),

$$\epsilon(\mathbf{k},\omega) = 1 - \frac{\omega_e^2}{k^2}\, P\int du\, \frac{d_u\, g(u)}{u - (\omega/k)} - \pi i\, \frac{\omega_e^2}{k^2}\, d_u\, g(u)|_{u=\omega/k} \tag{9.12}$$

Let us work out several examples.

EXAMPLE A VACUUM

Letting the plasma disappear, $\omega_e \to 0$, we have $\epsilon = 1$. Then

$$\varphi(\mathbf{k},t) = \frac{q_T}{2\pi^2 k^2}\, e^{-i\mathbf{k}\cdot\mathbf{v}_0 t} \tag{9.13}$$

Performing the inverse Fourier transform we have

$$\varphi(\mathbf{x},t) = \int d\mathbf{k}\, e^{i\mathbf{k}\cdot\mathbf{x}}\varphi(\mathbf{k},t)$$

$$= \int d\mathbf{k}\, e^{i\mathbf{k}\cdot\mathbf{x} - i\mathbf{k}\cdot\mathbf{v}_0 t}\, \frac{q_T}{2\pi^2 k^2} \tag{9.14}$$

This integral can be performed in spherical coordinates, letting the k_z-axis be in the direction of $\mathbf{x} - \mathbf{v}_0 t$. Then

$$\mathbf{k} \cdot (\mathbf{x} - \mathbf{v}_0 t) = k|\mathbf{x} - \mathbf{v}_0 t|\cos\theta \tag{9.15}$$

and

$$\varphi(\mathbf{x},t) = \frac{q_T}{2\pi^2}\int_0^\infty dk\, 2\pi \int_0^\pi d\theta \sin\theta\, e^{ik|\mathbf{x} - \mathbf{v}_0 t|\cos\theta} \tag{9.16}$$

With $u = \cos\theta$, $du = -\sin\theta\, d\theta$, we have

$$\varphi(\mathbf{x},t) = \frac{q_T}{\pi}\int_0^\infty dk \int_{-1}^{1} du\, e^{ik|\mathbf{x} - \mathbf{v}_0 t|u}$$

$$= \frac{2q_T}{\pi}\frac{1}{|\mathbf{x} - \mathbf{v}_0 t|}\underbrace{\int_0^\infty dk \frac{\sin(k|\mathbf{x} - \mathbf{v}_0 t|)}{k}}_{\underbrace{\int_0^\infty \frac{dz}{z}\sin z}_{\pi/2}} \tag{9.17}$$

or

$$\varphi(\mathbf{x},t) = \frac{q_T}{|\mathbf{x} - \mathbf{v}_0 t|} \tag{9.18}$$

Thus, we regain the potential due to a point charge in vacuum, moving with velocity $\mathbf{v}_0$.

EXAMPLE B TEST CHARGE AT REST IN PLASMA ($v_0 << v_e$)

For a test charge at rest, or moving very slowly ($|\mathbf{v}_0| << v_e$), we expect to regain the Debye shielding of Chapter 1 (for motionless ions). Setting $\omega = \mathbf{k} \cdot \mathbf{v}_0 \approx 0$, we have, taking the electrons Maxwellian [see Eq. (6.24)],

$$\epsilon(\mathbf{k},\omega = 0) = 1 - \frac{\omega_e^2}{k^2}\underbrace{\int du \frac{d_u g}{u}}_{\underbrace{-\int du \frac{g(u)}{v_e^2}}_{1/v_e^2}} \tag{9.19}$$

or

$$\epsilon(\mathbf{k},\omega = 0) = 1 + \frac{\omega_e^2}{k^2 v_e^2} = 1 + \frac{1}{k^2\lambda_e^2} \tag{9.20}$$

which is the "static dielectric function" with fixed ions. The potential is then

$$\varphi(\mathbf{k},t) = \frac{(2\pi^2)^{-1} q_T}{k^2 \epsilon(\mathbf{k},\omega = 0)} = \frac{(2\pi^2)^{-1} q_T}{k^2 + k_e^2} \tag{9.21}$$

where we have defined the Debye wave number

$$k_e \equiv \lambda_e^{-1} \tag{9.22}$$

without any factor of 2π. Then

$$\begin{aligned}
\varphi(\mathbf{x},t) &= \int d\mathbf{k}\, e^{i\mathbf{k}\cdot\mathbf{x}}\varphi(\mathbf{k},t) \\
&= \frac{q_T}{2\pi^2}\int_0^\infty 2\pi k^2\, dk \int_0^\pi d\theta \sin\theta\, \frac{e^{ikx\cos\theta}}{k^2 + k_e^2} \\
&= \frac{q_T}{\pi}\int_0^\infty dk\, \frac{k^2}{k^2 + k_e^2}\int_{-1}^1 du\, e^{ikxu} \\
&= \frac{2}{\pi x}\, q_T \int_0^\infty dk\, \frac{k \sin kx}{k^2 + k_e^2}
\end{aligned} \tag{9.23}$$

where the substitution $u' = \cos\theta$ has been used. Because the integrand is even in k, we can extend the integration to $-\infty$ and use contour techniques. We find $(x \equiv |\mathbf{x}| > 0)$

$$\begin{aligned}
\frac{1}{2}\int_{-\infty}^\infty dk\, \frac{k \sin kx}{k^2 + k_e^2} &= \frac{1}{2}\int_{-\infty}^\infty dk\, \frac{k}{(k + ik_e)(k - ik_e)}\left(\frac{e^{ikx} - e^{-ikx}}{2i}\right) \\
&= \frac{2\pi i}{2(2i)}\left[\frac{ik_e}{2ik_e}\, e^{i(ik_e)x} + \frac{-ik_e}{-2ik_e}\, e^{-i(-ik_e)x}\right] \\
&= \frac{\pi}{2}\, e^{-x/\lambda_e}
\end{aligned} \tag{9.24}$$

so that

$$\boxed{\varphi(\mathbf{x},t) = \frac{q_T}{x}\, e^{-x/\lambda_e}} \qquad v_0 << v_e \tag{9.25}$$

which is exactly what we expect for a Debye shielded test particle. Note that this formula is valid not only for motionless particles, but also for moving particles as long as $v_0 << v_e$. (See Refs. [1]–[9].)

EXAMPLE C VERY FAST TEST CHARGE ($v_0 >> v_e$)

For a very fast test charge, the dielectric function is

$$\begin{aligned}
\epsilon(\mathbf{k},\omega = \mathbf{k}\cdot\mathbf{v}_0) &\approx 1 - \frac{\omega_e^2}{k^2}\, P\int_{-\infty}^\infty du\, \frac{d_u g(u)}{u - \mathbf{k}\cdot\mathbf{v}_0/k} \\
&\approx 1 + \frac{\omega_e^2}{k\mathbf{k}\cdot\mathbf{v}_0}\underbrace{\int_{-\infty}^\infty du\, d_u g}_{0} \\
&\approx 1
\end{aligned} \tag{9.26}$$

where we have ignored u compared to $\mathbf{k}\cdot\mathbf{v}_0/k$ in the denominator. But this is the same result as for a test charge in vacuum (Example A), so we find

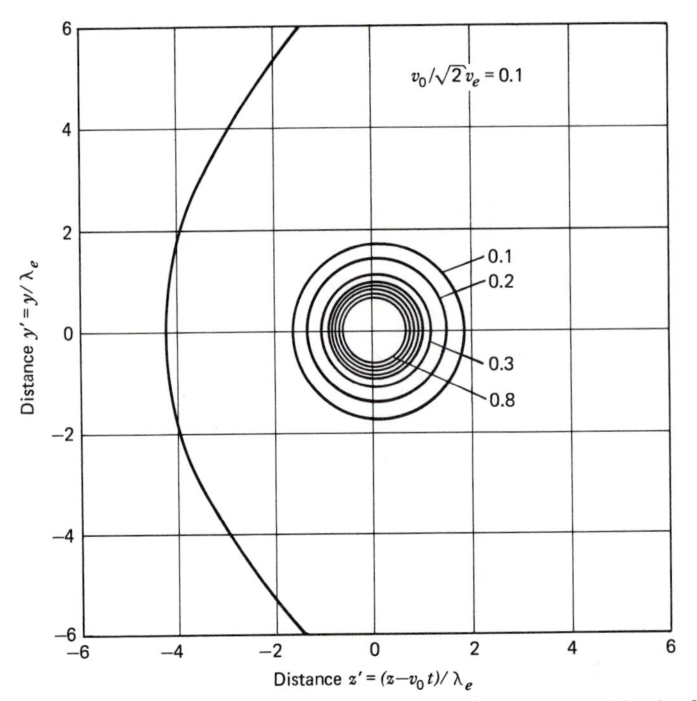

Fig. 9.2 Steady-state equipotential contours, as measured in the frame of a moving test charge at the origin, with charge q_T and speed $v_0/2^{1/2}\, v_e = 0.1$. Contour labels indicate the value of $\varphi\lambda_e/q_T$. Unclosed, unlabeled contour is at zero potential. From Ref. [9].

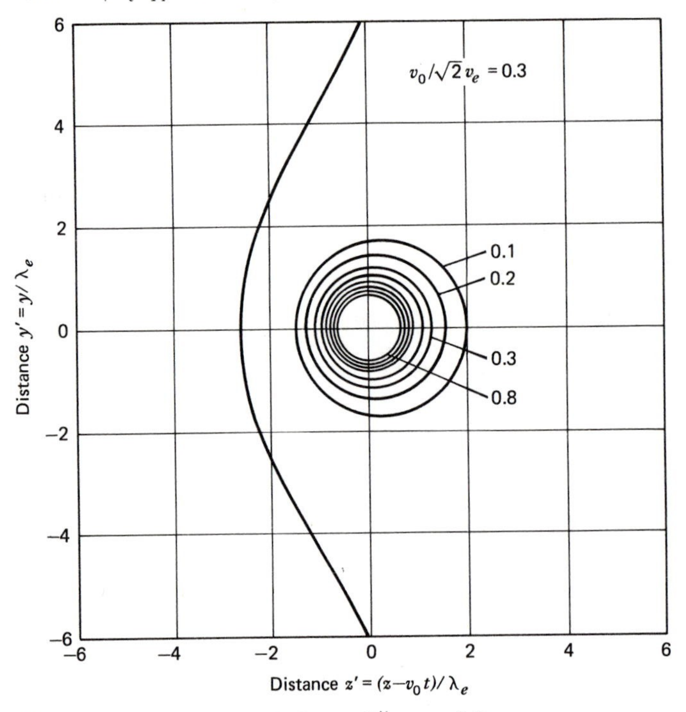

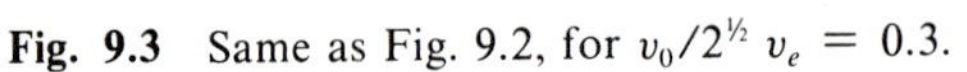
Fig. 9.3 Same as Fig. 9.2, for $v_0/2^{1/2}\, v_e = 0.3$.

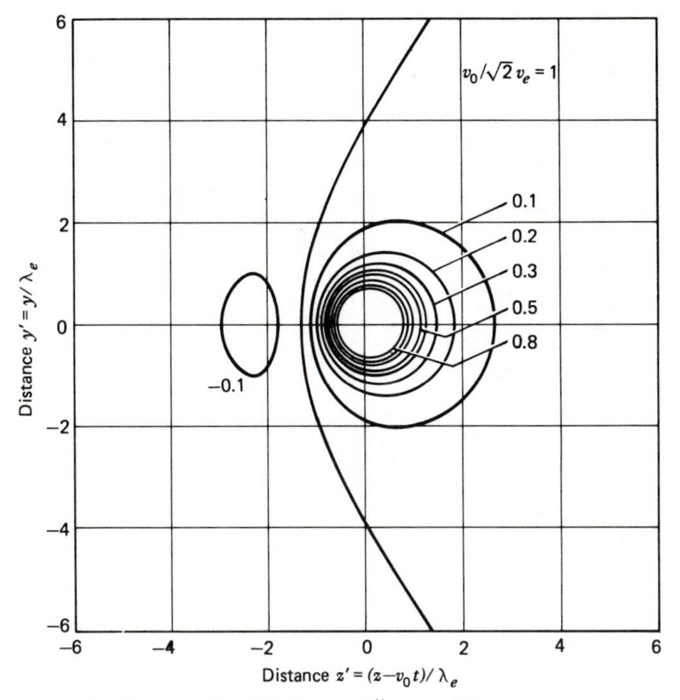

Fig. 9.4 Same as Fig. 9.2, for $v_0/2^{1/2}\, v_e = 1.0$.

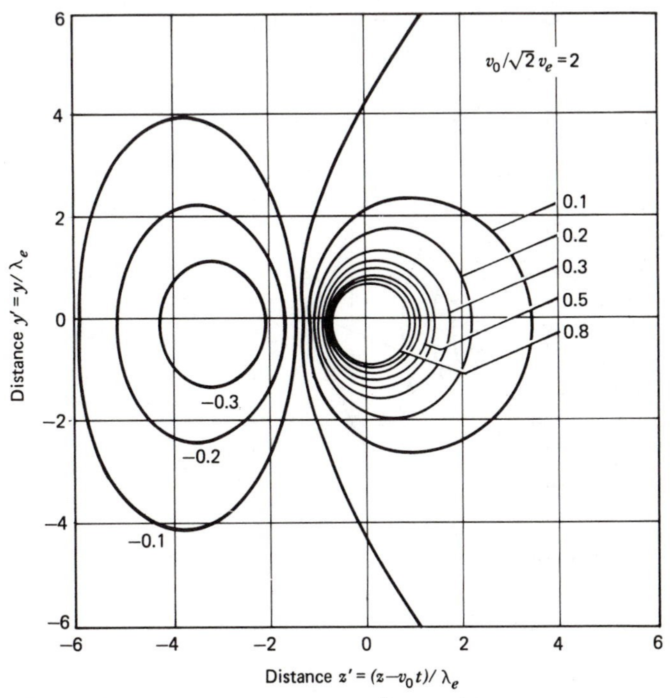

Fig. 9.5 Same as Fig. 9.2, for $v_0/2^{1/2}\, v_e = 2.0$.

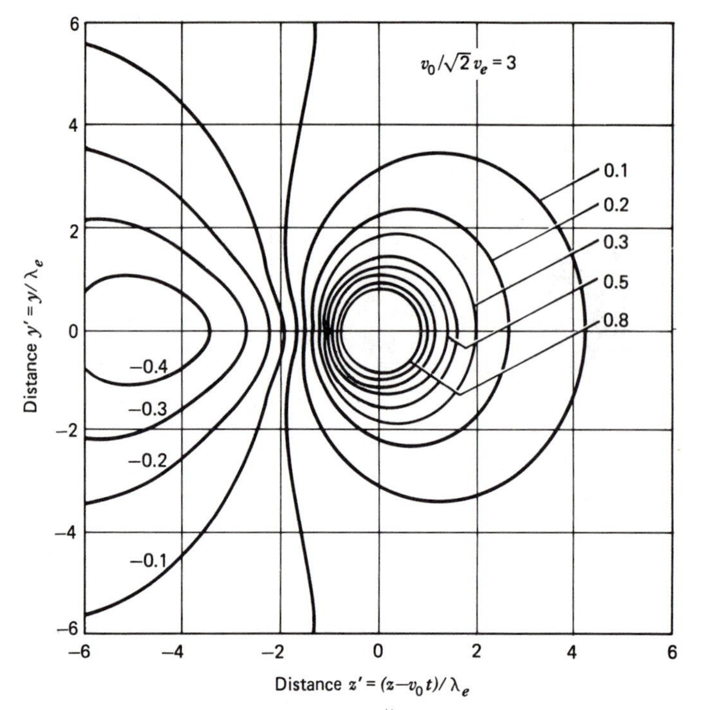

Fig. 9.6 Same as Fig. 9.2, for $v_0/2^{1/2}\, v_e = 3.0$.

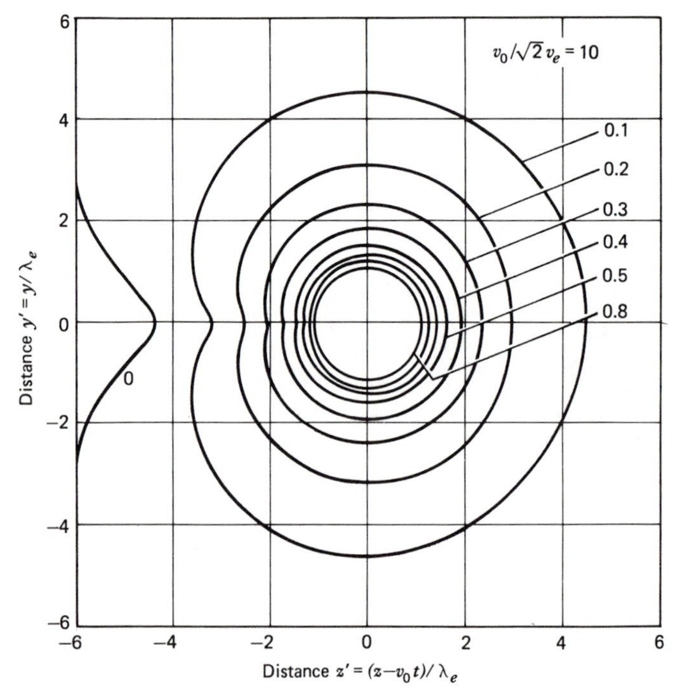

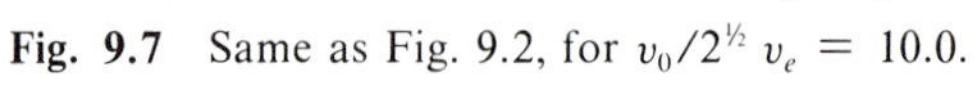

Fig. 9.7 Same as Fig. 9.2, for $v_0/2^{1/2}\, v_e = 10.0$.

$$\varphi(\mathbf{x},t) = \frac{q_T}{|\mathbf{x} - \mathbf{v}_0 t|}, \qquad v_0 >> v_e \tag{9.27}$$

The plasma does not have time to "see" a fast test charge, and thus does not have time to respond and shield. (See Refs. [1]–[9].)

We have seen that a fast particle has no shielding, while a slow particle is completely shielded. A particle moving at intermediate speeds will be partially shielded. The words fast, slow, and intermediate will depend on which plasma species we are talking about. Figures 9.2 to 9.7 show the transition from the Debye shielding of an almost motionless particle [$v_0/(2^{1/2} v_e) = 0.1$ in Fig. 9.2] to the almost unshielded behavior of a fast particle [$v_0/(2^{1/2} v_e) = 10.0$ in Fig. 9.7].

In the next section, we continue to exploit the test particle approach, calculating the equilibrium level of fluctuations in a plasma.

9.3 FLUCTUATIONS IN EQUILIBRIUM

In the previous section, we computed the electrostatic potential in a plasma due to a single test charge. We found that the rest of the plasma could be treated by the Vlasov equation.

In this section, we want to use the same ideas to compute the average level of electric field fluctuations in an equilibrium plasma. We do this by considering each and every plasma particle as a test charge. Each test charge sees the rest of the plasma as a Vlasov plasma, and it emits waves satisfying the *Cerenkov* condition $\omega = \mathbf{k} \cdot \mathbf{v}_0$, where $\mathbf{v}_0$ is the velocity of the test charge. Likewise, waves are damped via Landau damping, which again involves the *resonance condition* $\omega = \mathbf{k} \cdot \mathbf{v}_0 = ku$ where $\mathbf{v}_0$ is now the velocity of the particle that is doing the Landau damping, and $u \equiv \mathbf{k} \cdot \mathbf{v}_0/k$. Thus, we have a steady-state situation, with waves being emitted and absorbed. At each point in the plasma, the electric field is fluctuating wildly in space and time. However, the ensemble averaged electric field energy density is a constant in space and in time. It is this ensemble averaged electric field energy density that we wish to calculate.

The process considered here is an example of the *principle of detailed balance*, which states that to every emission process, there is a corresponding damping process, and vice versa. Cerenkov emission of Langmuir waves corresponds to Landau damping of Langmuir waves. In steady state, these two processes are balanced; the average rate of wave emission equals the average rate of wave damping.

Note that here we are using the term "wave" in its most general sense; we have fluctuations of all frequencies and all wave numbers, including the normal modes that we call "Langmuir waves."

The mathematics involved in this calculation is straightforward. Recall from Eq. (9.11) that the potential due to a single test charge is written in wave number space as

$$\varphi(\mathbf{k},t) = \frac{(2\pi^2)^{-1}\, q_T\, e^{-i\mathbf{k} \cdot \mathbf{x}_0(t)}}{k^2 \epsilon(\mathbf{k},\omega = \mathbf{k} \cdot \mathbf{v}_0)} \tag{9.28}$$

where in the exponent we have specified the orbit $\mathbf{v}_0 t$ by the expression $\mathbf{x}_0(t)$. Since

$$\mathbf{E}(\mathbf{x},t) = -\nabla\varphi(\mathbf{x},t) \tag{9.29}$$

we have

$$\mathbf{E}(\mathbf{k},t) = -i\mathbf{k}\varphi(\mathbf{k}) = -(2\pi^2)^{-1} i q_T \frac{\mathbf{k}\, e^{-i\mathbf{k}\cdot\mathbf{x}_0(t)}}{k^2\epsilon(\mathbf{k},\omega = \mathbf{k}\cdot\mathbf{v}_0)} \tag{9.30}$$

so that

$$\mathbf{E}(\mathbf{x},t) = \int d\mathbf{k}\, e^{i\mathbf{k}\cdot\mathbf{x}}\mathbf{E}(\mathbf{k},t)$$

$$= -(2\pi^2)^{-1} q_T i \int d\mathbf{k}\, \frac{\mathbf{k}\, e^{i\mathbf{k}\cdot\mathbf{x} - i\mathbf{k}\cdot\mathbf{x}_0(t)}}{k^2\epsilon(\mathbf{k},\omega = \mathbf{k}\cdot\mathbf{v}_0)} \tag{9.31}$$

Equation (9.31) gives the electric field at point $\mathbf{x}$ due to a particle with orbit $\mathbf{x}_0(t)$. If we add up the fields at $\mathbf{x}$ from all particles in the plasma, and take the ensemble average, we have

$$\langle \mathbf{E}(\mathbf{x},t)\rangle = \int d\mathbf{v}_0 \int d\mathbf{x}_0\, \mathbf{E}(\mathbf{x},t)\, f_0(\mathbf{x}_0,\mathbf{v}_0) \tag{9.32}$$

where f_0 is the zero order distribution function. The function f_0 is the probability density for plasma particles to have velocity $\mathbf{v}_0$ and position $\mathbf{x}_0$; thus, the ensemble average of any plasma property that is caused by the discrete plasma particles is given by an equation like (9.32).

In a uniform isotropic plasma, there is no preferred direction; therefore we expect (9.32) to vanish. This indeed happens.

EXERCISE By performing the $\mathbf{x}_0$ integration, convince yourself that (9.32) vanishes.

Next, consider the ensemble average electric field energy density in the plasma,

$$W \equiv \frac{1}{8\pi}\langle \mathbf{E}(\mathbf{x})\cdot\mathbf{E}(\mathbf{x})\rangle \tag{9.33}$$

Since this is a positive definite quantity, we expect a nonzero result. Taking the ensemble average in the same way as in (9.32), we have

$$W = \frac{1}{8\pi}\int d\mathbf{v}_0 \int d\mathbf{x}_0 \left(\frac{q_T}{2\pi^2}\right)^2 f_0(\mathbf{v}_0)$$

$$\times \left[\int d\mathbf{k}\, \frac{\mathbf{k}\, e^{i\mathbf{k}\cdot\mathbf{x} - i\mathbf{k}\cdot\mathbf{x}_0(t)}}{k^2\epsilon(\mathbf{k},\omega = \mathbf{k}\cdot\mathbf{v}_0)}\right]\cdot\left[\int d\mathbf{k}'\, \frac{\mathbf{k}'\, e^{-i\mathbf{k}'\cdot\mathbf{x} + i\mathbf{k}'\cdot\mathbf{x}_0(t)}}{(k')^2\,\epsilon^*(\mathbf{k}',\omega = \mathbf{k}'\cdot\mathbf{v}_0)}\right] \tag{9.34}$$

where we have used $\mathbf{E}^*(\mathbf{x}) = \mathbf{E}(\mathbf{x})$ for the real electric field in the second set of square brackets. The $\mathbf{x}_0$ integration yields $(2\pi)^3\delta(\mathbf{k} - \mathbf{k}')$ which facilitates the $\mathbf{k}'$ integration; we find

$$W = \frac{q_T^2}{4\pi^2} \int d\mathbf{v}_0 f_0(\mathbf{v}_0) \int d\mathbf{k} \; \frac{1}{k^2 |\epsilon(\mathbf{k}, \omega = \mathbf{k} \cdot \mathbf{v}_0)|^2} \tag{9.35}$$

We can perform the two velocity integrations in the directions perpendicular to **k**, and extract a factor of n_0, to obtain in the usual fashion

$$W = \frac{e^2}{4\pi^2} n_0 \int du \, g(u) \int d\mathbf{k} \; \frac{1}{k^2 |\epsilon(\mathbf{k}, \omega = ku)|^2} \tag{9.36}$$

Defining $\omega \equiv ku$, we find

$$W = \frac{n_0 e^2}{2\pi} \int \frac{d\omega}{2\pi} \int d\mathbf{k} \; \frac{g(\omega/k)}{k^3 |\epsilon(\mathbf{k},\omega)|^2} \tag{9.37}$$

Thus, we can define an energy density $W(\mathbf{k},\omega)$ such that

$$W \equiv \frac{\langle E^2 \rangle}{8\pi} = \int \frac{d\omega}{2\pi} \int d\mathbf{k} \; W(\mathbf{k},\omega) \tag{9.38}$$

with

$$W(\mathbf{k},\omega) = \frac{n_0 e^2 \, g(\omega/k)}{2\pi \, k^3 |\epsilon(\mathbf{k},\omega)|^2} \tag{9.39}$$

Since ω is purely real, we have for the Vlasov–Poisson system the exact expression (9.12). Thus,

$$|\epsilon(\mathbf{k},\omega)|^2 = \left[1 - \frac{\omega_e^2}{k^2} P \int du \; \frac{d_u \, g(u)}{u - (\omega/k)}\right]^2 + \left[\frac{\omega_e^2}{k^2} \, \pi \, d_u g(u)|_{u=\omega/k}\right]^2 \tag{9.40}$$

We can easily perform the frequency integration in Eq. (9.38) in two simple limiting cases.

CASE A: $k\lambda_e << 1$

This criterion is exactly the one for the existence of Langmuir waves. In this wave number regime we expect all of the energy in fluctuations to be concentrated in Langmuir waves with frequencies $\omega \approx \pm \omega_e$. Then in the denominator of the real part of ϵ, we have $\omega/k \approx \omega_e/k$ but $k << \lambda_e^{-1}$; therefore $\omega/k \approx \omega_e/k >> \omega_e \lambda_e = v_e$, which means that $\omega/k >> u$ for almost all the range of u integration. Thus, we integrate by parts to obtain

$$P \int du \; \frac{d_u g(u)}{u - (\omega/k)} = P \int du \; \frac{g(u)}{[u - (\omega/k)]^2} \approx \frac{k^2}{\omega^2} \int du \, g(u) = \frac{k^2}{\omega^2} \tag{9.41}$$

Thus

$$|\epsilon(\mathbf{k},\omega)|^2 = \left(1 - \frac{\omega_e^2}{\omega^2}\right)^2 + \left[\frac{\omega_e^2}{k^2} \, \pi \, d_u g|_{\omega/k}\right]^2 \tag{9.42}$$

Since this expression goes into the denominator of (9.38), we see that there will be a large contribution when $\omega \approx \pm \omega_e$. Thus, we evaluate the numerator ~ $g(\omega/k) \sim g(\omega_e/k)$ and the term $d_u g|_{\omega/k} \approx d_u g|_{\omega_e/k}$ in the denominator at $\omega \approx \omega_e$. We are left with the integration in the form

$$I \equiv \int_{\omega_e-\Delta}^{\omega_e+\Delta} \frac{d\omega}{2\pi} \frac{1}{[1-(\omega_e^2/\omega^2)]^2 + c^2} \approx \omega_e^4 \int_{\omega_e-\Delta}^{\omega_e+\Delta} \frac{d\omega}{2\pi} \frac{1}{(\omega^2-\omega_e^2)^2 + c^2\omega_e^4} \tag{9.43}$$

where c^2 is small; since the main part of the integral comes from $\omega \approx \omega_e$, we can let the limits of integration $\to \pm \infty$. With $y = \omega^2$, $dy = 2\omega d\omega$ we have

$$\begin{aligned} I &= \frac{1}{4\pi}\int_{-\infty}^{\infty} dy \frac{\omega_e^3}{(y-\omega_e^2)^2 + c^2\omega_e^4} \\ &= \frac{\omega_e^3}{4\pi}\int_{-\infty}^{\infty} dy \frac{1}{[(y-\omega_e^2)+ic\omega_e^2][(y-\omega_e^2)-ic\omega_e^2]} \end{aligned} \tag{9.44}$$

which can be done by contour integration; closing either up or down we find (multiplying by a factor of 2 to take into account the frequency regime near $\omega \approx -\omega_e$)

$$2I = \frac{2(2\pi i)\omega_e^3}{4\pi} \frac{1}{2ic\omega_e^2} = \frac{\omega_e}{2c} \tag{9.45}$$

so that

$$W = \int d\mathbf{k} \frac{n_0 e^2}{2\pi k^3} g(\omega_e/k) \frac{\omega_e}{2\left[\frac{\omega_e^2}{k^2} \pi d_u g|_{\omega_e/k}\right]} \tag{9.46}$$

For a Maxwellian, $g/d_u g|_{\omega_e/k} = v_e^2 k/\omega_e$; therefore we obtain

$$\boxed{W = \int_{k<<k_e} \frac{d\mathbf{k}}{(2\pi)^3} \frac{mv_e^2}{2} = \int_{k<<k_e} \frac{d\mathbf{k}}{(2\pi)^3} T_e/2} \tag{9.47}$$

so that in the regime $k\lambda_e << 1$, we find $T_e/2$ energy per unit k-space per unit real space.

Let us crudely evaluate the total amount of energy, per unit real space, in long wavelength ($k\lambda_e << 1$) fluctuations. To do this, we perform the integration in (9.47) over a spherical volume from $k = 0$ to $k = k_e \equiv \lambda_e^{-1}$. We obtain

$$W = \left(\frac{4}{3}\pi k_e^3\right) \frac{1}{(2\pi)^3} \frac{T_e}{2} \tag{9.48}$$

Multiplying numerator and denominator by n_0, and dropping all numerical factors, we crudely obtain

$$W \approx \frac{n_0 T_e}{n_0 \lambda_e^3} \tag{9.49}$$

or

$$\boxed{W \approx \frac{n_0 T_e}{\Lambda}} \tag{9.50}$$

Thus, the average long wavelength fluctuation energy density is very small; it is the average electron kinetic energy density divided by the number of particles in a Debye cube.

CASE B: $k\lambda_e >> 1$

When k is very large, we have from (9.12) that

$$\begin{aligned} \epsilon(\mathbf{k},\omega) &= 1 - \frac{\omega_e^2}{k^2}\int_L du\, \frac{d_u g(u)}{u - (\omega/k)} \\ &\approx 1 - \frac{\omega_e^2}{k^2}\int_L du\, \frac{d_u g(u)}{u} \\ &\approx 1 \end{aligned} \tag{9.51}$$

Then from (9.38),

$$\begin{aligned} W &= \int d\mathbf{k}\, \frac{d\omega}{2\pi}\, \frac{n_0 e^2}{2\pi k^3}\, g(\omega/k) \\ &= \pi(4\pi n_0 e^2)\int \frac{d\mathbf{k}}{(2\pi)^3}\, \frac{1}{2\pi k^2}\, \underbrace{\int d\left(\frac{\omega}{k}\right) g\left(\frac{\omega}{k}\right)}_{1} \\ &= \frac{m_e\omega_e^2}{2}\int \frac{d\mathbf{k}}{(2\pi)^3}\, \frac{1}{k^2} \\ &= \frac{T_e}{2}\int \frac{d\mathbf{k}}{(2\pi)^3}\, \frac{1}{k^2\lambda_e^2} \end{aligned} \tag{9.52}$$

Writing this in the form

$$W = \int \frac{d\mathbf{k}}{(2\pi)^3}\, W(\mathbf{k}) \tag{9.53}$$

we have

$$\boxed{W(\mathbf{k}) = \frac{T_e}{2}\, \frac{1}{k^2\lambda_e^2}}, \qquad k\lambda_e >> 1 \tag{9.54}$$

which can be compared with (9.47) where

$$\boxed{W(\mathbf{k}) = \frac{T_e}{2}}, \qquad k\lambda_e << 1 \tag{9.55}$$

Thus, the fluctuation level is much smaller in the short wavelength region $k\lambda_e >> 1$ than in the long wavelength region $k\lambda_e << 1$. This agrees with our intuition, which would predict a high fluctuation level for the weakly damped long wavelength normal modes (Langmuir waves).

It has been shown in an elegant calculation by Rostoker [10] that for a Maxwellian $g(u)$, the exact expression for all wave numbers is

$$W(\mathbf{k}) = \frac{T_e}{2} \frac{1}{1 + k^2\lambda_e^2} \tag{9.56}$$

so that

$$\boxed{W = \frac{T_e}{2} \int \frac{d\mathbf{k}}{(2\pi)^3} \frac{1}{1 + k^2\lambda_e^2}} \tag{9.57}$$

EXERCISE Does (9.56) give the correct limits (9.54) and (9.55)? Is the integral in (9.57) convergent or divergent? What is your physical interpretation of this?

This brings us to the end of this brief chapter on fluctuations and shielding. The test particle technique used here can be used to study the Cerenkov emission of electrostatic waves and their absorption via Landau damping. This is an illustration of the principle of detailed balance. The net result is the steady-state level of electric field fluctuations.

Another application of the principle of detailed balance, in the presence of a background magnetic field, involves *synchrotron emission* and *cyclotron damping*. The principle of detailed balance also applies to the emission of electromagnetic radiation via *bremsstrahlung*, and its absorption via *collisional damping*. Since these processes involve the collision of *two* charged particles, it is not sufficient to use the simple test charge theory that we used for electrostatic fluctuations. Another way to say this is that Langmuir waves are emitted via Cerenkov emission, with $\omega = \mathbf{k} \cdot \mathbf{v}_0$. Electromagnetic waves cannot be emitted in this way in a plasma because their phase speed in a plasma is always greater than the speed of light. For further discussion of these topics, see Refs. [11]–[17].

REFERENCES

[1] P. Chenevier, J. M. Dolique, and H. Perès, *J. Plasma Phys.*, *10*, 185 (1973).

[2] G. Cooper, *Phys. Fluids*, *12*, 2707 (1969).

[3] G. Joyce and D. Montgomery, *Phys. Fluids*, *10*, 2017 (1967).

[4] E. W. Laing, A. Lamont, and P. J. Fielding, *J. Plasma Phys.*, *5*, 441 (1971).

[5] D. Montgomery, G. Joyce, and R. Sugihara, *Plasma Phys.*, *10*, 681 (1968).

[6] J. Neufeld and R. H. Ritchie, *Phys. Rev.*, *98*, 1632 (1955).

[7] I. Oppenheim and N. G. van Kampen, *Phys. Fluids*, *7*, 813 (1964).

[8] L. Stenflo, M. Y. Yu, and P. K. Shukla, *Phys. Fluids*, *16*, 450 (1973).

[9] C.-L. Wang, G. Joyce, and D. R. Nicholson, *J. Plasma Phys.*, *25*, 225 (1981).

[10] N. Rostoker, *Nucl. Fusion*, *1*, 101 (1961).

[11] G. Bekefi, *Radiation Processes in Plasmas*, Wiley, New York, 1966.

[12] M. Mitchner, ed., *Radiation and Waves in Plasmas*, Stanford University Press, Stanford, Calif., 1961.

[13] Yu. L. Klimontovich, *The Statistical Theory of Non-Equilibrium Processes in a Plasma*, The M.I.T. Press, Cambridge, Mass., 1967.

[14] J. D. Jackson, *Classical Electrodynamics*, 2nd ed., Wiley, New York, 1975.

[15] P. C. Clemmow and J. P. Dougherty, *Electrodynamics of Particles and Plasmas*, Addison-Wesley, Reading, Mass., 1969.

[16] N. A. Krall and A. W. Trivelpiece, *Principles of Plasma Physics*, McGraw-Hill, New York, 1973.

[17] D. C. Montgomery and D. A. Tidman, *Plasma Kinetic Theory*, McGraw-Hill, New York, 1964.

CHAPTER 10

Weak Turbulence Theory

10.1 INTRODUCTION

Most of the wave phenomena treated in earlier chapters apply to linear waves so small in amplitude that terms which are quadratic in the wave quantities can be discarded. When the wave amplitudes are large enough, the nonlinear terms cannot be discarded. Examples of these nonlinear effects include the particle trapping discussed in Section 6.8, the nonlinear wave equations with soliton solutions introduced in Sections 7.15 and 7.16, and the parametric instabilities treated in Section 7.17. All of these nonlinear effects can be called *coherent*, since each solution applies to one realization of a plasma and each solution has a unique spatial dependence.

In this chapter, we discuss the solution of nonlinear equations under conditions that can be called *turbulent*. By this we mean that we seek the time evolution of certain spatially averaged or ensemble-averaged quantities. The averaging process means that we lose information on the detailed spatial variations in each realization of the plasma.

We treat three topics that commonly come under the heading of weak turbulence theory (Refs. [1] to [5]): quasilinear theory, induced scattering, and wave-wave interactions. Other topics such as resonance broadening and strong turbulence theories like the direct interaction approximation are topics of vigorous current research [6], and are beyond the scope of this book.

10.2 QUASILINEAR THEORY

Although it is possible to develop a general framework that includes all aspects of weak turbulence theory, we will take the alternative approach of considering each kind of interaction separately, in order to understand the physics involved. Crudely, the concept of weak turbulence means that the nonlinearities are small and yield small corrections to linear behavior, whereas strong turbulence means that the nonlinearities are as strong as the linear terms. In *quasilinear theory* [7, 8], the wave amplitudes are considered to be so small that the wave propagation can be treated by the linear theory of Chapter 7. The nonlinear part of quasilinear theory concerns the long-term effect of many waves on the background distribution function.

Consider the bump-on-tail situation. We have seen in Section 6.5 that Langmuir waves are unstable when their phase speeds correspond to regions of the one-dimensional distribution function $g(u)$ with a positive slope. The growth rate as given by (6.52) is proportional to the slope of the distribution function $d_u g(u)|_{\omega_r/k}$. Thus, at early times when linear theory is valid we have the growth situation displayed in Fig. 10.1. Here we have only indicated regions of positive imaginary frequency $\omega_i(k)$, graphed as a function of the phase speed of a Langmuir wave $v_\varphi \equiv \omega_r/k \approx \omega_e/k$. We know that linear instability cannot go on forever. For one thing, there is only so much energy in the bump, and the waves certainly cannot grow to levels such that the wave energy is larger than the initial particle energy.

As the waves grow, the particles find themselves in a turbulent situation. Thus, a typical particle's orbit, in both real space and velocity space, will be affected by the turbulent electric fields. In particular, particles will diffuse in both real space and velocity space. Consider all of the particles at position x_0 with speed u_0 at the initial time $t = 0$. In the absence of any electric fields, these particles would have orbits

$$x(t) = x_0 + u_0 t \tag{10.1}$$

and

$$u(t) = u_0 \tag{10.2}$$

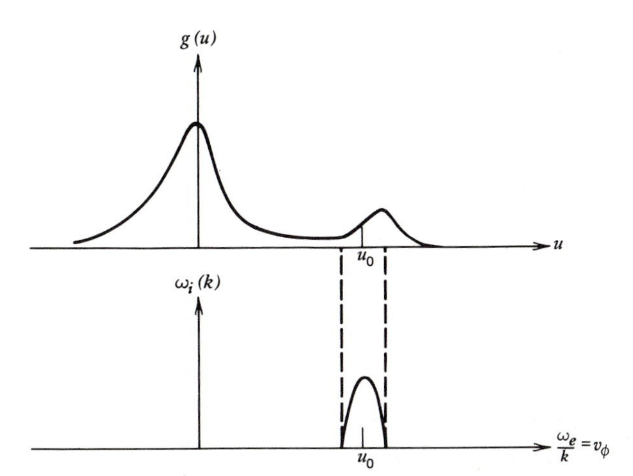

Fig. 10.1 Initial bump-on-tail distribution function, and corresponding region of positive growth rate $\omega_i(k)$.

for all times according to Vlasov theory. In the presence of the turbulent electric fields, the particles are accelerated. Because the fields are turbulent, the acceleration is not constant but is random, being alternately positive and negative. Thus, a typical particle will perform a random walk in velocity space. As with all random walk processes, this implies a diffusion of particles in velocity space (see Appendix B). While all the particles described by (10.1) and (10.2) will experience the same fields and thus will have the same speed $u(t)$ at time t, particles starting at a neighboring point $x_0 + \Delta$ with speed u_0 will have a different time history and thus a different speed at time t.

Diffusion tends to spread out the particles in velocity space. Thus, after some time the waves have grown and the particle distribution $g(u)$ and wave intensity distribution $I(v_\varphi)$ might look as shown in Fig. 10.2. The slope on the distribution function has changed and, thus, the linear growth rate of each wave has changed; since the maximum positive slope is smaller at time t than at $t = 0$, the maximum growth rate is smaller.

Eventually, the particles diffuse so much that the hole between the bump and the background has filled in, and there is no longer a region of positive slope. Then the linear growth rate of the waves is zero, and we have the steady-state situation shown in Fig. 10.3. Of course, this situation is only a steady state in the context of the Vlasov equation. In practice, the waves will eventually decay away due to collisions, and the distribution will eventually become Maxwellian due to collisions.

Let us develop a mathematical framework for these ideas. To do this, we must be a little more clever than if we were simply doing linear theory. This is because we wish to follow changes in the background distribution function on a very long time scale. It would not do to write $f = f_0 + f_1$ where f_0 is the initial spatially averaged or ensemble averaged background distribution. Then we would have to

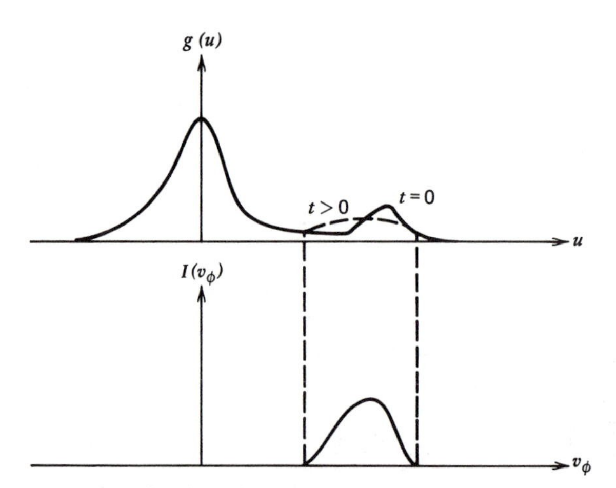

Fig. 10.2 Particle distribution and wave intensity distribution after the waves have grown for some time.

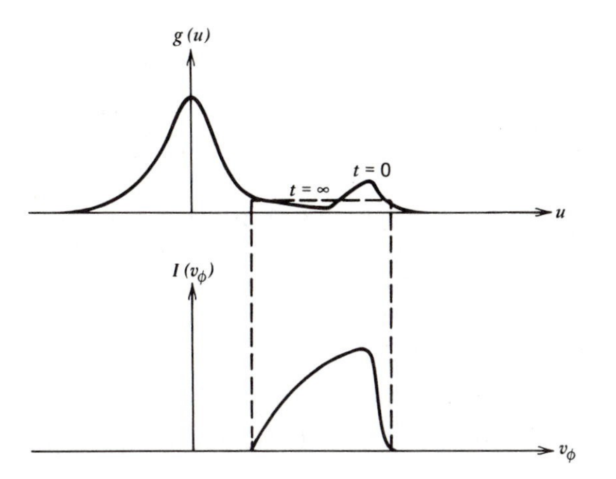

Fig. 10.3 Particle and wave distributions at a late time.

include in f_1 the difference between the final distribution and the initial distribution. But since such differences are as large as or larger than f_0 in the vicinity of the bump, f_1 would be as large or larger than f_0, and the expansion would break down.

A more successful approach is to separate

$$f(x,v,t) = f_0(v,t) + f_1(x,v,t) \tag{10.3}$$

where

$$f_0(v,t) = \langle f(x,v,t)\rangle_x \tag{10.4}$$

and the y and z dependencies of f have been integrated out. Thus, f_0 is the spatially averaged distribution function, and it changes on the slow time scale as the bump diffuses away. The waves are represented by $f_1(x,v,t)$ and, thus, $\langle f_1\rangle_x = 0$ as given by (10.3) and (10.4). With this separation, we take $f_1 << f_0$ for all (x,v,t).

Let us now write the Vlasov equation for electrons, assuming the ions are fixed. We have, in one dimension,

$$\partial_t f + v\,\partial_x f - \frac{e}{m} E\,\partial_v f = 0 \tag{10.5}$$

and

$$\partial_x E = -\,4\pi e \int_{-\infty}^{\infty} dv\, f_1(x,v,t) \tag{10.6}$$

where the spatially averaged electron distribution function f_0 cancels the ion contribution to Poisson's equation (10.6). Next, average (10.5) over space, to obtain

$$\partial_t\langle f\rangle_x + v\left\langle\frac{\partial f}{\partial x}\right\rangle_x - \frac{e}{m}\left\langle E\,\frac{\partial f}{\partial v}\right\rangle_x = 0 \tag{10.7}$$

Our spatial averaging procedure is to integrate over a finite length $-L/2 \le x \le L/2$, divide by L, and take the limit as $L \to \infty$. The first term in (10.7) is clearly

$\partial_t f_0$. The second term in (10.7) is

$$\left\langle \frac{\partial f}{\partial x} \right\rangle_x = \frac{1}{L}\int_{-L/2}^{L/2} dx\, \frac{\partial f}{\partial x} = \frac{1}{L}\left[f\left(x = \frac{L}{2}\right) - f\left(x = -\frac{L}{2}\right)\right] \xrightarrow[L\to\infty]{} 0 \tag{10.8}$$

where $\lim_{L\to\infty}$ is always implied when we write $1/L$, and where we have assumed that $f(x,v,t)$ is a smooth, well-behaved function. The third term in (10.7) is simplified by using the assumption that $\langle E\rangle_x = 0$; this is because we take the electric field to be produced by f_1 inside the plasma volume, with no component of E being produced by capacitor plates at $x \to \pm\infty$. Then

$$\begin{aligned} \left\langle E\,\frac{\partial f}{\partial v} \right\rangle_x &= \left\langle E\,\frac{\partial f_0}{\partial v} \right\rangle_x + \left\langle E\,\frac{\partial f_1}{\partial v} \right\rangle_x \\ &= \langle E\rangle_x \frac{\partial f_0}{\partial v} + \left\langle E\,\frac{\partial f_1}{\partial v} \right\rangle_x \\ &= \left\langle E\,\frac{\partial f_1}{\partial v} \right\rangle_x \end{aligned} \tag{10.9}$$

where $f_0(v,t)$ can be removed from the averaging bracket because it is not a function of x. Equation (10.7) then reads

$$\boxed{\partial_t f_0(v,t) = \frac{e}{m}\left\langle E\,\frac{\partial f_1}{\partial v} \right\rangle_x} \tag{10.10}$$

Equation (10.10) is the nonlinear part of quasilinear theory; f_0 is changing because of the product of E and f_1, which is a second order quantity.

The remainder of quasilinear theory is completely linear. The Vlasov equation (10.5) is linearized using $f_1 << f_0$, and with the help of Poisson's equation (10.6) a dispersion relation is obtained. This development is precisely as in Section 6.4, and we obtain the normal mode Langmuir wave frequencies for $k > 0$, adapted from (6.52),

$$\boxed{\omega(k,t) = \omega_e\left(1 + \frac{3}{2}\, k^2\lambda_e^{\,2}\right) + i\,\frac{\pi}{2}\,\frac{\omega_e^{\,3}}{k^2}\, d_u g(u,t)\big|_{\omega_r/k}} \tag{10.11}$$

where $g(u,t) = (1/n_0) f_0(u,t)$. Thus, the linear Langmuir waves evolve with complex frequency as given by (10.11), while the background distribution evolves according to the nonlinear equation (10.10).

Consistent with this ordering, we wish to evaluate the nonlinear term on the right of (10.10) by inserting the linear form of f_1. We consider the linear waves to consist of a spectrum of right-going waves with different wave numbers; each of these waves has its normal mode frequency $\omega(k)$ given by (10.11). Thus, the real electric field is

$$E(x,t) = \int_{-\infty}^{\infty} dk\, E(k,t)e^{ikx} = \int_{-\infty}^{\infty} dk\, \tilde{E}(k,t)e^{-i\omega_r(k)t + ikx} \tag{10.12}$$

where the Fourier transform conventions are given by (5.11) and (5.12). Now $E(x,t)$ must be real, and in order for it to be real, the component at each wave number (the part at $k = |k|$ plus the part at $k = -|k|$) must be real. Thus, the elementary properties of Fourier transforms imply

$$\omega_r(-k) = -\omega_r(k) \tag{10.13}$$

and

$$\tilde{E}(k,t) = \tilde{E}^*(-k,t) \tag{10.14}$$

The latter implies that $E(k,t) = E^*(-k,t)$ and $\omega_i(-k,t) = \omega_i(k,t)$.

EXERCISE Demonstrate these results.

The perturbed distribution function $f_1(x,v,t)$ can also be written in the form (10.12), as

$$f_1(x,v,t) = \int_{-\infty}^{\infty} dk\, f_1(k,v,t)e^{ikx}$$

$$= \int_{-\infty}^{\infty} dk\, \tilde{f}_1(k,v,t)e^{-i\omega_r(k)t+ikx} \tag{10.15}$$

The relation between $f_1(k,v,t)$ and $E(k,t)$ is as usual obtained by linearizing the Vlasov equation (10.5), with the assumed dependence $\exp[-i\omega(k)t + ikx]$, to obtain

$$f_1(k,v,t) = \frac{-e/m}{i[\omega(k,t) - kv]}\, \partial_v f_0(v,t)E(k,t) \tag{10.16}$$

In the right side of (10.10), we can move the velocity derivative outside the brackets since $E(x,t)$ does not depend on velocity. We are thus interested in the quantity

$$\langle E f_1\rangle_x = \frac{1}{L}\int_{-L/2}^{L/2} dx\, E(x,t)\, f_1(x,v,t)$$

$$= \frac{1}{L}\int_{-L/2}^{L/2} dx \int dk\, f_1(k)e^{ikx}\int dk'\, E(k')e^{ik'x}$$

$$= \frac{1}{L}\int dk\, f_1(k)\int dk'\, E(k')\int_{-L/2}^{L/2} dx\, e^{i(k+k')x} \tag{10.17}$$

Equation (10.17) can be simplified by using the standard formula

$$\lim_{L\to\infty}\int_{-L/2}^{L/2} dx\, e^{iax} = 2\pi\delta(a) \tag{10.18}$$

The right-most integral in (10.17) thus becomes $2\pi\delta(k + k')$, upon which the k' integration is trivial, and we obtain

$$\langle E f_1\rangle_x = \frac{2\pi}{L}\int dk\, E(-k)\, f_1(k)$$

$$= \frac{2\pi}{L}\int dk\, E(-k)\left[\frac{ie}{m}\,\frac{1}{\omega - kv}\,\partial_v f_0\right]E(k) \tag{10.19}$$

where (10.16) has been used. Inserting (10.19) in (10.10) we have

$$\partial_t f_0(v,t) = -\frac{e^2}{m^2}\frac{2\pi}{L}\partial_v \int_{-\infty}^{\infty} dk\, E(-k)E(k)\,\frac{\partial_v f_0(v,t)}{i(\omega - kv)} \qquad (10.20)$$

With the result below (10.14) this becomes

$$\partial_t f_0(v,t) = -\frac{e^2}{m^2}\frac{2\pi}{L}\partial_v \left\{\left[\int_{-\infty}^{\infty} dk\, \frac{|E(k)|^2}{i(\omega - kv)}\right]\partial_v f_0(v,t)\right\} \qquad (10.21)$$

Equation (10.21) can be simplified by defining the so-called *spectral density* $\epsilon(k)$ of the electric field. The natural definition of this quantity is

$$\epsilon(k) \equiv \frac{1}{4L}\,|E(k)|^2 \qquad (10.22)$$

Then the average electric field energy density is

$$\begin{aligned}
\left\langle \frac{E^2}{8\pi}\right\rangle_x &= \frac{1}{8\pi L}\int_{-L/2}^{L/2} dx\, E^2(x)\\
&= \frac{1}{8\pi L}\int_{-L/2}^{L/2} dx \int dk\, E(k)e^{ikx}\int dk'\, E(k')e^{ik'x}\\
&= \frac{1}{8\pi L}\int dk\, E(k)\int dk'\, E(k')\int_{-L/2}^{L/2} dx\, e^{i(k+k')x}\\
&= \frac{1}{4L}\int dk\, E(k)E(-k)\\
&= \frac{1}{4L}\int dk\, |E(k)|^2\\
&= \int_{-\infty}^{\infty} dk\, \epsilon(k)
\end{aligned} \qquad (10.23)$$

which shows that $\epsilon(k)$ is the wave energy density, per unit interval of wave number space.

With the definition (10.22), Eq. (10.21) reads

$$\partial_t f_0(v,t) = -\frac{8\pi e^2}{m^2}\partial_v \left\{\left[\int_{-\infty}^{\infty} dk\, \frac{\epsilon(k)}{i(\omega - kv)}\right]\partial_v f_0(v,t)\right\} \qquad (10.24)$$

This equation is in the form of a diffusion equation,

$$\boxed{\partial_t f_0(v,t) = \partial_v[D(v,t)\,\partial_v f_0(v,t)]} \qquad (10.25)$$

with

$$\boxed{D(v,t) \equiv -\frac{8\pi e^2}{m^2}\int_{-\infty}^{\infty} dk\, \frac{\epsilon(k,t)}{i[\omega(k,t) - kv]}} \qquad (10.26)$$

Since $\partial_t \tilde{E}(k) = \omega_i \tilde{E}(k)$, the spectral density (10.22) must satisfy

$$\partial_t \epsilon(k,t) = 2\omega_i(k,t)\epsilon(k,t) \qquad (10.27)$$

or

$$\epsilon(k,t) = \epsilon(k,t=0) \exp\left[2\int_0^t \omega_i(k,t')\,dt'\right] \tag{10.28}$$

where $\omega_i(k,t)$ is determined from the normal mode frequency (10.11). Given initial conditions $f_0(v,t = 0)$ and $\epsilon(k,t = 0)$, Eqs. (10.25), (10.26) and (10.28) provide a complete description of the time evolution of the system. For the bump-on-tail problem, the evolution is as described in the beginning of this section, with diffusion eventually resulting in a flat distribution function $f_0(v,t \to \infty)$.

There are several useful forms of the diffusion coefficient $D(v,t)$ in (10.26). Since f_0 in (10.25) is real, it must be true that $D(v,t)$ is real. The integrand of (10.26) is of the form

$$\frac{-i\epsilon(k)}{(\omega - kv)} = \frac{-i\epsilon(k)}{\omega_r + i\omega_i - kv} = \frac{-i\epsilon(k)[\omega_r - kv - i\omega_i]}{(\omega_r - kv)^2 + \omega_i^2} \tag{10.29}$$

Using the symmetries (10.13), and the fact that $\epsilon(-k) = \epsilon(k)$, we see that the imaginary part of (10.29) is odd in k and, thus, vanishes upon integration in (10.26).

EXERCISE Verify the last statement.

We have left

$$D(v,t) = \frac{8\pi e^2}{m^2}\int_{-\infty}^{\infty} dk\, \frac{\epsilon(k,t)\omega_i(k,t)}{[\omega_r(k,t) - kv]^2 + [\omega_i(k,t)]^2} \tag{10.30}$$

In the limit of very tiny ω_i, the integrand in (10.30) takes a form $\sim \delta(\omega_r - kv)$, and we find

$$D(v,t) = \frac{16\pi^2 e^2}{m^2}\,\frac{1}{v}\,\epsilon(k = \omega_r/v,t) \tag{10.31}$$

EXERCISE Verify (10.31) by going back to (10.26) and using the formula

$$\lim_{\epsilon\to 0^+} \frac{1}{x - i\epsilon - a} = P\left(\frac{1}{x - a}\right) + \pi i\delta(x - a) \tag{10.32}$$

To show that the $P(\quad)$ part vanishes, and the constant in (10.31) comes out right, be careful to count all of the places where $x = a$. The plus sign is chosen because we use the equivalent of a Landau contour in (10.26), as demanded by a proper treatment of the initial value problem.

Thus, (10.31) shows that the diffusion of particles with speed v is caused by waves with phase speeds $\omega_r/k = v$. This is *resonant* behavior, where particles interact strongly with those waves with which they are resonant, with $\omega_r = kv$. In the linear theory, it is the resonant particles that cause linear Landau growth or damping, as shown in (10.11). In the quasilinear theory, it is the resonant particles that are diffused because of the wave fields.

Quasilinear theory, with its simplicity and straightforward application to magnetized plasmas, has seen and continues to see a great deal of use in both fusion and astrophysical applications. In the next section, we shall proceed to consider another aspect of weak turbulence theory.

10.3 INDUCED SCATTERING

In the previous section, we considered one aspect of weak turbulence theory, the nonlinear diffusion of particles due to the presence of many waves. We now wish to consider another aspect of weak turbulence theory, the nonlinear coupling of one wave to another wave through the background particles. This is called *induced scattering* [9], induced scattering off ions, induced scattering off the polarization clouds of ions, and nonlinear Landau damping. All of these terms refer to the same process; the last expression is somewhat unfortunate, as this process has nothing to do with the nonlinear stage of linear Landau damping, as discussed in Section 6.8.

In linear Landau damping or growth, the important concept is a resonance between one wave and one particle, satisfying the linear resonance condition

$$\omega = kv \tag{10.33}$$

such that the particle velocity v equals the wave phase velocity ω/k. In induced scattering, the important concept is a resonance between two waves and one particle, satisfying the resonance condition

$$\omega_1 - \omega_2 = (k_1 - k_2)v \tag{10.34}$$

such that

$$v = \frac{\omega_1 - \omega_2}{k_1 - k_2} \tag{10.35}$$

that is, the particle velocity equals the velocity of the *beat* of two waves.

Consider a particle in the presence of two real waves. Then Newton's force law says

$$m\ddot{x} = qE_1 \exp(-i\omega_1 t + ik_1 x) + qE_2 \exp(-i\omega_2 t + ik_2 x) + \text{c.c.} \tag{10.36}$$

where all fields are real so that the complex conjugate must be added to the right side of (10.36). Equation (10.36) is a very nonlinear equation for the particle orbit $x(t)$. If the fields E_1, E_2 are not too large, we can solve (10.36) perturbatively. We have

$$x = x_0 + x_1 \tag{10.37}$$

where

$$x_0 = v_0 t \tag{10.38}$$

is the orbit of the particle in the absence of the fields. Inserting (10.37) into (10.36) and expanding the exponents, we find

$$\begin{aligned} m\ddot{x}_1 &= qE_1 \exp(-i\omega_1 t + ik_1 v_0 t + ik_1 x_1) \\ &+ qE_2 \exp(-i\omega_2 t + ik_2 v_0 t + ik_2 x_1) + \text{c.c.} \\ &= qE_1 \exp(-i\omega_1 t + ik_1 v_0 t)(1 + ik_1 x_1) \\ &+ qE_2 \exp(-i\omega_2 t + ik_2 v_0 t)(1 + ik_2 x_1) + \text{c.c.} \end{aligned} \tag{10.39}$$

First, ignore x_1 on the right of (10.39). Then the lowest order solution to x_1 comes from integrating (10.39) twice with respect to time. We find

$$x_1(t) \approx -\frac{q}{m}\frac{E_1 \exp(-i\omega_1 t + ik_1 v_0 t)}{(\omega_1 - k_1 v_0)^2} - \frac{q}{m}\frac{E_2 \exp(-i\omega_2 t + ik_2 v_0 t)}{(\omega_2 - k_2 v_0)^2} + \text{c.c.} \tag{10.40}$$

Inserting this lowest order solution to x_1 in the right side of (10.39), we obtain 20 terms, or

$$\ddot{x}_1 = -\frac{(q^2/m^2)ik_1E_1E_2^*}{(\omega_2 - k_2v_0)^2}\exp[-i(\omega_1 - \omega_2)t + i(k_1 - k_2)v_0t] + (19\ \text{terms}) \tag{10.41}$$

If the resonance condition (10.34),

$$(\omega_1 - \omega_2) - (k_1 - k_2)v_0 = 0 \tag{10.42}$$

is satisfied, we see that (10.41) will have at least one force term that is constant in time. Thus, a particle with initial velocity v_0 as given by (10.35) can interact very strongly with two waves [3].

A plasma contains not one but many particles. As a specific problem, we consider the interaction of two Langmuir waves E_1 and E_2, with a low frequency disturbance. We assume that the low frequency disturbance is dominated by details of the ion distribution, with the electrons simply supplying the charge to almost neutralize the low frequency disturbance; that is, we invoke quasineutrality of the low frequency disturbance. Then we can use fluid theory to describe the electrons; from Chapter 7 we have

$$m_en_e\,\partial_tV_e + m_en_eV_e\,\partial_xV_e = -\gamma_eT_e\,\partial_xn_e - en_eE \tag{10.43}$$

$$\partial_tn_e + \partial_x(n_eV_e) = 0 \tag{10.44}$$

and

$$\partial_xE = -4\pi en_e \tag{10.45}$$

where we shall only use Poisson's equation to describe high frequency electron oscillations. The ions are described by the Vlasov equation with $f_i(v) = f_0(v) + f_1(v)$,

$$\partial_tf_1 + v\,\partial_xf_1 + \frac{e}{m_i}E\,\partial_vf_0 = 0 \tag{10.46}$$

where $v \equiv v_x$ and f_1 has already been integrated over v_y and v_z. Although the theory we are performing is not a linear one, we have ignored the term involving ∂_vf_1 in (10.46). Thus, the theory is limited to those regimes where $\partial_vf_1 << \partial_vf_0$ for all ion velocities of interest.

The high frequency fields are given by $E_1\exp(-i\omega_1t + ik_1x) + E_2\exp(-i\omega_2t + ik_2x) + \text{c.c.}$ We consider E_1 to be a relatively large wave with a real frequency ω_1. The field E_2 is considered to be a tiny wave, which can use the ions to drain energy from the large field E_1; hence the term *induced scattering*. The frequency ω_2 is therefore complex, with a positive imaginary part describing its growth. The low frequency response is described by a density perturbation $n_3 \exp(-i\omega_3 t + ik_3 x)$ + c.c. We assume that the frequency matching condition

$$\omega_1 = \omega_2 + \omega_3^* \tag{10.47}$$

and the wave number matching condition

$$k_1 = k_2 + k_3 \tag{10.48}$$

are both satisfied. Here, ω_1, k_1, k_2, and k_3 are real while ω_2 and ω_3 are complex with the same imaginary part. We note that

$$\omega_1 \approx \omega_{2r} >> \omega_{3r} \tag{10.49}$$

For the high frequency modes, we have in mind waves that would be normal modes (Langmuir waves) of the plasma in the absence of the coupling. The low frequency mode, however, is a disturbance that would be strongly damped in the absence of the coupling. For example, it could be an ion-acoustic mode in an equal temperature plasma, which is strongly Landau damped by the ions. By contrast, the parametric instability theory of Section 7.17 is valid when $T_e >> T_i$ so that ion-acoustic waves are not strongly damped.

We first treat the behavior at frequency ω_2. From Eq. (10.43), we write all terms that vary as $\exp(-i\omega_2 t + ik_2 x)$, given the matching conditions (10.47) and (10.48). We keep linear terms and second order terms, but we discard terms that are third order in small quantities. Dividing (10.43) by $m_e n_e$, we have for high frequency motions

$$\partial_t V_e + V_e \partial_x V_e = -\gamma_e \frac{T_e}{m_e n_0} \partial_x n_e - \frac{e}{m_e} E \tag{10.50}$$

where nonlinearities in the pressure term (which would multiply the small $k^2\lambda_e^2$ correction to the Langmuir dispersion relation) have been ignored. The appropriate terms are

$$-i\omega_2 v_2 - ik_3 v_3^* v_1 + ik_1 v_1 v_3^* = -3ik_2 \frac{T_e}{m_e n_0} n_2 - \frac{e}{m_e} E_2 \tag{10.51}$$

where v_2 is the component of V_e with time dependence $\sim \exp(-i\omega_2 t)$, etc. Similarly, the terms in the continuity equation (10.44) that vary as $\exp(-i\omega_2 t + ik_2 x)$ are

$$-i\omega_2 n_2 + ik_2(n_0 v_2 + n_1 v_3^* + n_3^* v_1) = 0 \tag{10.52}$$

EXERCISE Verify that all the appropriate terms appear in (10.51) and (10.52).

Equation (10.51) can be solved for v_2, and the result inserted in (10.52). We find

$$-i\omega_2 n_2 = -ik_2 \Big[\underset{①}{n_1 v_3^*} + \underset{④}{n_3^* v_1} - \underset{②}{n_0 \frac{k_3}{\omega_2} v_3^* v_1} + \underset{③}{\frac{n_0 k_1}{\omega_2} v_1 v_3^*} + \frac{3k_2(T_e/m_e)n_2}{\omega_2} + \frac{n_0 e}{i\omega_2 m_e} E_2 \Big] \tag{10.53}$$

We next wish to discard the sum of terms ①, ②, and ③. This sum is of the form

$$①+②+③ = v_3^*\left(n_1 - n_0 \frac{k_3}{\omega_2} v_1 + n_0 \frac{k_1}{\omega_2} v_1\right) \tag{10.54}$$

To lowest order, the continuity equation (10.44) gives

$$n_1 = n_0 \frac{k_1}{\omega_1} v_1 \approx n_0 \frac{k_1}{\omega_2} v_1 \tag{10.55}$$

where the last form is valid since $\omega_1 \approx \omega_2$. Equation (10.54) becomes

$$①+②+③ = \frac{v_3^* v_1 n_0}{\omega_2}(k_1 - k_3 + k_1)$$

$$= \frac{v_3^* v_1 n_0}{\omega_2}(k_1 + k_2) \tag{10.56}$$

However, this is much smaller than term ④,

$$④ = n_3^* v_1 = n_0 v_3^* v_1 \frac{k_3}{\omega_3^*} \tag{10.57}$$

where the last form follows from the lowest order continuity equation for n_3^*. Equation (10.57) is much larger than Eq. (10.56) because $|\omega_3^*| << |\omega_2|$, provided the wave numbers are of the same order. Thus, in (10.53) we neglect terms ①, ②, and ③ compared to term ④ to obtain

$$\left[-i\omega_2 + \frac{3ik_2^2(T_e/m_e)}{\omega_2} - \frac{\omega_e^2}{i\omega_2}\right] n_2 = -ik_2 n_3^* v_1 \tag{10.58}$$

where E_2 in (10.53) is eliminated using the component of Poisson's equation varying at the frequency ω_2. Multiplying by $i\omega_2$ one finds

$$(\omega_2^2 - \omega_e^2 - 3k_2^2 v_e^2) n_2 = \omega_2 k_2 n_3^* v_1 \tag{10.59}$$

If (10.59) were linearized by neglecting the right side, the remainder would yield the familiar Langmuir wave dispersion relation. Separating the factor $\omega_2^2 - 3k_2^2 v_e^2$ on the left, we obtain

$$\boxed{(\omega_2^2 - 3k_2^2 v_e^2)\epsilon(\omega_2, k_2) n_2 = \omega_2 k_2 n_3^* v_1} \tag{10.60}$$

where $\epsilon(\omega_2, k_2)$ is the high frequency dielectric function

$$\epsilon(\omega, k) = 1 - \frac{\omega_e^2}{\omega^2 - 3k^2 v_e^2} \tag{10.61}$$

Having obtained the high frequency equation (10.60), we are one-half done with our derivation. We must now obtain a low frequency equation for n_3^*. Recall that to lowest order, v_1 is obtained from the force equation

$$m_e \partial_t v_1 = -eE_1 \tag{10.62}$$

or

$$v_1 = \frac{eE_1}{i\omega_1 m_e} \tag{10.63}$$

so that v_1 in (10.60) is determined by (10.63), with ω_1 real as discussed above (10.49).

Let us derive the low frequency part. We intend to enforce quasineutrality, $n_3 \equiv n_{e3} \approx n_{i3}$. For the electrons, the components of the force equation (10.43) varying as $\exp(-i\omega_3 t + ik_3 x)$ are, dividing first by $m_e n_e$ and ignoring nonlinearities in the pressure term,

$$-i\omega_3 v_3 - v_1 ik_2 v_2^* + ik_1 v_1 v_2^* = -i\frac{T_e}{n_0 m_e} k_3 n_3 - \frac{e}{m_e} E_3 \tag{10.64}$$

where we have chosen $\gamma_e = 1$ for low frequency motions. We neglect $-i\omega_3 v_3$ because ω_3 is small, and use $k_1 - k_2 = k_3$ to obtain

$$E_3 = -\frac{T_e}{en_0} ik_3 n_3 - \frac{im_e k_3}{e} v_1 v_2^* \tag{10.65}$$

From the lowest order continuity equation, $v_2^* = (\omega_2^*/k_2)(n_2^*/n_0)$, therefore,

$$E_3 = -\frac{T_e}{en_0} ik_3 n_3 - \frac{im_e}{n_0 e}\frac{k_3}{k_2} v_1 \omega_2^* n_2^* \tag{10.66}$$

The low frequency ion response is given by the Vlasov equation (10.46), which yields

$$f_{13} = \frac{-(e/m_i)}{-i(\omega_3 - k_3 v)} \partial_v f_0 E_3 \tag{10.67}$$

where f_{13} is the perturbed ion distribution function at frequency ω_3 and f_0 is the zero order ion distribution function. The low frequency density disturbance is

$$\begin{aligned} n_3 &= \int_{-\infty}^{\infty} dv f_{13} = E_3 \int_{-\infty}^{\infty} dv \frac{(e/m_i)}{i(\omega_3 - k_3 v)} \partial_v f_0 \\ &= \left(\frac{-T_e ik_3 n_3}{en_0} - \frac{im_e k_3 v_1 \omega_2^* n_2^*}{n_0 e k_2}\right) \int_{-\infty}^{\infty} dv \frac{(e/m_i)}{i(\omega_3 - k_3 v)} \partial_v f_0 \end{aligned} \tag{10.68}$$

Defining

$$W \equiv \frac{k_3 T_e}{m_i n_0} \int_{-\infty}^{\infty} dv \frac{\partial_v f_0}{\omega_3 - k_3 v} \tag{10.69}$$

we solve (10.68) for n_3 to obtain

$$n_3 = -\frac{m_e v_1 \omega_2^* n_2^*}{k_2 T_e} \frac{W}{1 + W} \tag{10.70}$$

for the low frequency density perturbation n_3.

We finally insert the complex conjugate of (10.70) into the high frequency equation (10.60) to obtain an equation with n_2 on both sides. Cancelling n_2, we have

$$(\omega_2^2 - 3k_2^2 v_e^2)\epsilon(\omega_2, k_2) = -\frac{m_e}{T_e} |v_1|^2 \omega_2^2 \frac{W^*}{1 + W^*} \tag{10.71}$$

which is a nonlinear dispersion relation for the complex frequency ω_2. Ignoring the small thermal correction $3k_2^2 v_e^2$ on the left, we can divide out ω_2^2 on both sides.

Then defining the thermal speed $v_e^2 \equiv T_e/m_e$ we have

$$\boxed{\epsilon_{NL} \equiv \epsilon(\omega_2,k_2) + \frac{|v_1|^2}{v_e^2}\,\frac{W^*}{1 + W^*} = 0} \tag{10.72}$$

where we have introduced the nonlinear dielectric function ϵ_{NL}. If the fixed electric field $E_1 \rightarrow 0$, then $v_1 = eE_1/im_e\omega_1 \rightarrow 0$, and we regain the linear Langmuir dispersion relation $\omega_2(k_2)$.

With v_1 finite, we have the possibility of instability. We call ϵ_{NL} a nonlinear dispersion relation because it contains the square of the field $E_1 \sim v_1$. However, with v_1 considered a constant, we can treat ϵ_{NL} in the same fashion as we did linear dielectric functions in Section 6.5. Recall that for small instability $|\omega_i| << |\omega_r|$, we have from (6.42) and (6.43),

$$\epsilon_r(\omega_{2r}) = 0 \tag{10.73}$$

and

$$\omega_{2i} = \frac{-\epsilon_i(\omega_{2r})}{\partial\epsilon_r/\partial\omega|_{\omega_{2r}}} \tag{10.74}$$

The imaginary part of the dielectric function (10.72) comes from W, given in (10.69). Evaluating W using the Landau contour for $|\omega_{3i}| << |\omega_{3r}|$, we have

$$W = W_r - \frac{T_e}{m_i n_0}\,\pi i\,\partial_v f_0|_{\omega_{3r}/k_3} \tag{10.75}$$

where the real part of W, W_r, involves a principal value integral that we shall not bother to evaluate.

We require the imaginary part of ϵ_{NL} in (10.72), which is proportional to

$$\mathrm{Im}[W^*/(1 + W^*)] = -\,W_i/(1 + W_r)^2 \tag{10.76}$$

where W_i has been treated as a small quantity and terms quadratic in W_i have been ignored.

EXERCISE Verify (10.76).

Then

$$\mathrm{Im}(\epsilon_{NL}) = -\,\frac{|v_1|^2}{v_e^2}\,\frac{W_i}{(1 + W_r)^2}$$

$$= \frac{1}{(1 + W_r)^2}\,\frac{|v_1|^2}{v_e^2}\,\frac{T_e\,\pi}{m_i n_0}\,\partial_v f_0|_{\omega_{3r}/k_3} \tag{10.77}$$

For $\partial\epsilon_r/\partial\omega|_{\omega_{2r}}$ in (10.74), we can use the dielectric function $\epsilon = 1 - \omega_e^2/\omega^2$ to obtain

$$\left.\frac{\partial\epsilon_r}{\partial\omega}\right|_{\omega_{2r}} \approx \frac{2}{\omega_e} \tag{10.78}$$

Equation (10.78) ignores the thermal correction and the *nonlinear frequency shift* that would be given by (10.72). Then the growth rate (10.74) is

$$\boxed{\omega_{3i} = \omega_{2i} = -\frac{1}{(1 + W_r)^2}\frac{\omega_e}{2}\frac{|v_1|^2}{v_e^2}\frac{T_e\,\pi}{m_i n_0}\,\partial_v f_0|_{\omega_{3r}/k_3}} \tag{10.79}$$

Notice that the derivative $\partial_v f_0$ is evaluated at

$$v = \frac{\omega_{3r}}{k_3} = \frac{\omega_1 - \omega_{2r}}{k_1 - k_2} \tag{10.80}$$

which is the phase speed of the *beat* between ω_1 and ω_{2r}. This reinforces our notion of a nonlinear resonance between two waves and one particle.

The growth rate (10.79) is positive when the slope $\partial_v f_0$ is negative. Thus, the waves E_2 that grow fastest are the ones whose beat with E_1 falls near the ion thermal speed for Maxwellian ions (Fig. 10.4).

There is a very close relation between the *induced scattering* considered here, involving two high frequency waves and the ions, and the *parametric decay instability* discussed in Section7.17. The former is more appropriate when $T_e \approx T_i$ and the beat phase velocity is likely to fall in the body of the ion distribution. The latter is inappropriate when $T_e \approx T_i$ because the low frequency wave equation is an undamped ion-acoustic wave in the absence of nonlinear coupling; when $T_e \approx T_i$ ion-acoustic waves are strongly Landau damped by ions. The latter is more appropriate when $T_e >> T_i$ and ion-acoustic waves are undamped. In the next section, we introduce a statistical approach for the case $T_e >> T_i$.

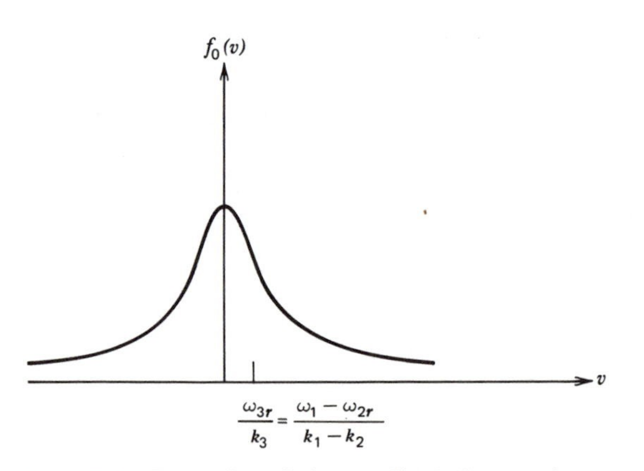

Fig. 10.4 For a given finite amplitude Langmuir wave with frequency ω_1 and wave number k_1, the fastest growing Langmuir waves due to induced scattering are those with frequency ω_{2r}, wave number k_2, such that $(\omega_{3r}/k_3) = (\omega_1 - \omega_{2r})/(k_1 - k_2) \approx v_i$.

10.4 WAVE-WAVE INTERACTIONS

In the preceding two sections, we have discussed two important aspects of weak turbulence theory. The first was *quasilinear* theory, which involves the *linear wave-particle interaction* characterized by the expression

$$\omega = \mathbf{k} \cdot \mathbf{v} \tag{10.81}$$

The second was *induced scattering*, which involves the *nonlinear wave-particle interaction* characterized by the expression

$$\omega_1 - \omega_{2r} = (\mathbf{k}_1 - \mathbf{k}_2) \cdot \mathbf{v} \tag{10.82}$$

In the first case, a particle is resonant with one wave; its speed is equal to the phase speed of the wave in the direction of the particle's velocity,

$$|\mathbf{v}| = \omega/k_{\|} \tag{10.83}$$

where $k_{\|} = \mathbf{k} \cdot \mathbf{v}/v$. In the second case, a particle is resonant with the *beat* between two waves; its speed is equal to the phase speed of the *beat* between the two waves in the direction of the particle's velocity,

$$|\mathbf{v}| = \frac{\omega_1 - \omega_{2r}}{(\mathbf{k}_1 - \mathbf{k}_2)_{\|}} \tag{10.84}$$

where

$$(\mathbf{k}_1 - \mathbf{k}_2)_{\|} \equiv [(\mathbf{k}_1 - \mathbf{k}_2) \cdot \mathbf{v}]/v \tag{10.85}$$

In this section, we wish to consider a third important aspect of weak turbulence theory, that of nonlinear *wave-wave interaction*. We shall find that this interaction is characterized by a *frequency matching* condition

$$\omega_1 = \omega_2 + \omega_3 \tag{10.86}$$

and by a *wave number matching* condition

$$\mathbf{k}_1 = \mathbf{k}_2 + \mathbf{k}_3 \tag{10.87}$$

For wave-wave interactions, we assume that resonant particle interactions (both linear and nonlinear) are not important, so that the conditions (10.86) and (10.87) do not involve a particle velocity.

We shall illustrate the ideas of wave-wave interaction by using the Zakharov equations discussed in Section 7.16. Recall that this equation describes the nonlinear coupling between linear Langmuir waves and linear ion-acoustic waves. Thus, this approach is valid for a plasma with $T_e >> T_i$, when we know that ion-acoustic waves exist. For the case $T_e \approx T_i$, there are no undamped ion-acoustic modes, and we cannot use the wave-wave interaction ideas to be developed here. Rather, when $T_e \approx T_i$ we would expect the nonlinear wave-particle interactions (induced scattering) of the previous section to be the dominant nonlinear interaction between high frequency electron waves and low frequency ion fluctuations.

The physical interaction discussed here is the same as the physical interaction that yielded the parametric decay instability of Section 7.17. However, in that case we considered a single finite-amplitude Langmuir wave that decays into one other Langmuir wave and one ion-acoustic wave. By contrast, here we consider a broad spectrum of Langmuir waves that evolves according to the nonlinear physics contained in the Zakharov equations. Our approach is a statistical one, so that we predict the evolution of the waves in an ensemble of systems rather than the evolution of the waves in a single system.

The Zakharov equations (7.329) and (7.330) are

$$i\,\partial_t E(x,t) + \partial_x^2 E = nE \tag{10.88}$$

and

$$\partial_t^2 n(x,t) - \partial_x^2 n = \partial_x^2 |E|^2 \tag{10.89}$$

The spatial Fourier transform of (10.88) is

$$\begin{aligned} i\,\partial_t E(k,t) - k^2 E &= \int_{-\infty}^{\infty} \frac{dx}{2\pi}\, e^{-ikx}\, n(x,t)E(x,t) \\ &= \int_{-\infty}^{\infty} \frac{dx}{2\pi}\, e^{-ikx} \left[\int_{-\infty}^{\infty} dk'\, n(k',t) e^{ik'x} \right] \\ &\quad \times \left[\int_{-\infty}^{\infty} dk''\, E(k'',t) e^{ik''x} \right] \\ &= \int_{-\infty}^{\infty} dk'\, n(k',t) \int_{-\infty}^{\infty} dk''\, E(k'',t) \underbrace{\int_{-\infty}^{\infty} \frac{dx}{2\pi}\, e^{i(-k+k'+k'')x}}_{\delta(k - k' - k'')} \end{aligned} \tag{10.90}$$

so

$$\boxed{i\,\partial_t E(k,t) - k^2 E(k,t) = \int_{-\infty}^{\infty} dk'\, n(k',t) E(k - k',t)} \tag{10.91}$$

Note that the wave number matching condition $k = k' + k''$ has already appeared in the argument of the δ-function in (10.90).

Suppose we look at the linear limit of (10.91). We ignore the nonlinear right side, and find

$$i\,\partial_t E(k,t) = k^2 E(k,t) \tag{10.92}$$

so

$$E(k,t) = E(k,t = 0) \exp(-ik^2 t) \tag{10.93}$$

Defining

$$\omega_l(k) = k^2 \tag{10.94}$$

we have

$$E(k,t) = E(k,t = 0) \exp[-i\omega_l(k)t] \tag{10.95}$$

where the subscript l in (10.94) means "Langmuir." Recall that the Zakharov equations are obtained by factoring out the high frequency time dependence exp

$(-i\omega_e t)$; if we put this time dependence back in and change back to dimensional variables, we would find that (10.94) becomes

$$\tilde{\omega}_l(\tilde{k}) = \omega_e \left(1 + \frac{3}{2}\tilde{k}^2\lambda_e^2\right) \tag{10.96}$$

where $\tilde{k}$ is the dimensional wave number and $\tilde{\omega}_l$ is the dimensional frequency; this is just our old friend the Langmuir wave dispersion relation.

EXERCISE Why is $\omega^2 = \omega_e^2 + 3k^2v_e^2$ the same as $\omega = \omega_e\left(1 + \frac{3}{2}k^2\lambda_e^2\right)$ for Langmuir waves?

The idea of weak turbulence theory is to assume that each Langmuir wave approximately obeys its linear solution (10.95). However, the amplitude is allowed to have a slow time variation because of the nonlinear term on the right of (10.91), rather than being an exact constant as in the linear solution (10.95).

Consider next the spatial Fourier transform of (10.89), which is

$$\begin{aligned}\partial_t^2 n(k,t) + k^2 n(k,t) &= -k^2 \int_{-\infty}^{\infty} \frac{dx}{2\pi} e^{-ikx} E(x,t)E^*(x,t) \\ &= -k^2 \int_{-\infty}^{\infty} \frac{dx}{2\pi} e^{-ikx}\left[\int_{-\infty}^{\infty} dk'\, e^{ik'x} E(k',t)\right] \\ &\quad \times \left[\int_{-\infty}^{\infty} dk''\, e^{-ik''x} E^*(k'',t)\right] \\ &= -k^2 \int_{-\infty}^{\infty} dk'\, E(k',t) \int_{-\infty}^{\infty} dk''\, E^*(k'',t) \underbrace{\int_{-\infty}^{\infty} \frac{dx}{2\pi} e^{i(-k+k'-k'')x}}_{\delta(k - k' + k'')}\end{aligned} \tag{10.97}$$

so

$$\boxed{\partial_t^2 n(k,t) + k^2 n(k,t) = -k^2 \int_{-\infty}^{\infty} dk'\, E(k',t)E^*(k'-k,t)} \tag{10.98}$$

The linear limit of (10.98) is

$$\partial_t^2 n(k,t) + k^2 n(k,t) = 0 \tag{10.99}$$

with solution

$$n(k,t) = A(k)\exp(-ikt) + B(k)\exp(ikt) \tag{10.100}$$

Since $n(x,t)$ is real, it must be true that

$$n(k,t) = n^*(-k,t) \tag{10.101}$$

EXERCISE Prove (10.101).

Then

$$A(k) = A^*(-k) \tag{10.102}$$

and

$$B(k) = B^*(-k) \tag{10.103}$$

Linear theory tells us that $A(k)$ and $B(k)$ are constants in time. Defining

$$\omega_s^+(k) \equiv k \tag{10.104}$$

and

$$\omega_s^-(k) \equiv -k \tag{10.105}$$

where s means "sound" (acoustic), we can write (10.100) as

$$n(k,t) = A(k) \exp[-i\omega_s^+(k)t] + B(k) \exp[-i\omega_s^-(k)t] \tag{10.106}$$

Note that $A(k)$ is the amplitude of the right-going ion-acoustic waves $[\omega_s^+(k)/k = 1 > 0]$ while $B(k)$ is the amplitude of the left-going ion-acoustic waves $[\omega_s^-(k)/k = -1 < 0]$. The idea of weak turbulence theory is to use the form (10.106) in (10.98), but to allow the coefficients $A(k,t)$ and $B(k,t)$ to be slowly varying functions of time. The word "slowly" in this context means slow compared to the terms $\exp[-i\omega_s^{\pm}(k)t]$. (Note that in physical units, the frequencies $\omega_s^{\pm}(k)$ are just the frequencies of our old friends the ion-acoustic waves, $\tilde{\omega}_s^{\pm}(k) = \pm \tilde{k}c_s$.)

Thus, we want to solve equations (10.91) and (10.98) with solutions of the form

$$E(k,t) = \tilde{E}(k,t) \exp[-i\omega_l(k)t] \tag{10.107}$$

and

$$n(k,t) = A(k,t) \exp[-i\omega_s^+(k)t] + B(k,t) \exp[-i\omega_s^-(k)t] \tag{10.108}$$

where $\tilde{E}(k,t)$, $A(k,t)$, and $B(k,t)$ are slowly varying. After we insert these forms into (10.91), the left side becomes

$$\begin{aligned} i\,\partial_t E(k,t) - k^2 E(k,t) &= i\,\partial_t\,\tilde{E}(k,t) \exp[-i\omega_l(k)t] \\ &+ \omega_l(k)\tilde{E}(k,t) \exp[-i\omega_l(k)t] - k^2\,\tilde{E}(k,t) \exp[-i\omega_l(k)t] \\ &= i\,\partial_t \tilde{E}(k,t) \exp[-i\omega_l(k)t] \end{aligned} \tag{10.109}$$

since by (10.94) the last two terms cancel. The entire equation (10.91) becomes

$$\begin{aligned} i\,\partial_t \tilde{E}(k,t) \exp[-i\omega_l(k)t] &= \int_{-\infty}^{\infty} dk' \,\{A(k',t) \exp[-i\omega_s^+(k')t] \\ &+ B(k',t) \exp[-i\omega_s^-(k')t]\} \cdot \tilde{E}(k-k',t) \exp[-i\omega_l(k-k')t] \end{aligned} \tag{10.110}$$

or

$$\boxed{\begin{aligned} \partial_t\,\tilde{E}(k,t) = -i \int_{-\infty}^{\infty} dk' \Big(& A(k',t)\,\tilde{E}(k-k',t) \\ & \times \exp\{i[\omega_l(k) - \omega_s^+(k') - \omega_l(k-k')]t\} \\ & + B(k',t)\tilde{E}(k-k',t) \exp\{i[\omega_l(k) - \omega_s^-(k') - \omega_l(k-k')]t\}\Big) \end{aligned}} \tag{10.111}$$

In the exponents on the right side of (10.111) we can already see the terms that will lead to the three-wave frequency matching conditions.

In order to put the "ion" equation (10.98) in the same form as the "electron" equation (10.111), it is useful to assume at this point that all of the terms with frequency $\exp[-i\omega_s^+(k)t]$ will behave independently of all of the terms with frequency $\exp[-i\omega_s^-(k)t]$. Then looking only for the terms on the left side of (10.98) with frequency $\exp[-i\omega_s^+(k)t]$, we find

$$(\partial_t^2 + k^2)A(k,t)\exp[-i\omega_s^+(k)t] = \{\partial_t^2 A(k,t) - 2i\omega_s^+(k)\partial_t A(k,t) - [\omega_s^+(k)]^2 A(k,t) + k^2 A(k,t)\}\exp[-i\omega_s^+(k)t] \tag{10.112}$$

The last two terms cancel by the definition (10.104) of $\omega_s^+(k)$, and we ignore $\partial_t^2 A$ compared to $-2i\omega_s^+(k)\,\partial_t A$ just as in the derivation of the "electron" equation (10.88) in Section 7.16. We obtain for the entire equation (10.98)

$$-2i\omega_s^+(k)\,\partial_t A(k,t)\exp[-i\omega_s^+(k)t] = -k^2\int_{-\infty}^{\infty} dk'\,\tilde{E}(k',t)\exp[-i\omega_l(k')t] \times \tilde{E}^*(k'-k,t)\exp[i\omega_l(k'-k)t] \tag{10.113}$$

or

$$\partial_t A(k,t) = \frac{-ik^2}{2\omega_s^+(k)}\int_{-\infty}^{\infty} dk'\,\tilde{E}(k',t)\tilde{E}^*(k'-k,t) \times \exp\{i[\omega_s^+(k) - \omega_l(k') + \omega_l(k'-k)]t\} \tag{10.114}$$

where the three-wave frequency matching conditions can again be seen popping up on the right side.

The equation for $B(k,t)$ is obtained in the same manner, leading to the same equation as (10.114) with A replaced by B and ω_s^+ replaced by ω_s^-. This is

$$\partial_t B(k,t) = \frac{-ik^2}{2\omega_s^-(k)}\int_{-\infty}^{\infty} dk'\,\tilde{E}(k',t)\tilde{E}^*(k'-k,t) \times \exp\{i[\omega_s^-(k) - \omega_l(k') + \omega_l(k'-k)]t\} \tag{10.115}$$

Equations (10.111), (10.114), and (10.115) are now a complete set of equations for the slowly varying amplitudes $\tilde{E}$, A, and B.

In order to see clearly the method we are about to develop, let us consider the model equation

$$\partial_t C(k,t) = \int_{-\infty}^{\infty} dk'\,V(k,k',k-k')C(k',t)C(k-k',t) \times \exp\{i[\omega(k) - \omega(k') - \omega(k-k')]t\} \tag{10.116}$$

This model equation is easily seen to have the same basic structure of our three equations (10.111), (10.114), and (10.115).

The derivation proceeds formally with an expansion of the amplitude $C(k,t)$. To clearly distinguish the different terms in the expansion, we treat the "vertex" V as the expansion parameter, even though it is really C itself that is small in some sense.

Thus, we expand

$$C(k,t) = C^{(0)}(k,t) + C^{(1)}(k,t) + C^{(2)}(k,t) + \cdots \tag{10.117}$$

Substituting this in the basic dynamical equation (10.116) we obtain, to zeroth order in V,

$$\partial_t C^{(0)}(k,t) = 0 \tag{10.118}$$

with solution

$$C^{(0)}(k,t) = C^{(0)}(k,t = 0) \tag{10.119}$$

which is just what linear physics would tell us; we choose this value to be $C(k,t = 0)$.

Next, the zeroth order solution (10.119) is substituted into the basic dynamical equation (10.116) to obtain

$$\partial_t C^{(1)}(k,t) = \int_{-\infty}^{\infty} dk'\, V(k,k',k - k')C^{(0)}(k')C^{(0)}(k - k') \times \exp\{i[\omega(k) - \omega(k') - \omega(k - k')]t\} \tag{10.120}$$

where we have dropped the time index on $C^{(0)}$ since it is a constant. The solution of (10.120) is

$$C^{(1)}(k,t) = \int_0^t dt' \int_{-\infty}^{\infty} dk' \exp\{i[\omega(k) - \omega(k') - \omega(k - k')]t\} \times V(k,k',k - k')C^{(0)}(k')C^{(0)}(k - k') \tag{10.121}$$

We can write this in a more symmetric form if we introduce

$$k'' \equiv k - k' \tag{10.122}$$

Then defining

$$F(k,k',k'',t) \equiv V(k,k',k'') \exp\{i[\omega(k) - \omega(k') - \omega(k'')]t\} \times \delta(k - k' - k'') \tag{10.123}$$

we can write (10.121) in the form

$$C^{(1)}(k,t) = \int_{-\infty}^{\infty} dk' \int_{-\infty}^{\infty} dk''\, C^{(0)}(k')C^{(0)}(k'') \int_0^t dt'\, F(k,k',k'',t') \tag{10.124}$$

In this form, we should now consider F as the expansion parameter.

The equation for $C^{(2)}$ is obtained by substituting $C = C^{(0)} + C^{(1)}$ on the right side of the basic dynamical equation (10.116) and picking out only those terms that are second order in F. We find

$$\partial_t C^{(2)}(k,t) = \int_{-\infty}^{\infty} dk' \int_{-\infty}^{\infty} dk'' \, F(k,k',k'',t)$$

$$\times [C^{(0)}(k')C^{(1)}(k'',t) + C^{(1)}(k',t)C^{(0)}(k'')]$$

$$= \int_{-\infty}^{\infty} dk' \int_{-\infty}^{\infty} dk'' \, F(k,k',k'',t)\{C^{(0)}(k') \int_{-\infty}^{\infty} dk''' \int_{-\infty}^{\infty} dk''''$$

$$\times \, C^{(0)}(k''')C^{(0)}(k'''') \int_0^t dt' \, F(k'',k''',k'''',t')$$

$$+ \, C^{(0)}(k'') \int_{-\infty}^{\infty} dk''' \int_{-\infty}^{\infty} dk'''' \, C^{(0)}(k''')C^{(0)}(k'''')$$

$$\times \int_0^t dt' \, F(k',k''',k'''',t')\} \tag{10.125}$$

which is integrated in time to yield

$$C^{(2)}(k,t) = \int dk' \, dk'' \, dk''' \, dk'''' \, C^{(0)}(k')C^{(0)}(k''')C^{(0)}(k'''')$$

$$\times \int_0^t dt' \int_0^{t'} dt'' \, F(k,k',k'',t')F(k'',k''',k'''',t'')$$

$$+ \int dk' \, dk'' \, dk''' \, dk'''' \, C^{(0)}(k'')C^{(0)}(k''')C^{(0)}(k'''')$$

$$\times \int_0^t dt' \int_0^{t'} dt'' \, F(k,k',k'',t') \, F(k',k''',k'''',t'') \tag{10.126}$$

This is the highest order term that we shall need for our theory.

At this point, we wish to introduce the idea of *random phases*. We want to develop a *statistical theory* of weak turbulence. One way to do this is to consider an ensemble of realizations, in each of which the absolute value of the amplitude $C^{(0)}$, at a given wave number k, is the same. However, the complex quantity

$$C^{(0)}(k) \equiv |C^{(0)}(k)|e^{i\theta(k)} \tag{10.127}$$

has a phase $\theta(k)$, which is a random number $0 \leq \theta(k) < 2\pi$. Thus, $|C^{(0)}(k)|$ is the same in each realization, but $\theta(k)$ varies randomly from realization to realization.

Consider the *two-point correlation function*

$$\langle C^{(0)}(k)C^{(0)}(k')\rangle = |C^{(0)}(k)||C^{(0)}(k')|\langle e^{i\theta(k)+i\theta(k')}\rangle$$

$$= |C^{(0)}(k)||C^{(0)}(k')|\langle \cos[\theta(k) + \theta(k')]$$

$$+ \, i \sin[\theta(k) + \theta(k')]\rangle \tag{10.128}$$

where $\langle \ \rangle$ indicates ensemble average. If $k \neq k'$, then $\theta(k)$ and $\theta(k')$ are statistically independent, which means (since any θ between 0 and 2π is equally likely)

$$\langle \cos [\theta(k) + \theta(k')] \rangle = \frac{\int_0^{2\pi} d\theta\,(k) \int_0^{2\pi} d\theta\,(k') \cos [\theta(k) + \theta(k')]}{\int_0^{2\pi} d\theta\,(k) \int_0^{2\pi} d\theta\,(k')} = 0 \quad (10.129)$$

The same thing happens for the $\sin [\theta(k) + \theta(k')]$ term. If $k = k'$, then

$$\begin{aligned} \langle e^{i\theta(k) + i\theta(k)} \rangle &= \langle e^{2i\theta(k)} \rangle \\ &= \langle \cos [2\theta(k)] \rangle + i \langle \sin [2\theta(k)] \rangle \\ &= 0 \end{aligned} \quad (10.130)$$

Thus, we have for all k and k' that

$$\langle C^{(0)}(k) C^{(0)}(k') \rangle = 0 \quad (10.131)$$

However, consider

$$\langle C^{(0)}(k) C^{(0)*}(k') \rangle = |C^{(0)}(k)| |C^{(0)}(k')| \langle e^{i\theta(k) - i\theta(k')} \rangle \quad (10.132)$$

When $k = k'$ we have

$$\langle e^{i\theta(k) - i\theta(k')} \rangle = \langle e^0 \rangle = \langle 1 \rangle = 1 \quad (10.133)$$

whereas when $k \neq k'$, Eq. (10.132) vanishes as before. Thus, the quantity in (10.132) is zero when $k \neq k'$ and is nonzero when $k = k'$; it must be a Dirac delta function. Thus, we write (10.132) as

$$\langle C^{(0)}(k) C^{(0)*}(k') \rangle = n^{(0)}(k) \delta(k - k') \quad (10.134)$$

where the quantity $n(k)$ is sometimes called the *mode occupation number* by analogy to quantum theories of atomic transitions. In this section, we shall call $n^{(0)}(k)$ the zeroth order *intensity*, since it is proportional to the square of the *amplitude* $C^{(0)}$.

[*Note:* In this model problem, we have taken the phase $\theta(k)$ to be statistically independent of the phase $\theta(-k)$. In some cases, this assumption may need to be modified. For example, in the physics contained in the nonlinear Schrodinger equation, this assumption is quite appropriate for the amplitude $E(k,t)$, but it is wrong for $A(k)$ and $B(k)$ as can be seen by (10.102) and (10.103). The result of treating these cases properly is to add more terms of the same form as we shall find for our model problem.]

The total intensity $n(k,t)$ is related to the total amplitude $C(k,t) = C^{(0)}(k,t) + C^{(1)}(k,t) + C^{(2)}(k,t)$ in the same way [Eq. (10.134)] that the zeroth order intensity $n^{(0)}$ is related to the zeroth order amplitude $C^{(0)}$:

$$\langle C(k,t) C^*(k',t) \rangle = n(k,t) \delta(k - k') \quad (10.135)$$

EXERCISE Show that the form (10.135) is a rigorous consequence of the assumption of statistical spatial homogeneity; that is, ensemble averages can only depend on spatial differences $x - x'$, not on absolute spatial location, x. See any introductory book on turbulence theory, for example, Leslie [10].

We look for an equation that describes the time evolution of $n(k,t)$. With the expansion (10.117) of the amplitude C, keeping only terms up to second order in F, we have (setting $k = k'$),

$$\begin{aligned}\langle|C(k,t)|^2\rangle &= \langle(C^{(0)} + C^{(1)} + C^{(2)})(C^{(0)*} + C^{(1)*} + C^{(2)*})\rangle \\ &= |C^{(0)}|^2 + \langle C^{(0)}C^{(1)*}\rangle + \langle C^{(1)}C^{(0)*}\rangle + \langle C^{(0)}C^{(2)*}\rangle \\ &+ \langle C^{(2)}C^{(0)*}\rangle + \langle C^{(1)}C^{(1)*}\rangle + \{\text{higher order terms}\} \end{aligned} \tag{10.136}$$

Many of the terms on the right of (10.136) vanish. For example, from (10.124) we see that all terms of the form $C^{(0)}C^{(1)*}$ look like

$$\begin{aligned}\langle C^{(0)}C^{(1)*}\rangle &\sim \langle C^{(0)}(k)C^{(0)*}(k')C^{(0)*}(k'')\rangle \\ &\sim \langle e^{i\theta(k)-i\theta(k')-i\theta(k'')}\rangle \\ &\sim 0\end{aligned} \tag{10.137}$$

EXERCISE Convince yourself of the last step.

Likewise, all terms of the form $C^{(0)}\ C^{(2)*}$ are of the form [see Eq. (10.126)]

$$\begin{aligned}\langle C^{(0)}C^{(2)*}\rangle &\sim \langle C^{(0)}(k)C^{(0)*}(k')C^{(0)*}(k''')C^{(0)*}(k'''')\rangle \\ &\sim \langle e^{i\theta(k)-i\theta(k')-i\theta(k''')-i\theta(k'''')}\rangle \\ &\sim 0\end{aligned} \tag{10.138}$$

EXERCISE Convince yourself of the last step.

[*Note:* When the conditions discussed in the previous note prevail, that is, $C(k)$ is correlated to $C(-k)$, then (10.137) still vanishes but (10.138) no longer vanishes in general.]

Thus, the only terms contributing in (10.136) are

$$\langle|C(k,t)|^2\rangle - |C^{(0)}(k)|^2 = \langle C^{(1)}(k,t)C^{(1)*}(k,t)\rangle \tag{10.139}$$

From (10.124), (10.134), and (10.135) we find

$$\begin{aligned}&[n(k,t) - n^{(0)}(k)]\delta(0) \\ &= \int dk'\, dk''\, dk'''\, dk'''' \int_0^t dt'\, F(k,k',k'',t') \\ &\cdot \int_0^t dt''\, F^*(k,k''',k'''',t'') \\ &\cdot \langle C^{(0)}(k')C^{(0)}(k'')C^{(0)*}(k''')C^{(0)*}(k'''')\rangle\end{aligned} \tag{10.140}$$

where $\delta(0)$ means $\delta(k - k')|_{k'=k}$; in order for (10.140) to make any sense we will need to find a similar factor on the right side.

Consider the factor

$$\langle C^{(0)}(k')C^{(0)}(k'')C^{(0)*}(k''')C^{(0)*}(k'''')\rangle \sim \langle e^{i\theta(k')+i\theta(k'')-i\theta(k''')-i\theta(k'''')}\rangle \tag{10.141}$$

There are two possible ways to obtain a nonzero result on the right. The first is when

$$k' = k''' \qquad \text{and} \qquad k'' = k'''' \tag{10.142}$$

and the second is when

$$k' = k'''' \qquad \text{and} \qquad k'' = k''' \tag{10.143}$$

Thus,

$$\begin{aligned} &\langle C^{(0)}(k')C^{(0)}(k'')C^{(0)*}(k''')C^{(0)*}(k'''')\rangle \\ &= n^{(0)}(k')n^{(0)}(k'')\delta(k' - k''')\delta(k'' - k'''') \\ &+ n^{(0)}(k')n^{(0)}(k'')\delta(k' - k'''')\delta(k'' - k''') \end{aligned} \tag{10.144}$$

Substituting this on the right of (10.140), and using the delta functions to perform the k''' and k'''' integrations, we obtain

$$\begin{aligned} [n(k,t) - n^{(0)}(k)]\delta(0) &= \int dk'\, dk'' \int_0^t dt' \int_0^t dt'' \\ &\times F(k,k',k'',t')n^{(0)}(k')n^{(0)}(k'') \\ &\times \{F^*(k,k',k'',t'') + F^*(k,k'',k',t'')\} \end{aligned} \tag{10.145}$$

To make life simpler, let us assume for this model problem that

$$V(k,k',k'') = V(k,k'',k') \tag{10.146}$$

which means that [see Eq. (10.123)]

$$F(k,k',k'',t'') = F(k,k'',k',t'') \tag{10.147}$$

Equation (10.145) then reads

$$\boxed{\begin{aligned} [n(k,t) - n^{(0)}(k)]\delta(0) &= 2\int dk'\, dk''\, n^{(0)}(k')n^{(0)}(k'') \\ &\times \left|\int_0^t dt'\, F(k,k',k'',t')\right|^2 \end{aligned}} \tag{10.148}$$

From the definition (10.123),

$$\left|\int_0^t dt'\, F\right|^2 \sim \left|\int_0^t dt'\, e^{i[\omega(k)-\omega(k')-\omega(k'')]t'}\right|^2 \tag{10.149}$$

The idea of weak turbulence theory is to consider changes on a time scale long compared to that of any of the characteristic frequencies. With this in mind, we wish to apply to (10.149) the formula

$$\lim_{t\to\infty} \left|\int_0^t dt'\, e^{i\Omega t'}\right|^2 = 2\pi t\delta(\Omega) \tag{10.150}$$

Equation (10.150) can be derived as follows:

$$\lim_{t\to\infty} \left| \int_0^t dt'\, e^{i\Omega t'} \right|^2 = \lim_{t\to\infty} \left| \frac{1}{i\Omega} e^{i\Omega t'} \Big|_0^t \right|^2$$

$$= \lim_{t\to\infty} \left| \frac{1}{i\Omega} (e^{i\Omega t} - 1) \right|^2$$

$$= \lim_{t\to\infty} \frac{1}{\Omega^2} (e^{i\Omega t} - 1)(e^{-i\Omega t} - 1)$$

$$= \lim_{t\to\infty} \frac{1}{\Omega^2} (e^{i\Omega t/2} - e^{-i\Omega t/2})(e^{-i\Omega t/2} - e^{i\Omega t/2})$$

$$= \lim_{t\to\infty} \frac{4}{\Omega^2} \sin^2 \left(\frac{\Omega t}{2} \right) \tag{10.151}$$

Now, multiply the argument by t^{-1}. Then, if $\Omega \neq 0$, we have

$$\lim_{t\to\infty} \frac{4}{\Omega^2 t} \sin^2 \left(\frac{\Omega t}{2} \right) \to 0 \tag{10.152}$$

If we first take the limit $\Omega \to 0$, we obtain

$$\lim_{t\to\infty} \lim_{\Omega\to 0} \frac{4}{\Omega^2 t} \sin^2 \left(\frac{\Omega t}{2} \right) = \lim_{t\to\infty} \frac{4}{\Omega^2 t} \left(\frac{\Omega t}{2} \right)^2$$

$$= \lim_{t\to\infty} t$$

$$= \infty \tag{10.153}$$

Thus, the argument of (10.152) vanishes if $\Omega \neq 0$, and becomes infinite if $\Omega = 0$. Because it must therefore be proportional to $\delta(\Omega)$, we write

$$\lim_{t\to\infty} \frac{4}{\Omega^2 t} \sin^2 \left(\frac{\Omega t}{2} \right) = \alpha \delta(\Omega) \tag{10.154}$$

To determine the constant α, we integrate both sides over all Ω, to obtain

$$\alpha = \int_{-\infty}^{\infty} d\Omega \lim_{t\to\infty} \frac{4}{\Omega^2 t} \sin^2 \left(\frac{\Omega t}{2} \right)$$

$$= 2 \int_{-\infty}^{\infty} \frac{dx}{x^2} \sin^2 x$$

$$= 2\pi \tag{10.155}$$

EXERCISE Obtain (10.155) by contour integration, moving the contour off of the nonexistent pole at $x = 0$.

EXERCISE By techniques similar to those used in the proof of (10.150), show that

$$\lim_{t\to\infty} \int_0^t dt'\, e^{i\Omega t'} \int_0^{t'} dt''\, e^{-i\Omega t''} = \pi t \delta(\Omega)$$

Finally, moving the factor t to the right in (10.154), we obtain the result (10.150).

Using this result and the definition (10.123), we find

$$\left| \int_0^t dt'\, F(k,k',k'',t') \right|^2$$
$$= |V(k,k',k'')|^2\, 2\pi t \delta[\omega(k) - \omega(k') - \omega(k'')]$$
$$\times\, \delta(k - k' - k'')\underbrace{\delta(k - k' - k'')}_{\delta(0)} \qquad (10.156)$$

where one may write $\delta(x)\delta(x) = \delta(x)\delta(0)$ since $x = 0$ is the only value that counts. Substituting this into (10.148) and canceling a $\delta(0)$ on each side, we find

$$n(k,t) - n^{(0)}(k) = 4\pi t \int dk'\, dk'' |V(k,k',k'')|^2$$
$$\times\, n^{(0)}(k')n^{(0)}(k'')\delta[\omega(k) - \omega(k') - \omega(k'')]\delta(k - k' - k'') \qquad (10.157)$$

This calculation started at $t = 0$ and went out a time t that was considered long. However, considering the right side of (10.157) as tiny, we may say that only a small change in $n(k,t)$ results. Thus, one can imagine performing this calculation over and over again, each time inserting the new value of $n(k,t)$ on the right of (10.157) instead of $n^{(0)}(k)$. Dividing (10.157) by t, we obtain a differential equation for $n(k,t)$; the left side becomes

$$\frac{n(k,t) - n^{(0)}(k)}{t} = \frac{\partial}{\partial t} n(k,t) \qquad (10.158)$$

Finally we have

$$\boxed{\frac{\partial}{\partial t} n(k,t) = 4\pi \int dk'\, dk'' |V(k,k',k'')|^2 n(k',t)n(k'',t) \times \delta[\omega(k) - \omega(k') - \omega(k'')]\delta(k - k' - k'')} \qquad (10.159)$$

The important quantities on the right are the two delta functions, which indicate that wave number matching ($k = k' + k''$) and frequency matching [$\omega(k) = \omega(k') + \omega(k'')$] are operative.

Equation (10.159) is the basic result of weak turbulence theory as applied to fluid equations of the form (10.88) and (10.89). We will not present any of the details of these calculations, but rather we refer the reader to the extensive discussions in other places. (See Refs. [1] to [5], [11] to [16].) Let us conclude this section by describing qualitatively the predictions of the weak turbulence theory as applied to the equations (10.111), (10.114), and (10.115), which contain the nonlinear

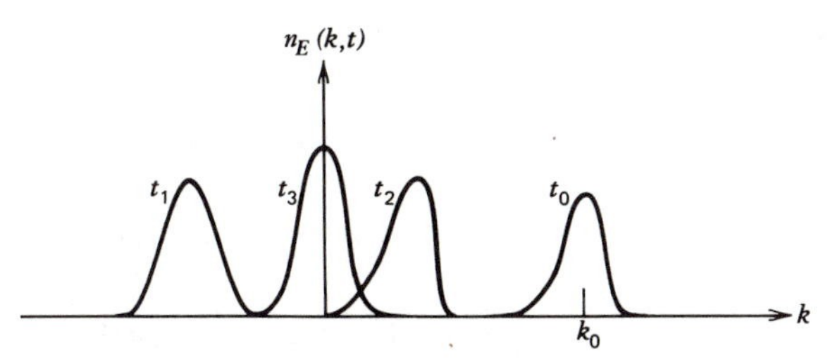

Fig. 10.5 Behavior of the intensity $n_E(k,t)$ predicted by weak turbulence theory; $t_3 > t_2 > t_1 > t_0$.

coupling of Langmuir waves to ion-acoustic waves. For a large range of parameters, it is found (Fig. 10.5) that the intensity n_E, defined by $n_E(k,t)\delta(k - k') = \langle E(k,t)E^*(k,t)\rangle$, initially localized about some k_0, migrates to wave numbers localized about a wave number opposite in sign and somewhat smaller in magnitude than k_0. This process is the same as the parametric decay instability of Section 7.17. The process continues until the intensity piles up about $k = 0$. For some time this phenomenon of condensation in wave number space was thought of as a paradox in plasma physics, because Landau damping is small at small wave numbers so that in the absence of collisional damping, there was no known dissipation mechanism. We now know (Section 7.17) that intense waves localized around $k = 0$ can drive the oscillating two-stream instability leading to soliton formation. Since solitons are localized in configuration space, their formation leads to a broadening of $n_E(k,t)$ in wave number space and thus leads to the possibility of Landau damping. Unfortunately, the four-wave modulational instabilities cannot be treated within the context of the weak turbulence theory of the present chapter. The development of a complete theory of Langmuir turbulence including modulational instability and soliton formation is only one of the many fascinating aspects of nonlinearity and turbulence that are being treated in current research in plasma physics ([6], [16] to [31]).

REFERENCES

[1] B. B. Kadomtsev, *Plasma Turbulence*, Academic, New York, 1965.

[2] R. C. Davidson, *Methods in Nonlinear Plasma Theory*, Academic, New York, 1972.

[3] V. N. Tsytovich, *Nonlinear Effects in Plasmas*, Plenum, New York, 1970.

[4] V. N. Tsytovich, *Theory of Turbulent Plasma*, Consultants Bureau, New York, 1977.

[5] R. Z. Sagdeev and A. A. Galeev, *Nonlinear Plasma Theory*, Benjamin, New York, 1969.

[6] J. A. Krommes, in *Handbook of Plasma Physics*, edited by R. Sudan and A. A. Galeev, North-Holland, Amsterdam, to be published.

[7] W. E. Drummond and D. Pines, *Suppl. Nucl. Fusion Part 3*, 1049 (1962).

[8] A. A. Vedenov, E. P. Velikhov, and R. Z. Sagdeev, *Suppl. Nucl. Fusion Part 2*, 465 (1962).

[9] F. W. Perkins, C. Oberman, and E. J. Valeo, *J. Geophys. Res.*, *79*, 1478 (1974).

[10] D. C. Leslie, *Developments in the Theory of Turbulence*, Clarendon, Oxford, 1973.

[11] V. E. Zakharov and E. A. Kuznetsov, *Sov. Phys.-JETP*, *48*, 458 (1978).

[12] A. A. Vedenov, *Theory of Turbulent Plasma*, American Elsevier, New York, 1968.

[13] V. N. Tsytovich, *An Introduction to the Theory of Plasma Turbulence*, Pergamon, Oxford, 1972.

[14] S. A. Kaplan and V. N. Tsytovich, *Plasma Astrophysics*, Pergamon, Oxford, 1973.

[15] A. Hasegawa, *Plasma Instabilities and Nonlinear Effects*, Springer-Verlag, New York, 1975.

[16] W. Horton and D. Choi, *Phys. Rep.*, *49*, 273 (1979).

[17] S. A. Orszag and R. H. Kraichnan, *Phys. Fluids, 10,* 1720 (1967).

[18] L. I. Rudakov and V. N. Tsytovich, *Plasma Phys.*, *13*, 213 (1971).

[19] T. H. Dupree, *Phys. Fluids*, *9*, 1733 (1966).

[20] J. Weinstock, *Phys. Fluids*, *11*, 1977 (1968).

[21] T. H. Dupree, *Phys. Fluids*, *15*, 334 (1972).

[22] B. H. Hui and T. H. Dupree, *Phys. Fluids*, *18*, 235 (1975).

[23] T. H. Dupree, *Phys. Fluids*, *21*, 783 (1978).

[24] H. A. Rose, *J. Stat. Phys.*, *20*, 415 (1979).

[25] D. F. DuBois and M. Espedal, *Plasma Phys.*, *20*, 1209 (1978).

[26] J. A. Krommes and R. G. Kleva, *Phys. Fluids*, *22*, 2168 (1979).

[27] J. C. Adam, G. Laval, and D. Pesme, *Phys. Rev. Lett.*, *43*, 1671 (1979).

[28] P. J. Hansen and D. R. Nicholson, *Phys. Fluids*, *24*, 615 (1981).

[29] D. F. DuBois, *Phys. Rev.*, *A23*, 865 (1981).

[30] T. Boutros-Ghali and T. H. Dupree, *Phys. Fluids*, *24*, 1839 (1981).

[31] D. F. DuBois and H. A. Rose, *Phys. Rev.*, *A24*, 1476 (1981).

PROBLEM

10.1 Quasilinear Theory

(a) Suppose two electron beams are incident on a Maxwellian plasma, so that the one-dimensional distribution function is as shown in Fig. 10.6. Using the ideas of quasilinear theory, consider an initial value problem consisting of

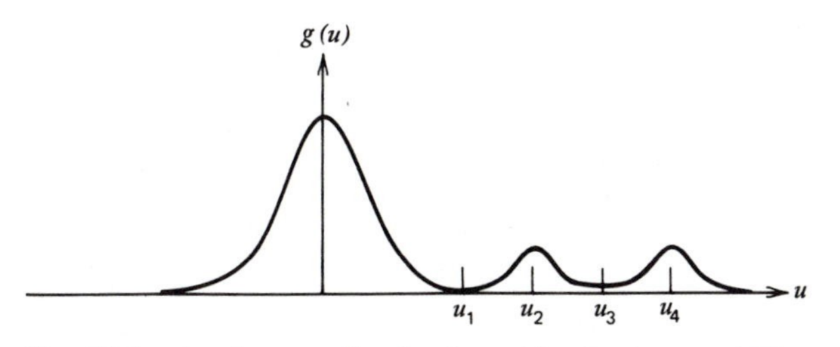

Fig. 10.6 An electron distribution with a background Maxwellian and two electron beams.

the distribution as shown plus a small noise level of electric field fluctuations. Draw a series of sketches, at several different times, of: the growth rate of Langmuir waves as a function of wave number; the intensity of Langmuir waves as a function of phase speed; and the distribution function. Use u_1, u_2, u_3, and u_4 as benchmarks. Make your final set of sketches correspond to $t \to$ "∞." Crudely estimate a time scale for this entire process, using the electron plasma frequency and the ratio n_b/n_0 where n_b is the density of beam particles.

(b) The derivation of quasilinear theory proceeded from the Vlasov equation and thus left out certain physics. Discuss this physics, and recall an equation from an earlier chapter that would allow us to include this physics. Draw a new set of sketches as in part (a), using the $t \to$ "∞" sketch of part (a) as the $t = 0$ sketch of part (b). Make your final set of sketches correspond to $t \to \infty$. Crudely estimate a time scale for this entire process.

In each of parts (a) and (b), state explicitly what is being assumed about the ions. Should this be the same in each part?

Derivation of the Lenard–Balescu Equation

In this appendix, we complete the derivation of the Lenard–Balescu equation (5.19) starting from Eqs. (5.1) and (5.4), which in turn are obtained from the BBGKY hierarchy by discarding three-particle correlations (Refs. [1] to [5]). From (5.1) and (5.4) to (5.8), we have

$$\partial_t f_1(\mathbf{v}_1,t) = -n_0 \int d\mathbf{x}_2\, d\mathbf{v}_2 \mathbf{a}_{12} \cdot \nabla_{\mathbf{v}_1} g(\mathbf{x}_1 - \mathbf{x}_2,\mathbf{v}_1,\mathbf{v}_2,t) \tag{A.1}$$

$$\frac{\partial}{\partial \tilde{t}} g(\mathbf{x}_1 - \mathbf{x}_2,\mathbf{v}_1,\mathbf{v}_2,\tilde{t}) + V_1 g + V_2 g = S(\mathbf{x}_1 - \mathbf{x}_2,\mathbf{v}_1,\mathbf{v}_2) \tag{A.2}$$

$$\begin{aligned} V_1 g(12) &= \mathbf{v}_1 \cdot \nabla_{\mathbf{x}_1} g(12) \\ &+ [n_0 \int d3\ \mathbf{a}_{13}\, g(32)] \cdot \nabla_{\mathbf{v}_1} f_1(\mathbf{v}_1) \end{aligned} \tag{A.3}$$

$$\begin{aligned} V_2 g(12) &= \mathbf{v}_2 \cdot \nabla_{\mathbf{x}_2} g(12) \\ &+ [n_0 \int d3\ \mathbf{a}_{23}\, g(13)] \cdot \nabla_{\mathbf{v}_2} f_1(\mathbf{v}_2) \end{aligned} \tag{A.4}$$

$$S(\mathbf{x}_1 - \mathbf{x}_2,\mathbf{v}_1,\mathbf{v}_2) = -(\mathbf{a}_{12} \cdot \nabla_{\mathbf{v}_1} + \mathbf{a}_{21} \cdot \nabla_{\mathbf{v}_2}) f_1(\mathbf{v}_1) f_1(\mathbf{v}_2) \tag{A.5}$$

where we have used $g(32) = g(23)$, we alternate between the notations (1) and $(\mathbf{x}_1,\mathbf{v}_1)$ depending on convenience, and we recall from Chapter 5 that we wish to solve for $g(\tilde{t} \to \infty)$ where $\tilde{t}$ is the fast time scale on which g relaxes. On this fast time scale, the functions f_1 and thus S are considered to be constants. We shall also need, from (5.9),

$$\mathbf{a}_{ij} = \frac{e^2}{m_e |\mathbf{x}_i - \mathbf{x}_j|^3} (\mathbf{x}_i - \mathbf{x}_j) \tag{A.6}$$

Using the Fourier transform conventions in Chapter 5, we spatially Fourier transform these equations with respect to $\mathbf{x}_1$ and $\mathbf{x}_2$. Because of the appearance of the combination $(\mathbf{x}_1 - \mathbf{x}_2)$, we obtain the factor $\delta(\mathbf{k}_1 + \mathbf{k}_2)$ in several places and, thus, can replace $\mathbf{k}_2$ by $-\mathbf{k}_1$.

EXERCISE For any function $f(\mathbf{x}_1 - \mathbf{x}_2)$, show that the double Fourier transform with respect to $\mathbf{x}_1$ and $\mathbf{x}_2$ is $\delta(\mathbf{k}_1 + \mathbf{k}_2)\, f(\mathbf{k}_1)$ where $f(\mathbf{k})$ is the Fourier transform of $f(\mathbf{x})$ with respect to $\mathbf{x}$.

EXERCISE Show that $\int d\mathbf{x}\, f_1(\mathbf{x}) f_2(\mathbf{x}) = (2\pi)^3 \int d\mathbf{k}\, f_1(-\mathbf{k}) f_2(\mathbf{k})$ for any functions f_1 and f_2; here, $f_1(\mathbf{k})$ is the Fourier transform of $f_1(\mathbf{x})$, etc., as usual.

EXERCISE Show that the double Fourier transform of $\int d\mathbf{x}_3\, f_1(\mathbf{x}_1 - \mathbf{x}_3) \times f_2(\mathbf{x}_2 - \mathbf{x}_3)$ is $(2\pi)^3\, \delta(\mathbf{k}_1 + \mathbf{k}_2)\, f_1(\mathbf{k}_1)\, f_2(-\mathbf{k}_1)$.

With the results of these exercises, and Eq. (5.16) for $\mathbf{a}_{12}(\mathbf{k})$, the Fourier transformed version of (A.1) to (A.6) is

$$\partial_t f_1(\mathbf{v}_1,t) = -\frac{in_0(2\pi)^3}{m_e} \nabla_{\mathbf{v}_1} \cdot \int d\mathbf{v}_2\, d\mathbf{k}_1\, \mathbf{k}_1\, \varphi(k_1) g(\mathbf{k}_1,\mathbf{v}_1,\mathbf{v}_2,\tilde{t} = \infty) \tag{A.7}$$

$$\frac{\partial}{\partial \tilde{t}} g(\mathbf{k}_1,\mathbf{v}_1,\mathbf{v}_2,\tilde{t}) + V_1 g + V_2 g = S(\mathbf{k}_1,\mathbf{v}_1,\mathbf{v}_2) \tag{A.8}$$

$$V_1\, g(12) = i\mathbf{k}_1 \cdot \mathbf{v}_1 g(12) - \frac{n_0(2\pi)^3}{m_e} i\mathbf{k}_1 \cdot \nabla_{\mathbf{v}_1} f_1(\mathbf{v}_1)\varphi(k_1) \int d\mathbf{v}_3\, g(\mathbf{k}_1,\mathbf{v}_3,\mathbf{v}_2,\tilde{t}) \tag{A.9}$$

$$V_2 g(12) = -i\mathbf{k}_1 \cdot \mathbf{v}_2\, g(12) + \frac{n_0(2\pi)^3}{m_e} i\mathbf{k}_1 \cdot \nabla_{\mathbf{v}_2} f_1(\mathbf{v}_2)\varphi(k_1) \int d\mathbf{v}_3\, g(\mathbf{k}_1,\mathbf{v}_1,\mathbf{v}_3,\tilde{t}) \tag{A.10}$$

$$S(\mathbf{k}_1,\mathbf{v}_1,\mathbf{v}_2) = \frac{\varphi(k_1)}{m_e} i\mathbf{k}_1 \cdot (\nabla_{\mathbf{v}_1} - \nabla_{\mathbf{v}_2}) f_1(\mathbf{v}_1) f_1(\mathbf{v}_2) \tag{A.11}$$

Our goal is to express the right side of (A.7) in terms of f_1 by solving (A.8) for g. With (A.7) in its present form, the remainder of the calculation can be performed in wave number space; because of the factor i on the right of (A.7) and the fact that the right of (A.7) must be real, we need only calculate the imaginary part of $g(\mathbf{k}_1,\mathbf{v}_1,\mathbf{v}_2,\tilde{t} = \infty)$.

The solution of (A.8) for $g(\mathbf{k}_1,\mathbf{v}_1,\mathbf{v}_2,\tilde{t} = \infty)$ is accomplished by Laplace transforming with respect to the fast time $\tilde{t}$.

EXERCISE For any function $g(t)$, show that the Laplace transform of dg/dt is $-g(t = 0) - i\omega\, g(\omega)$.

With the result of this exercise, the Laplace transform of (A.8) is

$$- g(\mathbf{k}_1,\mathbf{v}_1,\mathbf{v}_2,\tilde{t} = 0) - i\omega\, g(\mathbf{k}_1,\mathbf{v}_1,\mathbf{v}_2,\omega) + V_1\, g(12\omega)$$

$$+ V_2\, g(12\omega) = - \frac{1}{i\omega} S(\mathbf{k}_1,\mathbf{v}_1,\mathbf{v}_2) \tag{A.12}$$

where $g(\omega)$ is defined only for $\omega_i \equiv \mathrm{Im}(\omega)$ sufficiently large, and where the operators V_1 and V_2 can be regarded as numbers since they have no time dependence in (A.9) and (A.10). Solving (A.12) for $g(\omega)$ we find

$$g(\omega) = \frac{g(\tilde{t} = 0) - (S/i\omega)}{-i\omega + V_1 + V_2} \tag{A.13}$$

We require $g(\tilde{t} = \infty)$. This can be obtained from the inverse Laplace transform of (A.13). It turns out that distribution functions $f_1(\mathbf{v})$ that are stable in the Vlasov sense (Chapter 6) are such that the zeros of $-i\omega + V_1 + V_2$ always occur in the lower half ω-plane. We consider only such stable distribution functions $f_1(\mathbf{v})$. Thus, the inverse Laplace transform

$$g(\tilde{t}) = \int_L \frac{d\omega}{2\pi} \frac{g(\tilde{t} = 0) - S/i\omega}{-i\omega + V_1 + V_2} e^{-i\omega t} \tag{A.14}$$

can be performed by deforming the Laplace contour as shown in Fig. A.1. Since poles in the lower half plane contribute only damped functions of time, $\sim$ exp $(\omega_i t)$, the only pole that contributes to $g(\tilde{t} = \infty)$ is the one at $\omega = 0$; therefore,

$$g(\tilde{t} = \infty) = \lim_{\omega \to 0} \frac{S}{-i\omega + V_1 + V_2} \tag{A.15}$$

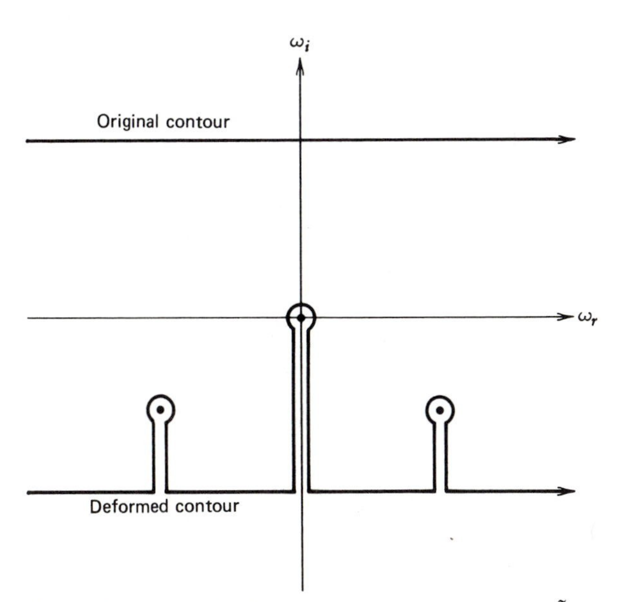

Fig. A.1 Inverse Laplace contour for calculating $g(\tilde{t} = \infty)$.

where we retain the $\lim_{\omega \to 0}$ to help us interpret other contour integrations that occur in the calculation.

At this point, we introduce a trick that allows us to treat the operators V_1 and V_2 separately, rather than in the combination $V_1 + V_2$. Consider

$$\frac{1}{-i\omega + V_1 + V_2} = \int_0^{\infty} dt\, e^{(i\omega - V_1 - V_2)t}$$

$$= \int_0^{\infty} dt\, e^{i\omega t} \int_{C1} \frac{d\omega_1}{2\pi} \frac{e^{-i\omega_1 t}}{-i\omega_1 + V_1} \int_{C2} \frac{d\omega_2}{2\pi} \frac{e^{-i\omega_2 t}}{-i\omega_2 + V_2}$$

$$= \int_{C1} \frac{d\omega_1}{2\pi} \int_{C2} \frac{d\omega_2}{2\pi} \frac{1}{-i\omega_1 + V_1} \frac{1}{-i\omega_2 + V_2} \frac{1}{-i(\omega - \omega_1 - \omega_2)} \tag{A.16}$$

where the contours C_1 and C_2 must be chosen so that $\omega_i > \omega_{1i} + \omega_{2i}$. Then (A.15) becomes

$$g(\mathbf{k}_1, \mathbf{v}_1, \mathbf{v}_2, \tilde{t} = \infty)$$

$$= \lim_{\omega \to 0} \int_{C1} \frac{d\omega_1}{2\pi} \int_{C2} \frac{d\omega_2}{2\pi} \frac{1}{-i\omega_2 + V_2} \frac{1}{-i\omega_1 + V_1} \frac{S(\mathbf{k}_1, \mathbf{v}_1, \mathbf{v}_2)}{-i(\omega - \omega_1 - \omega_2)} \tag{A.17}$$

In expressions (A.13) to (A.17), we interpret the meaning of an inverse operator $(-i\omega_1 + V_1)^{-1}F$ acting on a function F to be that function G such that $F = (-i\omega_1 + V_1)G$.

We first need

$$\alpha(\mathbf{k}_1, \mathbf{v}_1, \mathbf{v}_2) = \frac{1}{-i\omega_1 + V_1} S(\mathbf{k}_1, \mathbf{v}_1, \mathbf{v}_2) \tag{A.18}$$

such that

$$S(\mathbf{k}_1, \mathbf{v}_1, \mathbf{v}_2) = (-i\omega_1 + V_1)\alpha(\mathbf{k}_1, \mathbf{v}_1, \mathbf{v}_2)$$

$$= (-i\omega_1 + i\mathbf{k}_1 \cdot \mathbf{v}_1)\alpha(\mathbf{k}_1, \mathbf{v}_1, \mathbf{v}_2) - \frac{i(2\pi)^3 n_0}{m_e} \mathbf{k}_1 \cdot \nabla_{\mathbf{v}_1}$$

$$\times f_1(\mathbf{v}_1)\varphi(k_1) \int d\mathbf{v}_3\, \alpha(\mathbf{k}_1, \mathbf{v}_3, \mathbf{v}_2) \tag{A.19}$$

In order to solve this for α we must first eliminate $\int d\mathbf{v}_3\, \alpha$; we express (A.19) as

$$\alpha(\mathbf{k}_1, \mathbf{v}_1, \mathbf{v}_2) = \frac{1}{-i\omega_1 + i\mathbf{k}_1 \cdot \mathbf{v}_1} [S(\mathbf{k}_1, \mathbf{v}_1, \mathbf{v}_2)$$

$$+ \frac{i(2\pi)^3 n_0}{m_e} \mathbf{k}_1 \cdot \nabla_{\mathbf{v}_1} f_1(\mathbf{v}_1)\varphi(k_1) \int d\mathbf{v}_3\, \alpha(\mathbf{k}_1, \mathbf{v}_3, \mathbf{v}_2)] \tag{A.20}$$

and integrate over all $\mathbf{v}_1$ to find

$$\int d\mathbf{v}_1\, \alpha(\mathbf{k}_1,\mathbf{v}_1,\mathbf{v}_2)$$

$$= \int d\mathbf{v}_1\, \frac{S(\mathbf{k}_1,\mathbf{v}_1,\mathbf{v}_2)}{-i\omega_1 + i\mathbf{k}_1\cdot\mathbf{v}_1} + [\int d\mathbf{v}_3\, \alpha(\mathbf{k}_1,\mathbf{v}_3,\mathbf{v}_2)]$$

$$\times \frac{i(2\pi)^3 n_0}{m_e}\, \varphi(k_1) \int d\mathbf{v}_1\, \frac{\mathbf{k}_1\cdot\nabla_{\mathbf{v}_1} f_1(\mathbf{v}_1)}{-i\omega_1 + i\mathbf{k}_1\cdot\mathbf{v}_1} \tag{A.21}$$

Realizing that $\mathbf{v}_3$ on the right is merely a dummy variable of integration, we find

$$\int d\mathbf{v}_3\, \alpha(\mathbf{k}_1,\mathbf{v}_3,\mathbf{v}_2) = \frac{1}{\epsilon(\mathbf{k}_1,\omega_1)} \int d\mathbf{v}_3\, \frac{S(\mathbf{k}_1,\mathbf{v}_3,\mathbf{v}_2)}{-i\omega_1 + i\mathbf{k}_1\cdot\mathbf{v}_3} \tag{A.22}$$

where

$$\epsilon(\mathbf{k}_1,\omega_1) = 1 + \frac{\omega_e^2}{k_1^2} \int d\mathbf{v}_1\, \frac{\mathbf{k}_1\cdot\nabla_{\mathbf{v}_1} f_1(\mathbf{v}_1)}{\omega_1 - \mathbf{k}_1\cdot\mathbf{v}_1} \tag{A.23}$$

is the dielectric function encountered in Chapter 6. Thus, (A.20) becomes

$$\alpha(\mathbf{k}_1,\mathbf{v}_1,\mathbf{v}_2) = \frac{1}{-i\omega_1 + i\mathbf{k}_1\cdot\mathbf{v}_1} \Bigg[S(\mathbf{k}_1,\mathbf{v}_1,\mathbf{v}_2)$$

$$+ \frac{i(2\pi^3) n_0 \mathbf{k}_1\cdot\nabla_{\mathbf{v}_1} f_1(\mathbf{v}_1)\varphi(k_1)/m_e}{\epsilon(\mathbf{k}_1,\omega_1)} \int d\mathbf{v}_3\, \frac{S(\mathbf{k}_1,\mathbf{v}_3,\mathbf{v}_2)}{-i\omega_1 + i\mathbf{k}_1\cdot\mathbf{v}_3} \Bigg] \tag{A.24}$$

which completes the inversion of the operator $(-i\omega_1 + V_1)^{-1}$

Next, we need

$$\int d\mathbf{v}_2\, \beta(\mathbf{k}_1,\mathbf{v}_1,\mathbf{v}_2) = \int d\mathbf{v}_2\, \frac{1}{-i\omega_2 + V_2}\, \alpha(\mathbf{k}_1,\mathbf{v}_1,\mathbf{v}_2) \tag{A.25}$$

where we have noted from (A.7) that we need $\int d\mathbf{v}_2\, g$ rather than g, allowing us to use the compact analogue of (A.22). Noting that V_2 is the same as V_1 if the sign of $\mathbf{k}_1$ is changed and if $\mathbf{v}_1$ and $\mathbf{v}_2$ are interchanged appropriately, we find

$$\int d\mathbf{v}_2\, \beta(\mathbf{k}_1,\mathbf{v}_1,\mathbf{v}_2) = \frac{1}{\epsilon(-\mathbf{k}_1,\omega_2)} \int d\mathbf{v}_2\, \frac{\alpha(\mathbf{k}_1,\mathbf{v}_1,\mathbf{v}_2)}{-i\omega_2 - i\mathbf{k}_1\cdot\mathbf{v}_2} \tag{A.26}$$

With the result (A.26) we have from (A.17)

$$\int d\mathbf{v}_2\, g(\mathbf{k}_1,\mathbf{v}_1,\mathbf{v}_2,\tilde{t} = \infty)$$

$$= \lim_{\omega\to 0} \int_{C1} \frac{d\omega_1}{2\pi} \int_{C2} \frac{d\omega_2}{2\pi}\, \frac{1}{\epsilon(-\mathbf{k}_1,\omega_2)}\, \frac{1}{-i(\omega - \omega_1 - \omega_2)}$$

$$\times \int d\mathbf{v}_2\, \frac{1}{-i\omega_2 - i\mathbf{k}_1\cdot\mathbf{v}_2}\, \frac{1}{-i\omega_1 + V_1}\, S(\mathbf{k}_1,\mathbf{v}_1,\mathbf{v}_2) \tag{A.27}$$

We perform the ω_2 integration first, along the contour C_2 shown in Fig. (A.2). Since the integrand behaves like ω_2^{-2} for large ω_2, we can close the contour upward

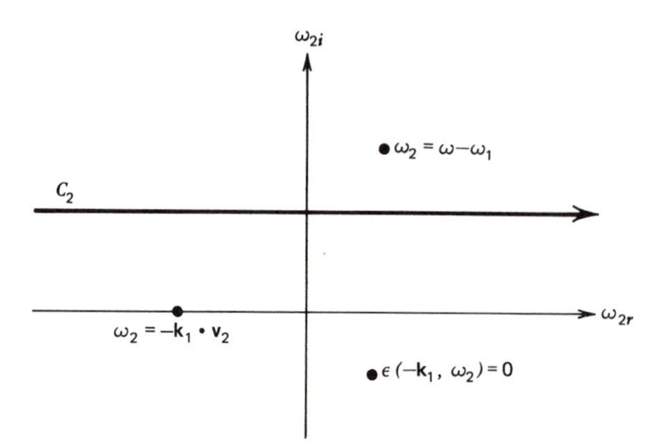

Fig. A.2 Contour C_2 used in evaluating (A.27).

and pick up only the pole at $\omega_2 = \omega - \omega_1$, yielding

$$\int d\mathbf{v}_2\, g(\mathbf{k}_1,\mathbf{v}_1,\mathbf{v}_2,\tilde{t} = \infty) = \lim_{\omega\to 0} \int d\mathbf{v}_2 \int_{C_1} \frac{d\omega_1}{2\pi}$$

$$\times \frac{1}{\epsilon(-\mathbf{k}_1,\omega - \omega_1)} \frac{1}{-i(\omega - \omega_1) - i\mathbf{k}_1 \cdot \mathbf{v}_2} \frac{1}{-i\omega_1 + V_1} S(\mathbf{k}_1,\mathbf{v}_1,\mathbf{v}_2) \qquad \text{(A.28)}$$

Inserting the results (A.11) and (A.24) we have

$$\int d\mathbf{v}_2\, g(\mathbf{k}_1,\mathbf{v}_1,\mathbf{v}_2,\tilde{t} = \infty) = \lim_{\omega\to 0} \int d\mathbf{v}_2 \int_{C_1} \frac{d\omega_1}{2\pi}$$

$$\times \frac{1}{\epsilon(-\mathbf{k}_1,\omega - \omega_1)} \frac{1}{-i(\omega - \omega_1) - i\mathbf{k}_1 \cdot \mathbf{v}_2} \frac{1}{-i\omega_1 + i\mathbf{k}_1 \cdot \mathbf{v}_1}$$

$$\times \left[\frac{i\mathbf{k}_1}{m_e} \cdot \varphi(k_1)(\overset{①}{\nabla_{\mathbf{v}_1}} - \overset{②}{\nabla_{\mathbf{v}_2}}) f_1(\mathbf{v}_1) f_1(\mathbf{v}_2) \right.$$

$$- \frac{i(2\pi)^3 n_0 \mathbf{k}_1 \cdot \nabla_{\mathbf{v}_1} f_1(\mathbf{v}_1) \varphi^2(k_1)/m_e^2}{\epsilon(\mathbf{k}_1,\omega_1)}$$

$$\left. \times \int d\mathbf{v}_3 \frac{\mathbf{k}_1 \cdot (\overset{③}{\nabla_{\mathbf{v}_3}} - \overset{④}{\nabla_{\mathbf{v}_2}})}{\omega_1 - \mathbf{k}_1 \cdot \mathbf{v}_3} f_1(\mathbf{v}_3) f_1(\mathbf{v}_2) \right] \qquad \text{(A.29)}$$

There are four numbered terms in the square brackets. Including the $\mathbf{v}_2$ integration and the denominator containing $\mathbf{v}_2$, we have

$$② = \frac{-i\mathbf{k}_1 \cdot \varphi(k_1)}{m_e} f_1(\mathbf{v}_1) \int d\mathbf{v}_2 \frac{\nabla_{\mathbf{v}_2} f_1(\mathbf{v}_2)}{-i(\omega - \omega_1) - i\mathbf{k}_1 \cdot \mathbf{v}_2}$$

$$= \frac{-f_1(\mathbf{v}_1)}{n_0(2\pi)^3} [\epsilon(-\mathbf{k}_1, \omega - \omega_1) - 1] \tag{A.30}$$

where (A.23) has been used. Similarly,

$$③ = \frac{-i(2\pi)^3 n_0}{\epsilon(\mathbf{k}_1, \omega_1)} \left[\int d\mathbf{v}_2 \frac{f_1(\mathbf{v}_2)}{-i(\omega - \omega_1) - i\mathbf{k}_1 \cdot \mathbf{v}_2}\right]$$

$$\times\ \mathbf{k}_1 \cdot \nabla_{\mathbf{v}_1} f_1(\mathbf{v}_1)\varphi(k_1)/m_e \left[\frac{\varphi(k_1)}{m_e} \int d\mathbf{v}_3 \frac{\mathbf{k}_1 \cdot \nabla_{\mathbf{v}_3} f_1(\mathbf{v}_3)}{\omega_1 - \mathbf{k}_1 \cdot \mathbf{v}_3}\right]$$

$$= \left[1 - \frac{1}{\epsilon(\mathbf{k}_1, \omega_1)}\right] (-i)\mathbf{k}_1 \cdot \nabla_{\mathbf{v}_1} f_1(\mathbf{v}_1) \frac{\varphi(k_1)}{m_e}$$

$$\times \int d\mathbf{v}_2 \frac{f_1(\mathbf{v}_2)}{-i(\omega - \omega_1) - i\mathbf{k}_1 \cdot \mathbf{v}_2} \tag{A.31}$$

where (A.23) has been used again. Likewise,

$$④ = \left[i\mathbf{k}_1 \cdot \frac{\varphi(k_1)}{m_e} (2\pi)^3 n_0 \int d\mathbf{v}_2 \frac{\nabla_{\mathbf{v}_2} f_1(\mathbf{v}_2)}{-i(\omega - \omega_1) - i\mathbf{k}_1 \cdot \mathbf{v}_2}\right]$$

$$\times \frac{\mathbf{k}_1 \cdot \nabla_{\mathbf{v}_1} f_1(\mathbf{v}_1)\varphi(k_1)/m_e}{\epsilon(\mathbf{k}_1, \omega_1)} \int d\mathbf{v}_3 \frac{f_1(\mathbf{v}_3)}{\omega_1 - \mathbf{k}_1 \cdot \mathbf{v}_3}$$

$$= \frac{[\epsilon(-\mathbf{k}_1, \omega - \omega_1) - 1]}{\epsilon(\mathbf{k}_1, \omega_1)} \mathbf{k}_1 \cdot \nabla_{\mathbf{v}_1} f_1(\mathbf{v}_1) \frac{\varphi(k_1)}{m_e}$$

$$\times \int d\mathbf{v}_3 \frac{f_1(\mathbf{v}_3)}{\omega_1 - \mathbf{k}_1 \cdot \mathbf{v}_3} \tag{A.32}$$

Cancelling term ① with one of the terms in term ③, we combine the remaining terms to obtain

$$\int d\mathbf{v}_2 g(\mathbf{k}_1, \mathbf{v}_1, \mathbf{v}_2, \tilde{t} = \infty) = \lim_{\omega \to 0} \int_{C1} \frac{d\omega_1}{2\pi} \frac{1}{-i\omega_1 + i\mathbf{k}_1 \cdot \mathbf{v}_1}$$

$$\times \left\{ \overset{ⓐ}{\left[1 - \frac{1}{\epsilon(-\mathbf{k}_1, \omega - \omega_1)}\right]} \left[\overset{ⓑ}{-\frac{f_1(\mathbf{v}_1)}{n_0(2\pi)^3}} + \overset{ⓒ}{\frac{\mathbf{k}_1 \cdot \nabla_{\mathbf{v}_1} f_1(\mathbf{v}_1)\varphi(k_1)/m_e}{\epsilon(\mathbf{k}_1, \omega_1)}}\right.\right.$$

$$\left.\left.\times \overset{ⓓ}{\int d\mathbf{v}_2 \frac{f_1(\mathbf{v}_2)}{\omega_1 - \mathbf{k}_1 \cdot \mathbf{v}_2}}\right] + \overset{ⓔ}{\frac{i\mathbf{k}_1 \cdot \nabla_{\mathbf{v}_1} f_1(\mathbf{v}_1)\varphi(k_1)/m_e}{\epsilon(\mathbf{k}_1, \omega_1)\epsilon(-\mathbf{k}_1, \omega - \omega_1)}}\right.$$

$$\left.\times \int d\mathbf{v}_2 \frac{f_1(\mathbf{v}_2)}{-i(\omega - \omega_1) - i\mathbf{k}_1 \cdot \mathbf{v}_2}\right\} \tag{A.33}$$

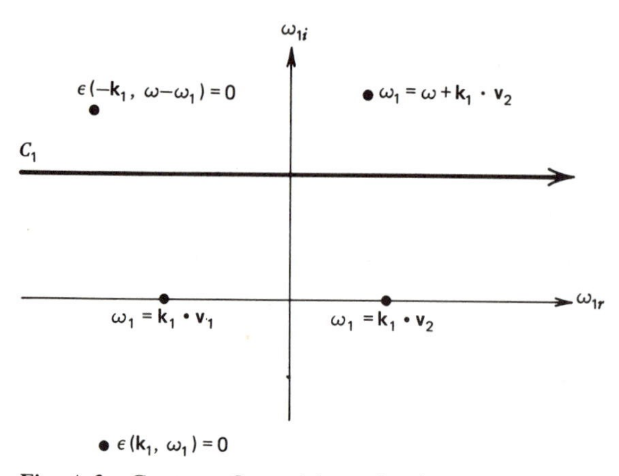

Fig. A.3 Contour C_1 used in evaluating (A.33).

where a change of dummy variable has occurred in term ⓓ. Recalling from below (A.16) that we still have $\omega_i > \omega_{1i}$ along the C_1 contour, and recalling that stable distributions $f_1(\mathbf{v})$ imply that the zeroes of $\epsilon(\mathbf{k},\omega)$ occur only for $\omega_i < 0$, the C_1 contour and the poles of the integrand on the right of (A.33) are as shown in Fig. A.3. Note that not all of the poles occur for each of the terms in (A.33).

Term ⓒ is evaluated by closing the contour C_1 downward, yielding a contribution only from the pole at $\omega_1 = \mathbf{k}_1 \cdot \mathbf{v}_1$, which gives

$$ⓒ = -\frac{f_1(\mathbf{v}_1)}{n_0(2\pi)^3}\left[1 - \frac{1}{\epsilon(-\mathbf{k}_1, \omega - \mathbf{k}_1 \cdot \mathbf{v}_1)}\right] \tag{A.34}$$

EXERCISE Convince yourself that the integrand falls off fast enough at large ω_1 to allow the contour to be closed downward.

Term ⓐ × ⓓ vanishes when the contour is closed upward. Finally, we consider the remaining two pieces together; these are

$$ⓑ \times ⓓ + ⓔ = \lim_{\omega \to 0} \int_{C_1} \frac{d\omega_1}{2\pi} \frac{1}{-i\omega_1 + i\mathbf{k}_1 \cdot \mathbf{v}_1} \frac{1}{\epsilon(-\mathbf{k}_1, \omega - \omega_1)}$$
$$\times \frac{1}{\epsilon(\mathbf{k}_1, \omega_1)} \mathbf{k}_1 \cdot \nabla_{\mathbf{v}_1} f_1(\mathbf{v}_1) \frac{\varphi(k_1)}{m_e} \int d\mathbf{v}_2 f_1(\mathbf{v}_2)$$
$$\times \left[\frac{-1}{\omega_1 - \mathbf{k}_1 \cdot \mathbf{v}_2} + \frac{1}{\omega_1 - \omega - \mathbf{k}_1 \cdot \mathbf{v}_2}\right] \tag{A.35}$$

At this point it is convenient to use the fact that in (A.7) we only need the imaginary part of $\int d\mathbf{v}_2\, g$. As we move the contour in Fig. A.3 down to the real axis (let $\omega \to 0$), the integrand of (A.35) appears to vanish. However, we must be careful at the pole $\omega_1 = \mathbf{k}_1 \cdot \mathbf{v}_1$ and at the two poles that pinch the contour at $\omega_1 = \mathbf{k}_1 \cdot \mathbf{v}_2$ and $\omega_1 = \omega + \mathbf{k}_1 \cdot \mathbf{v}_2$. Recall the Plemelj formulas,

$$\lim_{\eta \to 0} \frac{1}{\omega - a \pm i\eta} = P\left(\frac{1}{\omega - a}\right) \mp i\pi\delta(\omega - a) \tag{A.36}$$

where the upper sign is used when a contour passes above a pole, and the lower sign is used when a contour passes below a pole. Then

$$\begin{aligned}
&\operatorname{Re} \lim_{\omega \to 0} \frac{1}{\omega_1 - \mathbf{k}_1 \cdot \mathbf{v}_1} \left[\frac{-1}{\omega_1 - \mathbf{k}_1 \cdot \mathbf{v}_2} + \frac{1}{\omega_1 - \omega - \mathbf{k}_1 \cdot \mathbf{v}_2}\right] \\
&\quad = \operatorname{Re}\left[P\left(\frac{1}{\omega_1 - \mathbf{k}_1 \cdot \mathbf{v}_1}\right) - i\pi\delta(\omega_1 - \mathbf{k}_1 \cdot \mathbf{v}_1)\right] \\
&\quad \times \left[-P\left(\frac{1}{\omega_1 - \mathbf{k}_1 \cdot \mathbf{v}_2}\right) + i\pi\delta(\omega_1 - \mathbf{k}_1 \cdot \mathbf{v}_2)\right. \\
&\quad \left. + P\left(\frac{1}{\omega_1 - \mathbf{k}_1 \cdot \mathbf{v}_2}\right) + i\pi\delta(\omega_1 - \mathbf{k}_1 \cdot \mathbf{v}_2)\right] \\
&\quad = 2\pi^2\delta(\omega_1 - \mathbf{k}_1 \cdot \mathbf{v}_1)\delta(\omega_1 - \mathbf{k}_1 \cdot \mathbf{v}_2)
\end{aligned} \tag{A.37}$$

where Re indicates the real part. If we use one of the δ-functions to perform the ω_1 integration, (A.35) yields

$$\begin{aligned}
\operatorname{Im}[ⓑ \times ⓓ + ⓔ] &= i\pi\mathbf{k}_1 \cdot \nabla_{\mathbf{v}_1} f_1(\mathbf{v}_1) \frac{\varphi(k_1)}{m_e} \\
&\times \int d\mathbf{v}_2 \frac{\delta[\mathbf{k}_1 \cdot (\mathbf{v}_1 - \mathbf{v}_2)] f_1(\mathbf{v}_2)}{|\epsilon(\mathbf{k}_1, \mathbf{k}_1 \cdot \mathbf{v}_1)|^2}
\end{aligned} \tag{A.38}$$

where we have used the fact that $\epsilon(-\mathbf{k}, -\omega) = \epsilon^*(\mathbf{k}, \omega)$ when ω is real.

EXERCISE Demonstrate this fact from the definition (A.23) of $\epsilon(\mathbf{k}, \omega)$. Show that for ω real, $\operatorname{Im}[\epsilon(\mathbf{k}, \omega)] = -i\pi(\omega_e^2/k^2) \int d\mathbf{v}\, [\mathbf{k} \cdot \nabla_{\mathbf{v}} f_1(\mathbf{v})]\delta(\omega - \mathbf{k} \cdot \mathbf{v})$.

Similarly, if one uses the results of the exercise,

$$\begin{aligned}
\operatorname{Im}[ⓒ] &= \frac{f_1(\mathbf{v}_1)}{n_0(2\pi)^3} \frac{\operatorname{Im}[\epsilon(\mathbf{k}_1, \mathbf{k}_1 \cdot \mathbf{v}_1)]}{|\epsilon(\mathbf{k}_1, \mathbf{k}_1 \cdot \mathbf{v}_1)|^2} = \frac{-i\pi f_1(\mathbf{v}_1)\varphi(k_1)/m_e}{|\epsilon(\mathbf{k}_1, \mathbf{k}_1 \cdot \mathbf{v}_1)|^2} \\
&\times \int d\mathbf{v}_2\, [\mathbf{k}_1 \cdot \nabla_{\mathbf{v}_2} f_1(\mathbf{v}_2)]\delta[\mathbf{k}_1 \cdot (\mathbf{v}_1 - \mathbf{v}_2)]
\end{aligned} \tag{A.39}$$

Finally, inserting (A.38) and (A.39) into (A.7), one obtains

$$\begin{aligned}
\partial_t f_1(\mathbf{v}_1, t) &= -\frac{8\pi^4 n_0}{m_e^2} \nabla_{\mathbf{v}_1} \cdot \int d\mathbf{k}_1\, d\mathbf{v}_2 \frac{\mathbf{k}_1\mathbf{k}_1 \cdot \varphi^2(k_1)}{|\epsilon(\mathbf{k}_1, \mathbf{k}_1 \cdot \mathbf{v}_1)|^2} \\
&\times \delta[\mathbf{k}_1 \cdot (\mathbf{v}_1 - \mathbf{v}_2)][f_1(\mathbf{v}_1)\nabla_{\mathbf{v}_2} f_1(\mathbf{v}_2) - f_1(\mathbf{v}_2)\nabla_{\mathbf{v}_1} f_1(\mathbf{v}_1)]
\end{aligned} \tag{A.40}$$

which with appropriate changes of variables is the Lenard–Balescu equation (5.19).

REFERENCES

[1] D. C. Montgomery and D. A. Tidman, *Plasma Kinetic Theory*, McGraw-Hill, New York, 1964.

[2] P. C. Clemmow and J. P. Dougherty, *Electrodynamics of Particles and Plasmas*, Addison-Wesley, Reading, Mass., 1969.

[3] D. C. Montgomery, *Theory of the Unmagnetized Plasma*, Gordon and Breach, New York, 1971.

[4] A. Lenard, *Ann. Phys. (New York)*, *10*, 390 (1960).

[5] R. Balescu, *Phys. Fluids*, *3*, 52 (1960).

APPENDIX **B**

Langevin Equation, Fluctuation–Dissipation Theorem, Markov Process, and Fokker–Planck Equation

B.1 LANGEVIN EQUATION AND FLUCTUATION-DISSIPATION THEOREM

The discussion of plasma kinetic theory, including collisions, in Chapters 3, 4, and 5, led to the Fokker–Planck form of the plasma kinetic equation in (5.31). This is not a coincidence. In this appendix, it is shown that the Fokker–Planck equation arises naturally whenever a probability distribution [i.e., the one particle distribution function $f_s(\mathbf{v},t)$] changes slowly in time because of huge numbers of small changes (i.e., small angle collisions).

In order to motivate the Fokker–Planck equation, we use a physical example that is simpler than a plasma; namely, the case of Brownian motion. This will lead us to the related topics of the *Langevin equation*, the *fluctuation–dissipation theorem*, and *Markov processes*. As we study the example of Brownian motion, ask yourself how each step corresponds to its analogue in the plasma case.

The Langevin equation arises whenever a variable experiences a slow time variation as a result of a rapidly varying force. The best known example of this is the case of Brownian motion. A large particle (mass $\sim 10^{-12}$ gram) exhibits Brownian motion when bombarded by the molecules in air (mass $\sim 10^{-22}$ gram). The path of the particle may look as shown in Fig. B.1. The human eye, looking through a microscope, cannot see the fine structure on the curve shown, and so instead [1, 2] sees the curve in Fig. B.2. The wandering motion is, essentially, a random walk due to the large number of collisions that the particle suffers per unit time with the gas molecules. Picking out one of the dimensions of the motion, we can write Newton's force law in one spatial dimension,

$$\frac{dv(t)}{dt} = F(t) \tag{B.1}$$

Fig. B.1 Path of a Brownian particle.

where $F(t)$ is the force per unit mass on the Brownian particle. Thus, $F(t)$ contains the sum of many collisions, each lasting an extremely short time.

To study the physics of (B.1), we can consider an ensemble of realizations, each having the same initial speed $v(t = 0) = v_0$ but different random functions $F(t)$. Our intuition tells us that the overall effect of the many collisions will be to slow the Brownian particle, so that $\langle v(t) \rangle = 0$ as $t \to \infty$.

Microscopically, the Brownian particle slows because it collides with more particles in the direction of motion than in the opposite direction. It thus gives up net kinetic energy to the gas molecules, which leave the collision with a net gain in right-going momentum.

This discussion leads to the conclusion that the ensemble average of the force on the right of (B.1) must contain a term that tends to slow the Brownian particle.

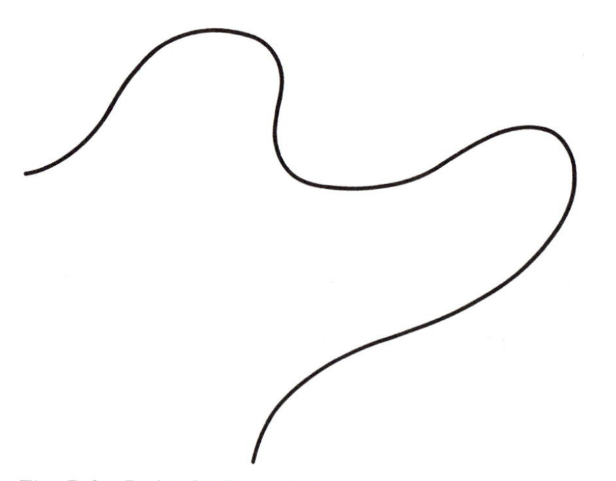

Fig. B.2 Path of a Brownian particle as seen by the human eye.

Thus, we split the force $F(t)$ into two terms,

$$F(t) = \langle F(t) \rangle + \delta F(t) \tag{B.2}$$

so that $\langle \delta F(t) \rangle = 0$. The ensemble averaged part of $F(t)$ will depend on the properties of the gas, and on the speed v of the Brownian particle. Suppose we Taylor expand this quantity in terms of the particle speed v:

$$\langle F(t) \rangle = c_1 + c_2 v + c_3 v^2 + \ldots \tag{B.3}$$

When $v = 0$, we want $\langle F \rangle = 0$, since there is then no preferred direction; thus, $c_1 = 0$. Let us then keep only the next term in (B.3). Because we expect this term to slow the particle, we introduce the minus sign explicitly through the introduction of a new constant ν such that $c_2 = -\nu$; our force equation (B.1) now reads

$$\boxed{\frac{dv(t)}{dt} = -\nu v(t) + \delta F(t)} \tag{B.4}$$

which is the famous *Langevin equation* (Refs. [3] to [7]).

The constant ν in (B.4) represents dissipation. This can be seen by taking the ensemble average of (B.4)

$$\frac{d}{dt}\langle v(t) \rangle = -\nu \langle v(t) \rangle \tag{B.5}$$

so that

$$\langle v(t) \rangle = v_0 e^{-\nu t} \tag{B.6}$$

(Recall that each realization of the ensemble has initial speed v_0). Thus, the characteristic slowing down time is ν^{-1}, and since the slowing down means a decrease in kinetic energy, ν represents dissipation.

Let us next investigate some of the statistical properties of (B.4). This equation is a linear inhomogeneous first order ordinary differential equation and thus is easy to solve. We have

$$\frac{dv(t)}{dt} + \nu v = \delta F(t) \tag{B.7}$$

Multiplying each side by $e^{\nu t}$ we have

$$\frac{d}{dt}\left[v(t)e^{\nu t}\right] = e^{\nu t}\,\delta F(t) \tag{B.8}$$

Thus

$$v(t)e^{\nu t} = v_0 + \int_0^t dt'\,\delta F(t')e^{\nu t'} \tag{B.9}$$

or

$$\boxed{v(t) = v_0\,e^{-\nu t} + e^{-\nu t}\int_0^t dt'\,\delta F(t')e^{\nu t'}} \tag{B.10}$$

The ensemble average of this equation reproduces (B.6),

$$\langle v(t) \rangle = v_0 e^{-\nu t} \tag{B.6}$$

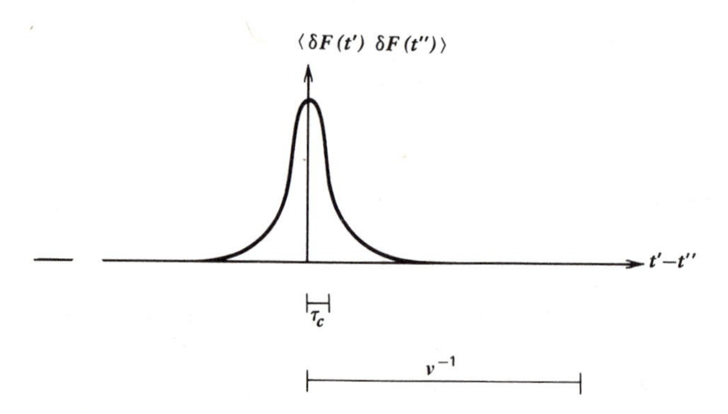

Fig. B.3 Autocorrelation function of the fluctuating force $\delta F(t)$, indicating the relative time scales τ_c and ν^{-1}.

Next, we square the velocity and ensemble average. Using (B.10), we have

$$\langle v^2(t)\rangle = \langle [v_0\, e^{-\nu t} + e^{-\nu t}\int_0^t dt'\ \delta F(t')e^{\nu t'}] \times [v_0 e^{-\nu t} + e^{-\nu t}\int_0^t dt''\ \delta F(t'')e^{\nu t''}]\rangle$$

$$= v_0{}^2 e^{-2\nu t} + e^{-2\nu t}\int_0^t dt'\ e^{\nu t'}\int_0^t dt''\ \langle \delta F(t')\delta F(t'')\rangle e^{\nu t''} \tag{B.11}$$

where two terms have disappeared in the ensemble average.

We now make the important assumption that δF is only correlated with itself over a time τ_c extremely short compared to the characteristic dissipation time ν^{-1} (Fig. B.3). We furthermore assume that δF is a stationary process, so that $\langle \delta F(t')\delta F(t'')\rangle$ is only a function of the time difference $t' - t''$. The correlation time τ_c is roughly the time of one molecular collision.

We are interested in the integral

$$I \equiv e^{-2\nu t}\int_0^t dt'\ e^{\nu t'}\int_0^t dt''\ \langle \delta F(t')\delta F(t'')\rangle e^{\nu t''} \tag{B.12}$$

The above arguments indicate that the integrand is only important (nonzero) for $t' \approx t''$, as shown in Fig. B.4. With the change of variable $y = t' - t''$, $dy =$

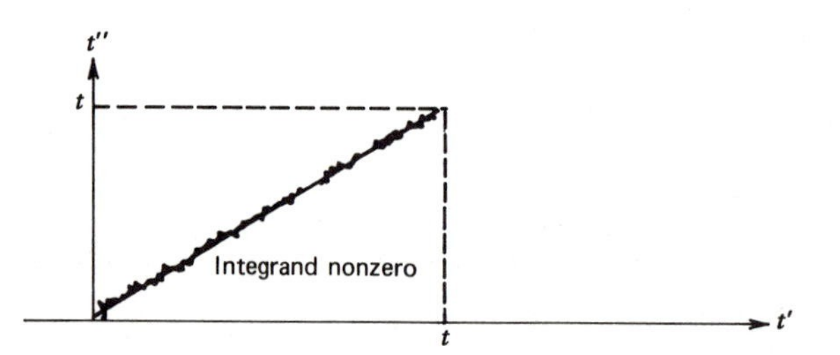

Fig. B.4 Region of the t'-t'' plane that contains a substantial contribution to the integral in (B.12).

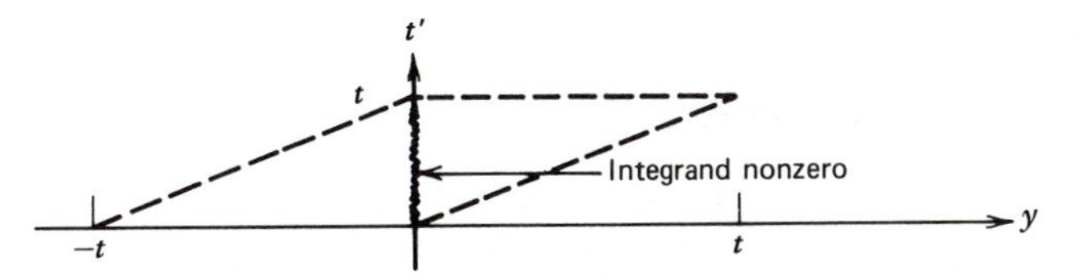

Fig. B.5 Region of the t'-y plane that contains a substantial contribution to the integral in (B.13).

$-\ dt''$, (B.12) becomes

$$I = e^{-2\nu t} \int_0^t dt'\ e^{\nu t'} \int_{t'-t}^{t'} dy\ e^{\nu t' - \nu y} \langle \delta F(t') \delta F(t' - y) \rangle \tag{B.13}$$

By stationarity, we can write

$$\begin{aligned} \langle \delta F(t') \delta F(t' - y) \rangle &= \langle \delta F(0) \delta F(-y) \rangle \\ &= \langle \delta F(0) \delta F(y) \rangle \end{aligned} \tag{B.14}$$

where the last equality is due to the evenness of the correlation function. The integral in (B.13) is now substantial in the region shown in Fig. B.5, where $y = t' - t'' \approx 0$. Since the integrand is only important near $y \approx 0$, we can replace the upper limit of y-integration by $+\infty$ and the lower limit of y-integration by $-\infty$. Then (B.13) becomes

$$I = e^{-2\nu t} \int_0^t dt'\ e^{2\nu t'} \int_{-\infty}^{\infty} dy\ \langle \delta F(0) \delta F(y) \rangle \tag{B.15}$$

where we have discarded the factor $e^{-\nu y}$ that is unity when $y \approx 0$ where the integrand is important. The t' integration can now be performed,

$$I = \frac{1}{2\nu} (1 - e^{-2\nu t}) \int_{-\infty}^{\infty} dy\ \langle \delta F(0) \delta F(y) \rangle \tag{B.16}$$

so that the full equation (B.11) now reads

$$\boxed{\langle v^2(t) \rangle = v_0^{\,2} e^{-2\nu t} + \frac{1}{2\nu} (1 - e^{-2\nu t}) \int_{-\infty}^{\infty} dy\ \langle \delta F(0) \delta F(y) \rangle} \tag{B.17}$$

If we allow the time to become very large compared to the dissipation time ν^{-1}, then we obtain an expression for the thermal fluctuations of v^2,

$$\boxed{\langle v^2(t) \rangle \overset{t \to \infty}{\longrightarrow} \frac{1}{2\nu} \int_{-\infty}^{\infty} dy\ \langle \delta F(0) \delta F(y) \rangle} \tag{B.18}$$

However, we know from elementary thermodynamics that in thermal equilibrium, the Brownian particle will have ½T of kinetic energy per degree of freedom (Boltzmann's constant is as usual absorbed into the temperature T). Thus, elementary thermodynamics predicts

$$\tfrac{1}{2} M \langle v^2(t) \rangle = \tfrac{1}{2} T \tag{B.19}$$

or

$$\langle v^2(t) \rangle = \frac{T}{M} \tag{B.20}$$

Equating (B.18) and (B.20) we have

$$\frac{T}{M} = \frac{1}{2\nu} \int_{-\infty}^{\infty} dy \, \langle \delta F(0) \delta F(y) \rangle \tag{B.21}$$

or

$$\boxed{\nu = \frac{M}{2T} \int_{-\infty}^{\infty} dy \, \langle \delta F(0) \delta F(y) \rangle} \tag{B.22}$$

which is the *fluctuation–dissipation* theorem.

Equation (B.22) expresses the amazing fact that the dissipation of a Brownian particle is directly related to the correlation function $\langle \delta F(0) \delta F(y) \rangle$ of the fluctuating force $F(t) = \langle F(t) \rangle + \delta F(t)$ whose ensemble average $\langle F \rangle$ produces the dissipation. This is a fundamental result of physics that applies in many situations; in the theory of electric circuits it is known as *Nyquist's theorem.*

This concludes our discussion of the Langevin equation and the fluctuation–dissipation theorem. In the next section, we shall consider the related topic of *Markov processes* and derive the *Fokker–Planck equation.*

B.2 MARKOV PROCESSES AND FOKKER-PLANCK EQUATION

In the previous section, we considered the behavior of a Brownian particle and derived the Langevin equation together with a fluctuation–dissipation theorem. In this section, we show how the behavior of a Brownian particle can be described by a Fokker–Planck equation. The Fokker–Planck equation is a very general equation in physics; it describes not only Brownian particles, but any phenomenon that in some approximate sense can be thought of as a *Markov process.*

A *Markov process* is one whose value at the next measuring time depends only on its value at the present measuring time, and not on any previous measuring time. Thus, if $x(t)$ is the random process, and $x_n \equiv x(t_n)$, with $t_n > t_{n-1} > \ldots > t_1 > t_0$, a Markov process has a probability density such that

$$\rho(x_n | x_{n-1} \, x_{n-2} \ldots x_1 x_0) = \rho(x_n | x_{n-1}) \tag{B.23}$$

where the notation $\rho(a|b)$ means "the probability density of a given that b was true." Thus, for a Markov process, the probability that $x_n = 5$ depends only on what the value of x_{n-1} was; it does not depend on what the values of x_{n-2}, x_{n-3}, etc. were.

There are both *discrete* and *continuous* Markov processes. An example of a discrete Markov process is given by flipping a coin. A trivial example comes if we give each toss a value $x(t_n) \equiv x_n = +1$ for a toss of "heads" and a value $x_n = -1$ for a toss of "tails." Then x is clearly a Markov process, since $\rho(x_n) = \frac{1}{2}\delta(x_n - 1) + \frac{1}{2}\delta(x_n + 1)$ does not depend on x_{n-1}, much less on x_{n-2}, x_{n-3}, etc.

A better example of a discrete Markov process is given by defining the random variable

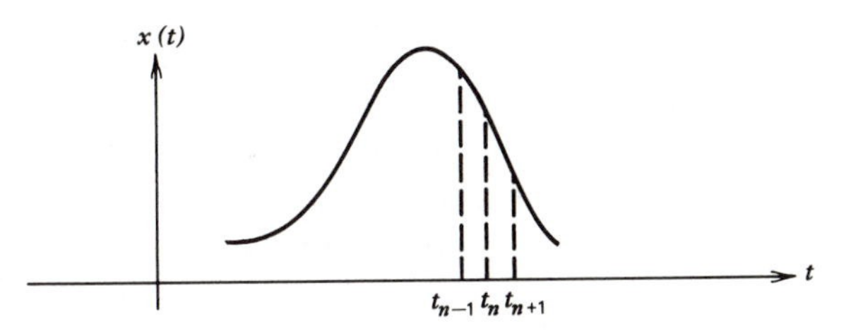

Fig. B.6 Any function in nature can be drawn as a smooth curve as shown.

$$X(t_n) \equiv X_n \equiv \sum_{i=1}^{n} x_i \tag{B.24}$$

where the x_i are given by the coin tosses of the previous paragraph. Now X is clearly a Markov process, whose probability density at t_n very definitely depends on the value of X_{n-1}, but on no previous value.

EXERCISE Calculate $\rho(X_n|X_{n-1})$ for this example.

To give an example of a continuous Markov process is more difficult, because a continuous Markov process cannot exist in nature. To see this, consider any random function that we can draw as a smooth curve, as in Fig. B.6. Now, on the time scale shown, it appears that x_{n+1} not only depends on x_n, but also on x_{n-1}. That is, x_{n+1} not only depends on x_n, but also on the derivative of the function $dx(t)/dt|_{t=t_n}$, which can be written

$$\left.\frac{dx(t)}{dt}\right|_{t=t_n} = \frac{x_n - x_{n-1}}{\Delta t} \tag{B.25}$$

Thus, this function is not a Markov process. In fact, no function that is a continuous curve and, therefore no physical function, can be a Markov process.

This does not mean that Markov processes cannot be a good approximation to a physical process. Consider the velocity function of the Brownian particle in the previous section (Fig. B.7). We have seen that the velocity consists of a rapid fluctuation due to each molecular collision, together with a slowing down or net friction force. Thus, on the time scale of molecular collisions, the process is not Markovian. However, on the much longer time scale of many collision times, the

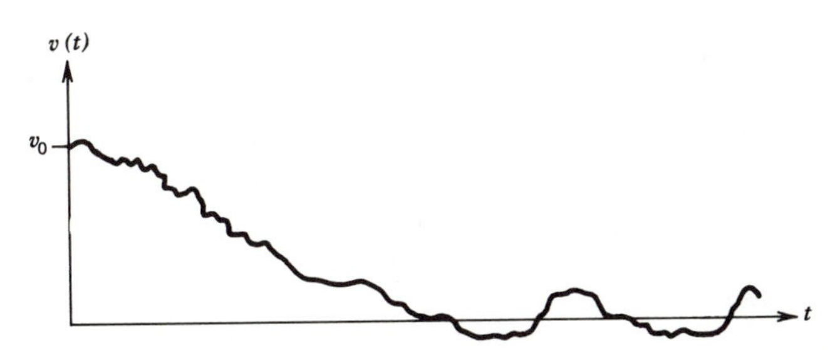

Fig. B.7 One realization of the velocity of a Brownian particle in a particular direction.

situation is very nearly Markovian. The Brownian particle is performing a random walk in velocity space, and soon forgets the details of its orbit near $t = 0$; it does, however, remember its velocity v_0 at $t = 0$.

Thus, we consider the process to have three time scales (Fig. B.8): the collision time τ_c, which is the autocorrelation time of the force $\delta F(t)$ in the Langevin equation; the time Δt after which we may assume to good approximation that the process is Markovian; and the dissipation time ν^{-1}. We must have $\Delta t >> \tau_c$; we shall further assume in this section that $\Delta t << \nu^{-1}$.

Let us develop some of the mathematical properties of Markov processes. This development will lead us to the Fokker–Planck equation.

Consider the probability of a sequence of values of the random function $x(t)$. This is

$\rho(x_n, x_{n-1}, \ldots, x_2, x_1, x_0) \equiv$ {probability that, at time t_0, the process $x(t)$ has the value x_0 *and* at time t_1, $x(t)$ has the value x_1, *and* . . . *and* at time t_n, $x(t)$ has the value x_n} where $t_n > t_{n-1} > t_{n-2} \ldots > t_1 > t_0$ (B.26)

By the definition (B.23) we can write

$$\rho(x_n, x_{n-1}, \ldots, x_0) = \rho(x_n | x_{n-1}, x_{n-2}, \ldots, x_0)$$

$$\times \rho(x_{n-1}, x_{n-2}, \ldots, x_0) = \rho(x_n | x_{n-1}) \rho(x_{n-1}, x_{n-2}, \ldots, x_0) \quad \text{(B.27)}$$

The same procedure can now be applied to the last factor on the right of (B.27), so that

$$\rho(x_{n-1}, x_{n-2}, \ldots, x_0) = \rho(x_{n-1} | x_{n-2}) \rho(x_{n-2}, \ldots, x_0) \quad \text{(B.28)}$$

and so on until we have finally, for a Markov process,

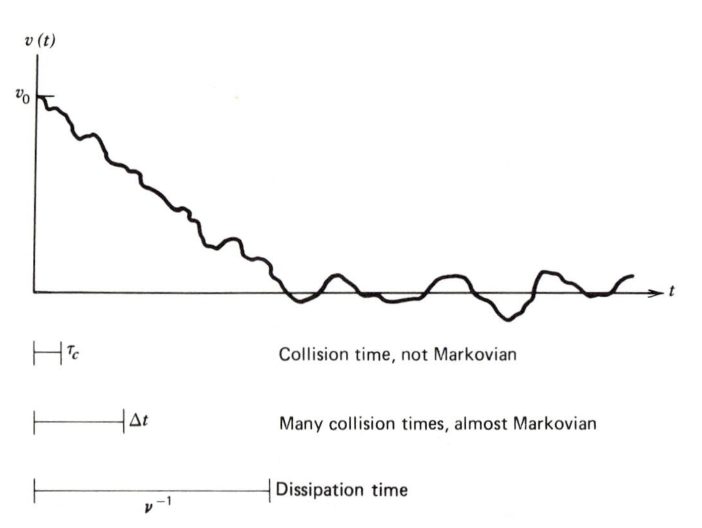

Fig. B.8 Three time scales of Brownian motion.

$$\boxed{\begin{aligned}\rho(x_n,x_{n-1},x_{n-2},\ldots,x_0) &= \rho(x_n|x_{n-1})\rho(x_{n-1}|x_{n-2})\\ &\ldots \rho(x_2|x_1)\rho(x_1|x_0)\rho(x_0)\end{aligned}} \tag{B.29}$$

By elementary considerations it must also be true that

$$\rho(x_n,x_{n-1},x_{n-2},\ldots,x_1,x_0) = \rho(x_n,x_{n-1},\ldots,x_1|x_0)\rho(x_0) \tag{B.30}$$

Comparing (B.30) and (B.29) we find

$$\begin{aligned}&\rho(x_n,x_{n-1},\ldots,x_1|x_0)\\ &= \rho(x_n|x_{n-1})\rho(x_{n-1}|x_{n-2})\ldots\rho(x_1|x_0)\end{aligned} \tag{B.31}$$

In particular, we can choose $n = 2$ to obtain

$$\rho(x_2,x_1|x_0) = \rho(x_2|x_1)\rho(x_1|x_0) \tag{B.32}$$

Let us now integrate this expression over all possible x_1 to obtain

$$\rho(x_2|x_0) = \int dx_1\ \rho(x_2,x_1|x_0) \tag{B.33}$$

or

$$\boxed{\rho(x_2|x_0) = \int dx_1\ \rho(x_2|x_1)\rho(x_1|x_0)} \tag{B.34}$$

which is the *Chapman–Kolmogorov equation*, or *Smoluchowsky equation* [8].

Suppose we identify x_1 with time t and x_2 as $x(t + \Delta t)$. Suppose we further assume that

$$\rho(x_0) \equiv \rho(x,t = t_0) = \delta(x - x_0) \tag{B.35}$$

Then we can drop the references to x_0 in (B.34), and write

$$\rho(x_2|x_0) = \rho(x,t + \Delta t) \tag{B.36}$$

that is, x_2 is now denoted by x, and

$$\rho(x_1|x_0) = \rho(x_1,t) \tag{B.37}$$

We can also change the notation of $\rho(x_2|x_1)$; with the definition

$$\Delta x \equiv x - x_1 \tag{B.38}$$

we can write

$$\begin{aligned}\rho(x_2|x_1) &= \rho(x,t + \Delta t|x - \Delta x,t)\\ &= \psi(\Delta x,t + \Delta t|x - \Delta x,t)\end{aligned} \tag{B.39}$$

where the transition probability ψ is defined by (B.39); ψ gives the probability that at time $t + \Delta t$, the random process has made a jump of Δx from its previous value $x - \Delta x$ at time t.

With these notational changes, we can rewrite (B.34) as

$$\rho(x,t + \Delta t) = \int d(\Delta x)\psi(\Delta x,t + \Delta t|x - \Delta x,t)\rho(x - \Delta x,t) \tag{B.40}$$

The value x appears on the right of (B.40) only in the combination $x - \Delta x$. Thus, if we assume that *all of the important physics happens for small* Δx, then we can make a Taylor series expansion on the right of (B.40), obtaining

$$\rho(x,t + \Delta t) = \int d(\Delta x) \sum_{l=0}^{\infty} \frac{(-\Delta x)^l}{l!}$$

$$\times \left\{ \frac{\partial^l}{\partial x^l} \left[\psi(\Delta x, t + \Delta t | x - \Delta x, t) \rho(x - \Delta x, t) \right]_{x-\Delta x = x} \right\}$$

or

$$\rho(x,t + \Delta t) = \int d(\Delta x) \sum_{l=0}^{\infty} \frac{(-\Delta x)^l}{l!} \frac{\partial^l}{\partial x^l}$$

$$\times [\psi(\Delta x, t + \Delta t | x,t) \rho(x,t)] \tag{B.41}$$

If the infinite sum converges, and if we can interchange the summation and integration, then we can write

$$\rho(x,t + \Delta t) = \sum_{l=0}^{\infty} \frac{(-1)^l}{l!} \frac{\partial^l}{\partial x^l}$$

$$\times [\rho(x,t) \int d(\Delta x) (\Delta x)^l \psi(\Delta x, t + \Delta t | x,t)] \tag{B.42}$$

The quantity given by the Δx integration is just the expectation value or ensemble average of $(\Delta x)^l$,

$$\langle (\Delta x)^l \rangle \equiv \int d(\Delta x) (\Delta x)^l \psi(\Delta x, t + \Delta t | x,t) \tag{B.43}$$

which is itself a function of x,t through ψ. Equation (B.42) becomes

$$\rho(x,t + \Delta t) = \sum_{l=0}^{\infty} \frac{(-1)^l}{l!} \frac{\partial^l}{\partial x^l} [\rho(x,t) \langle (\Delta x)^l \rangle (x,t)] \tag{B.44}$$

Moving the $l = 0$ term to the left side, and dividing by Δt, we have

$$\frac{\rho(x,t + \Delta t) - \rho(x,t)}{\Delta t} = \sum_{l=1}^{\infty} \frac{(-1)^l}{l! \Delta t} \frac{\partial^l}{\partial x^l} [\rho(x,t) \langle (\Delta x)^l \rangle (x,t)] \tag{B.45}$$

We next take the limit as $\Delta t \to$ "0". This means that we let Δt become very small, much smaller than any macroscopic time scale (e.g., ν^{-1}). However, Δt cannot really go to zero, because this development has assumed that Δt is large enough to justify the Markovian assumption. Thus, the left side of (B.45) becomes

$$\lim_{\Delta t \to \text{"0"}} \frac{\rho(x,t + \Delta t) - \rho(x,t)}{\Delta t} = \frac{\partial \rho(x,t)}{\partial t} \tag{B.46}$$

where the time derivative refers to macroscopic time. Equation (B.45) becomes

$$\frac{\partial \rho}{\partial t} = \sum_{l=1}^{\infty} (-1)^l \frac{\partial^l}{\partial x^l} \left[\lim_{\Delta t \to \text{“}0\text{”}} \frac{\langle (\Delta x)^l \rangle}{l!\Delta t} \rho(x,t) \right] \tag{B.47}$$

Defining the diffusion coefficients

$$D^{(l)}(x,t) \equiv \lim_{\Delta t \to \text{“}0\text{”}} \frac{\langle (\Delta x)^l \rangle}{l!\Delta t} \tag{B.48}$$

Equation (B.47) is

$$\frac{\partial \rho(x,t)}{\partial t} = \sum_{l=1}^{\infty} (-1)^l \frac{\partial^l}{\partial x^l} [D^{(l)}(x,t)\rho(x,t)] \tag{B.49}$$

If we keep only the first two terms on the right of (B.49), we have

$$\boxed{\frac{\partial \rho(x,t)}{\partial t} = - \frac{\partial}{\partial x} [D^{(1)}(x,t)\rho(x,t)] + \frac{\partial^2}{\partial x^2} [D^{(2)}(x,t)\rho(x,t)]} \tag{B.50}$$

which is the well-known *Fokker–Planck equation* [9].

For Brownian motion, the random variable x is replaced by the particle velocity $v(t)$. We shall leave it as an exercise to determine the diffusion coefficients $D^{(1)}(v,t)$ and $D^{(2)}(v,t)$.

EXERCISE Use the results of the previous section to evaluate the coefficients in Eq. (B.50) for Brownian motion. Show that $D^{(1)}(v,t) = -\nu v$, and $D^{(2)}(v,t) = \nu T/M$, so that the Fokker–Planck equation associated with the Langevin equation of Brownian motion is

$$\boxed{\frac{\partial \rho(v,t)}{\partial t} = \nu \frac{\partial}{\partial v} (v\rho) + \frac{\nu T}{M} \frac{\partial^2}{\partial v^2} \rho} \tag{B.51}$$

EXERCISE Use the results of the previous section to show that $D^{(3)}(v,t) \sim \Delta t$ and, thus, vanishes as $\Delta t \to$ “0”.

We can now understand why we are able to write the Lenard–Balescu equation in the form of a Fokker–Planck equation,

$$\frac{\partial f(\mathbf{v}_1,t)}{\partial t} = - \nabla_{\mathbf{v}_1} \cdot (\mathbf{A}f) + \frac{1}{2} \nabla_{\mathbf{v}_1} \nabla_{\mathbf{v}_1} : (\overleftrightarrow{\mathbf{B}}f) \tag{B.52}$$

Because the derivation of Lenard–Balescu assumed $g(1,2) << f_1(1)f_1(2)$, we have effectively limited ourselves to small angle two-body collisions. The quantity $f(\mathbf{v}_1,t)$ may be thought of as the probability density of particles in velocity space. Thus, $f(\mathbf{v}_1,t)$ is changing slowly on the time scale for a two-body collision. All of these features are precisely those assumed in the derivation of the Fokker–Planck equation. It should come as no surprise to us that the Lenard–Balescu equation can be written in the form of the Fokker–Planck equation. The coefficient $\mathbf{A}$ in (B.52) is called the coefficient of *dynamic friction*, and plays the same role as νv in the Fokker–Planck equation (B.51) for Brownian motion. It represents the slowing down of a particle due to many small angle Coulomb collisions. Likewise, the coefficient $\overleftrightarrow{\mathbf{B}}$ in (B.52) is called the *diffusion coefficient*, and plays the same role as $\nu T/M$ in (B.51). It represents the diffusion of the plasma particles in velocity space due to many small angle collisions.

In the steady state, a typical particle is suffering dynamic friction plus diffusion; the net effect is to produce a Maxwellian. This is just as true in a plasma as it is for a Brownian particle.

In addition to the stated references, sources for this appendix include the book by Stratonovich [10] and the ageless and excellent article by Chandrasekhar [11].

REFERENCES

[1] R. Brown, *Phil. Mag.*, *4*, 161 (1928).

[2] R. Brown, *Ann. Phys. Chem.*, *14*, 294 (1928).

[3] M. P. Langevin, *C. R. Acad. Sci. Paris*, *146*, 530 (1908).

[4] A. Einstein, *Ann. Phys.*, *17*, 549 (1905).

[5] A. Einstein, *Ann. Phys.*, *19*, 289 (1906).

[6] A. Einstein, *Ann. Phys.*, *19*, 371 (1906).

[7] D. K. C. MacDonald, *Noise and Fluctuations: An Introduction*, Wiley, New York, 1962.

[8] M. von Smoluchowsky, *Ann. Phys.*, *21*, 756 (1906).

[9] Lord Rayleigh, *Nature*, *72*, 318 (1905).

[10] R. L. Stratonovich, *Topics in the Theory of Random Noise*, Gordon and Breach, New York, 1963, Vol. 1.

[11] S. Chandrasekhar, *Rev. Mod. Phys.*, *15*, 1 (1943).

APPENDIX **C**

Pedestrian's Guide to Complex Variables

Many parts of this book make use of the basic results of the theory of complex variables [1, 2]. For the benefit of readers who have not yet studied this subject in detail, or who have studied it long enough ago to have forgotten it, these basic results are summarized here.

The most useful result is the *residue theorem*, which states that the integral in a counterclockwise direction around a closed curve is $2\pi i$ times the sum of the residues. If the integrand is of the form $f(z)(z - z_0)^{-1}$, the residue at the *simple pole* $z = z_0$ is $f(z_0)$. For example, consider the integral

$$I = \int_{-\infty}^{\infty} \frac{dz}{z^2 + a^2} \tag{C.1}$$

where the integration is along the real z-axis, and $a > 0$. The integration can be closed by a large semicircle at infinity, since the contribution from the semicircle is

$$\sim \lim_{R\to\infty} (\pi R/R^2) = \lim_{R\to\infty} (\pi/R) = 0$$

The semicircle can occur in either the upper-half z-plane or the lower-half z-plane (Fig. C.1). Writing (C.1) as

$$I = \oint \frac{dz}{(z + ia)(z - ia)} \tag{C.2}$$

we close the contour downward, changing the sign on the result because this is in the clockwise direction, to obtain (only the pole at $z = -ia$ is enclosed)

$$I = (-2\pi i) \left. \frac{1}{z - ia} \right|_{z=-ia} = \frac{\pi}{a} \tag{C.3}$$

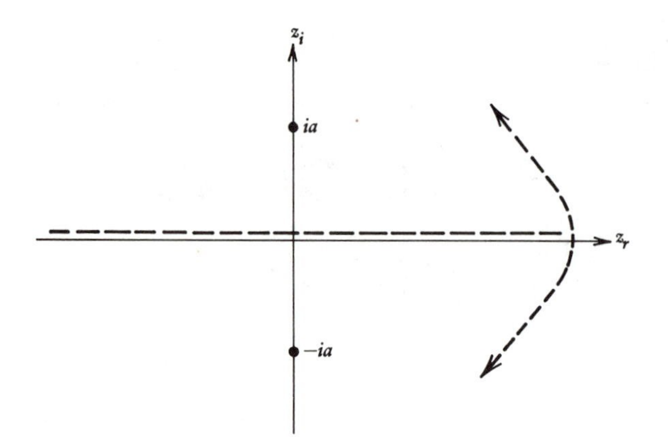

Fig. C.1 Integration contour in the z-plane.

EXERCISE Obtain the same result by closing upward.

The solution of many ordinary and partial differential equations is facilitated by *Fourier and Laplace transformation.* The Fourier transform conventions used in this book, stated in Chapter 5, are for functions of one spatial dimension x,

$$f(k) = \int_{-\infty}^{\infty} \frac{dx}{2\pi} f(x) \exp(-ikx) \tag{C.4}$$

with inverse transform

$$f(x) = \int_{-\infty}^{\infty} dk \quad f(k) \exp(ikx) \tag{C.5}$$

The x and k integrations are along the real axes. The Laplace transform conventions for functions of time t are

$$f(\omega) = \int_{0}^{\infty} dt\, f(t) \exp(i\omega t) \tag{C.6}$$

with inverse transform

$$f(t) = \int_{L} \frac{d\omega}{2\pi} f(\omega) \exp(-i\omega t) \tag{C.7}$$

where the Laplace contour L is a horizontal line in the complex ω-plane that must pass above all singularities of $f(\omega)$. The Laplace transform (C.6) can be considered as the Fourier transform of the function $\tilde{f}(t)$ such that $\tilde{f}(t) = f(t)$ for $t > 0$ and $\tilde{f}(t) = 0$ for $t < 0$. Then for $t < 0$ the inverse Laplace transform (C.7) can be closed upward [since $\exp(-i\omega t) \sim \exp(-|t|\omega_i) \to 0$ for $\omega_i \to +\infty$], yielding $\tilde{f}(t) = 0$ for $t < 0$ since the Laplace contour passes above all singularities of $f(\omega)$.

Consider the solution of the differential equation

$$\frac{df}{dt} = \alpha f \tag{C.8}$$

with $f(t = 0) = f_0$. The Laplace transform of the left side is

$$\int_0^\infty dt \, \frac{df(t)}{dt} \exp(i\omega t) = f(t) \exp(i\omega t)\Big|_0^\infty$$

$$- i\omega \int_0^\infty dt \, f(t) \exp(i\omega t) = f(t) \exp(i\omega t)\Big|_0^\infty - i\omega f(\omega) \tag{C.9}$$

Without knowing the function $f(t)$, we can only say that this integral is defined for ω_i large enough, for only then is $f(t) \exp(i\omega t)|_{t=\infty}$ equal to zero. How large is large enough remains to be seen. For large enough ω_i, Eq. (C.8) has the Laplace transform

$$- f(t = 0) - i\omega f(\omega) = \alpha f(\omega) \tag{C.10}$$

so that

$$f(\omega) = \frac{if(t = 0)}{\omega - i\alpha} \tag{C.11}$$

The inverse transform is

$$f(t) = \int_L \frac{d\omega}{2\pi} \frac{if(t = 0)}{\omega - i\alpha} \exp(-i\omega t) \tag{C.12}$$

where the contour must be placed high enough in the ω-plane so that $f(\omega)$ is defined (the shaded region in Fig. C.2). Once the contour is drawn in the shaded region of Fig. C.2, it can be moved around only if $f(\omega)$ is *analytically continued* to the remainder of the complex ω-plane. An *analytic* function is one that is *differentiable* (the derivative in the complex plane at the point z does not depend on which direction the point is approached from). The analytic continuation of a simple function like $f(\omega)$ in (C.11) from the shaded region in Fig. C.2 [the only region where (C.10) is defined] to the entire ω-plane is easy; it is just the function

$$f(\omega) = \frac{if(t = 0)}{\omega - i\alpha} \tag{C.13}$$

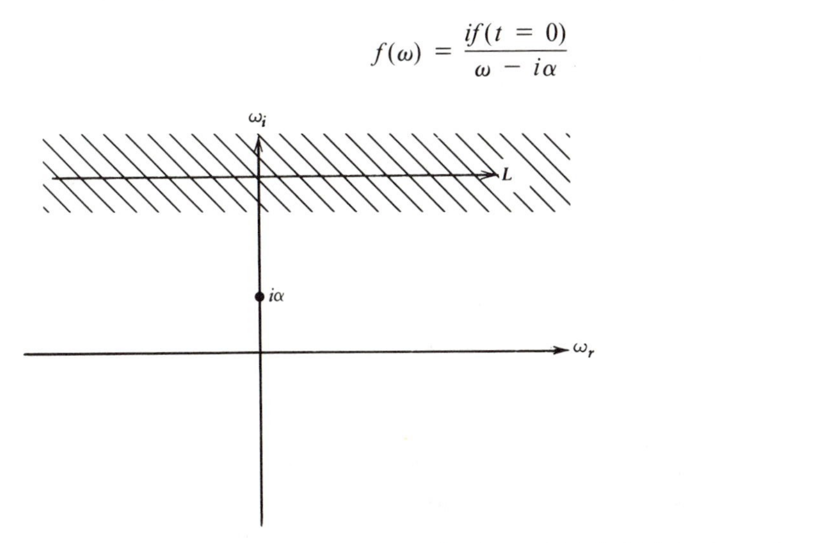

Fig. C.2 Inverse Laplace contour must be drawn initially high in the ω-plane.

itself. This function is now analytic in the entire ω-plane except at the point $\omega = i\alpha$. The contour in (C.12) can now be closed by a large semicircle in the lower-half ω-plane, since $f(\omega)$ is defined everywhere. The result is

$$f(t) = -\frac{2\pi i(i)}{2\pi} f(t = 0) \exp(\alpha t)$$

$$= f(t = 0) \exp(\alpha t) \tag{C.14}$$

which is the desired result. In retrospect, we can now see that (C.9) converges for $\omega_i > \alpha$. This is why the inverse Laplace contour must be drawn above all singularities of $f(\omega)$ in the ω-plane.

Analytic continuation is not always quite as simple as in (C.13). Consider

$$f(z) = \int_{-\infty}^{\infty} \frac{dx}{(x^2 + a^2)(x - z)} \tag{C.15}$$

defined for $z_i > 0$ with $a > 0$; the integration is along the real x-axis. Closing the contour either up or down (Fig. C.3), we find

$$f(z) = \frac{-\pi}{a} \frac{1}{z + ia} \tag{C.16}$$

for $z_i > 0$. We *cannot* use (C.15) as the analytic continuation of $f(z)$ for $z_i < 0$, for then (C.15) yields

$$f(z) = \frac{-\pi}{a} \frac{1}{z + ia} - \frac{2\pi i}{z^2 + a^2} \tag{C.17}$$

for $z_i < 0$. The function $f(z)$ defined by (C.16) and (C.17) is discontinuous at $z_i = 0$ and so is not analytic. In order to properly analytically continue (C.15), one must subtract the extra term in (C.17) that leads to the discontinuity, and write

$$f(z) = \int_{-\infty}^{\infty} \frac{dx}{(x^2 + a^2)(x - z)} + \frac{2\pi i}{z^2 + a^2} \tag{C.18}$$

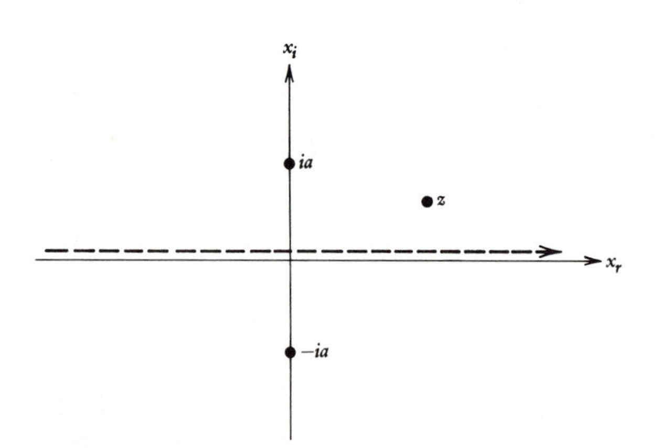

Fig. C.3 Evaluation of $f(z)$ for $z_i > 0$ in (C.15).

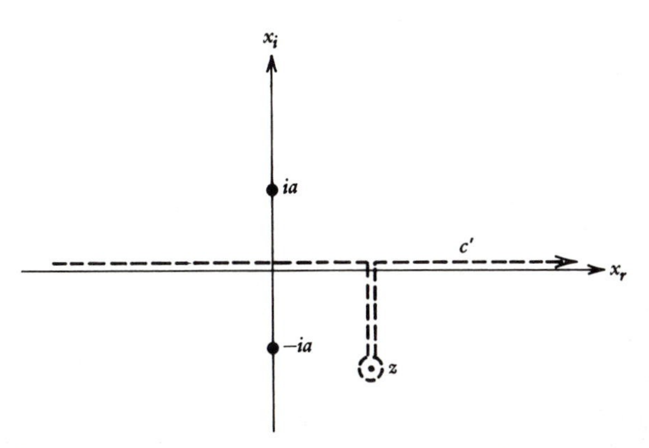

Fig. C.4 Integration contour for $z_i < 0$ that gives the proper analytic continuation of (C.15).

for $z_i < 0$. The combination (C.15) for $z_i > 0$ and (C.18) for $z_i < 0$ is now an analytic function everywhere except at the pole of (C.16), $z = -ia$. Alternatively, one can deform the contour in (C.15) as shown in Fig. C.4 for $z_i < 0$, and write

$$f(z) = \int_{C'} \frac{dx}{(x^2 + a^2)(x - z)} \tag{C.19}$$

The contour C' is as shown in Fig. C.4 for $z_i < 0$, and is along the real x-axis for $z_i > 0$. This gives the same result as the form (C.18) for $z_i < 0$.

A useful formula for Fourier transformation is

$$\int_{-\infty}^{\infty} \frac{dx}{2\pi} \exp(-ikx) = \delta(k) \tag{C.20}$$

This formula can be demonstrated by multiplying each side by an arbitrary function $f(k)$ and integrating over all k. The right side yields $f(k = 0)$, while the left side is

$$\begin{aligned} \int_{-\infty}^{\infty} dk\, f(k) \int_{-\infty}^{\infty} \frac{dx}{2\pi} \exp(-ikx) &= \int_{-\infty}^{\infty} \frac{dx}{2\pi} \int_{-\infty}^{\infty} dk\, f(k) \exp(-ikx) \\ &= \int_{-\infty}^{\infty} \frac{dx}{2\pi} f(x) \\ &= f(k = 0) \end{aligned} \tag{C.21}$$

where it has been assumed that the order of integrations can be reversed. Since the right and left sides of (C.20) yield the same result, the identification (C.20) must be correct.

Another useful formula concerns integrals of the form

$$I = \lim_{\eta \to 0} \int_{-\infty}^{\infty} dx \; \frac{1}{x - a \pm i|\eta|} \tag{C.22}$$

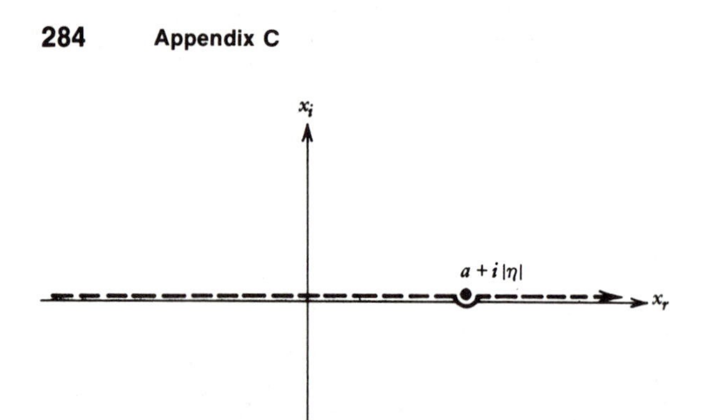

Fig. C.5 Integration contour leading to the Plemelj formulas.

where the integral is along the real x-axis, and $a > 0$. For the lower sign, the pole is at $x = a + i|\eta|$, and the integral can be performed by slightly deforming the contour as shown in Fig. C.5. This leads to

$$I = P\int_{-\infty}^{\infty} dx\, \frac{1}{x - a} + \pi i \tag{C.23}$$

where the semicircle in Fig. C.5 contributes half of $2\pi i$, and where

$$P\int_{-\infty}^{\infty} = \lim_{\epsilon \to 0} \left[\int_{-\infty}^{a-|\epsilon|} + \int_{a+|\epsilon|}^{\infty}\right] \tag{C.24}$$

Formally, one writes

$$\lim_{\eta \to 0} \quad \frac{1}{x - a - i|\eta|} = P\left(\frac{1}{x - a}\right) + \pi i \delta(x - a) \tag{C.25}$$

which when integrated over x yields (C.23). For the upper sign in (C.22), the pole approaches the integration contour from below, the integration contour must be deformed upward rather than downward, and the sign of the imaginary contribution changes. The general formula is finally

$$\lim_{\eta \to 0} \quad \frac{1}{x - a \pm i|\eta|} = P\left(\frac{1}{x - a}\right) \mp \pi i \delta(x - a) \tag{C.26}$$

Other properties of complex variables are explored throughout the book.

REFERENCES

[1] I. S. Sokolnikoff and R. M. Redheffer, *Mathematics of Physics and Modern Engineering*, 2nd edition, McGraw-Hill, New York, 1966.

[2] G. Arfken, *Mathematical Methods for Physicists*, Academic, New York, 1970.

APPENDIX D

Vector and Tensor Identities

The following vector and tensor identities are useful in the study of plasma physics [1].

$$\mathbf{A} \cdot (\mathbf{B} \times \mathbf{C}) = (\mathbf{A} \times \mathbf{B}) \cdot \mathbf{C} = \mathbf{B} \cdot (\mathbf{C} \times \mathbf{A}) = (\mathbf{B} \times \mathbf{C}) \cdot \mathbf{A} = \mathbf{C} \cdot (\mathbf{A} \times \mathbf{B}) = (\mathbf{C} \times \mathbf{A}) \cdot \mathbf{B} \tag{D.1}$$

$$\mathbf{A} \times (\mathbf{B} \times \mathbf{C}) = (\mathbf{A} \cdot \mathbf{C})\mathbf{B} - (\mathbf{A} \cdot \mathbf{B})\mathbf{C} \tag{D.2}$$

$$(\mathbf{A} \times \mathbf{B}) \cdot (\mathbf{C} \times \mathbf{D}) = (\mathbf{A} \cdot \mathbf{C})(\mathbf{B} \cdot \mathbf{D}) - (\mathbf{A} \cdot \mathbf{D})(\mathbf{B} \cdot \mathbf{C}) \tag{D.3}$$

$$\nabla(fg) = f\nabla g + g\nabla f \tag{D.4}$$

$$\nabla \cdot (f\mathbf{A}) = f\nabla \cdot \mathbf{A} + \mathbf{A} \cdot \nabla f \tag{D.5}$$

$$\nabla \times (f\mathbf{A}) = f\nabla \times \mathbf{A} + \nabla f \times \mathbf{A} \tag{D.6}$$

$$\nabla \cdot (\mathbf{A} \times \mathbf{B}) = \mathbf{B} \cdot (\nabla \times \mathbf{A}) - \mathbf{A} \cdot (\nabla \times \mathbf{B}) \tag{D.7}$$

$$\nabla \times (\mathbf{A} \times \mathbf{B}) = \mathbf{A}(\nabla \cdot \mathbf{B}) - \mathbf{B}(\nabla \cdot \mathbf{A}) + (\mathbf{B} \cdot \nabla)\mathbf{A} - (\mathbf{A} \cdot \nabla)\mathbf{B} \tag{D.8}$$

$$\nabla(\mathbf{A} \cdot \mathbf{B}) = \mathbf{A} \times (\nabla \times \mathbf{B}) + \mathbf{B} \times (\nabla \times \mathbf{A}) + (\mathbf{A} \cdot \nabla)\mathbf{B} + (\mathbf{B} \cdot \nabla)\mathbf{A} \tag{D.9}$$

$$\nabla^2\mathbf{A} = \nabla(\nabla \cdot \mathbf{A}) - \nabla \times (\nabla \times \mathbf{A}) \tag{D.10}$$

$$\nabla \times \nabla f = 0 \tag{D.11}$$

$$\nabla \cdot (\nabla \times \mathbf{A}) = 0 \tag{D.12}$$

$$\nabla \cdot (\mathbf{B}\,\mathbf{A}) = \mathbf{A}(\nabla \cdot \mathbf{B}) + (\mathbf{B} \cdot \nabla)\mathbf{A} \tag{D.13}$$

$$\nabla \cdot (f\overleftrightarrow{\mathbf{T}}) = (\nabla f) \cdot \overleftrightarrow{\mathbf{T}} + f\nabla \cdot \overleftrightarrow{\mathbf{T}} \tag{D.14}$$

REFERENCE

[1] D. L. Book, *Plasma Formulary*, Naval Research Laboratory, Washington, D.C., 1980.

Index